A FIRST COURSE IN THE THEORY OF LINEAR STATISTICAL MODELS

THE DUXBURY ADVANCED SERIES IN STATISTICS AND DECISION SCIENCES

Applied Nonparametric Statistics, Second Edition, Daniel
Applied Regression Analysis and Other Multivariate Methods, Second Edition, Kleinbaum/Kupper/Muller
Classical and Modern Regression with Applications, Second Edition, Myers
Elementary Survey Sampling, Fourth Edition, Scheaffer/Mendenhall/Ott
Introduction to Contemporary Statistical Methods, Second Edition, Koopmans
Introduction to Probability and Its Applications, Scheaffer
Introduction to Probability and Mathematical Statistics, Bain/Englehardt
A First Course in the Theory of Linear Statistical Models, Myers/Milton
Linear Statistical Models, Second Edition, Bowerman/O'Connell
Probability Modeling and Computer Simulation, Matloff
Quantitative Forecasting Methods, Farnum/Stanton
Time Series Analysis, Cryer
Time Series Forecasting: Unified Concepts and Computer Implementation, Second Edition, Bowerman/O'Connell

THE DUXBURY SERIES IN STATISTICS AND DECISION SCIENCES

Applications, Basics and Computing of Exploratory Data Analysis, Velleman/Hoaglin
Applied Regression Analysis for Business and Economics, Dielman
Elementary Statistics, Fifth Edition, Johnson
Elementary Statistics for Business, Second Edition, Johnson/Siskin
Essential Business Statistics: A Minitab Framework, Bond/Scott
Introduction to Mathematical Programming: Applications and Algorithms, Winston
Fundamental Statistics for the Behavioral Sciences, Second Edition, Howell
Fundamentals of Biostatistics, Third Edition, Rosner
Introduction to Probability and Statistics, Eighth Edition, Mendenhall/Beaver
An Introduction to Statistical Methods and Data Analysis, Third Edition, Ott
Introductory Business Statistics with Microcomputer Applications, Shiffler/Adams
Introductory Statistics for Management and Economics, Third Edition, Kenkel
Making Hard Decisions: An Introduction to Decision Analysis, Clemen
Mathematical Statistics with Applications, , Fourth Edition, Mendenhall/Wackerly/Scheaffer
Minitab Handbook, Second Edition, Ryan/Joiner/Ryan
Minitab Handbook for Business and Economics, Miller
Operations Research: Applications and Algorithms, Second Edition, Winston
Probability and Statistics for Engineers, Third Edition, Scheaffer/McClave
Probability and Statistics for Modern Engineering, Second Edition, Lapin
SAS Applications Programming: A Gentle Introduction, DiIorio
Statistical Methods for Psychology, Second Edition, Howell
Statistical Thinking for Managers, Third Edition, Hildebrand/Ott
Statistics: A Tool for the Social Sciences, Fourth Edition, Ott/Larson/Mendenhall
Statistics for Business and Economics, Bechtold/Johnson
Statistics for Business: Data Analysis and Modeling, Cryer/Miller
Statistics for Management and Economics, Sixth Edition, Mendenhall/Reinmuth/Beaver
Student Edition of Exec-U-Stat, The Authors of Statgraphics
Understanding Statistics, Fifth Edition, Ott/Mendenhall

A FIRST COURSE IN THE THEORY OF LINEAR STATISTICAL MODELS

Raymond H. Myers
Virginia Polytechnic Institute and State University
Janet S. Milton
Radford University

THE DUXBURY ADVANCED SERIES IN STATISTICS AND DECISION SCIENCES
PWS-KENT Publishing Company
Boston

PWS-KENT
Publishing Company

20 Park Plaza
Boston, Massachusetts 02116

To Our Families
Enid, Joan, Tom, Deborah, David
Sharon, Julie, Billy

PWS-KENT Publishing Company is a division of Wadsworth, Inc.

Library of Congress Cataloging-in-Publication Data

Myers, Raymond H.
A first course in the theory of linear statistical models
Raymond H. Myers, Janet S. Milton.
p. cm.
Includes bibliographical references and index.
ISBN 0-534-91645-7
1. Linear models (Statistics) I. Milton, J. Susan (Janet Susan)
II. Title.
QA279.M937 1991 90-39233
519.5′3—dc20 CIP

Sponsoring Editor: Michael Payne
Assistant Editor: Marcia Cole
Editorial Assistant: Patricia Schumacher
Production Editor: Helen Walden
Manufacturing Coordinator: Peter Leatherwood
Interior and Cover Design: Catherine L. Johnson
Composition: H. Charlesworth & Co. Ltd.
Printing and Binding: Maple-Vail Book Mfr. Group

Printed in the United States of America

91 92 93 94 95—10 9 8 7 6 5 4 3 2 1

CONTENTS

CHAPTER 6

HYPOTHESIS TESTING IN THE LESS THAN FULL RANK MODEL 238

CHAPTER 7

ADDITIONAL TOPICS 316

PREFACE

Students studying statistics for the first time typically begin with an introductory course in applied statistics that includes some discussion of analysis of variance and regression. Such courses usually rely on intuitive arguments to justify the statistics presented and emphasize their proper use. We believe it is important for the beginning statistician to see the theoretical development of these statistics early in his or her academic career. Most of these derivations have their genesis in the field of linear statistical models.

A First Course in the Theory of Linear Statistical Models is a teaching text for the advanced statistics undergraduate or the beginning graduate student of statistics. We provide a readable introduction to the theory of linear statistical models that links the theory to the applications that are already familiar to the beginner. Thus it is assumed that the user of this text has had at least a one-year course in applied or mathematical statistics. This text should serve as a bridge to the more advanced texts on the subject.

We believe that for a text to be appropriate in this setting it should be theoretically sound but as streamlined and self-contained as possible. For this reason, we have included a thorough introductory chapter on matrix algebra. This chapter can and should be skipped by students with a strong and recent background in linear algebra.

We also believe that a result derived by the student is more easily remembered than one for which a proof is supplied by the author. For this reason, many of the theoretical exercises entail leading the student through the proof of a lemma, theorem, or corollary. This is done even for some very important results. These exercises are starred (*). They are especially important in the work to come and should not be skipped. A complete solutions manual is available for instructors or students using the text for independent study. When a proof is given in the text, it is given in detail. These proofs are usually ones that are intricate and that would be difficult for a beginner to generate alone. They can be skipped or assigned as outside reading if so desired.

Exercises for each section of each chapter are found at the end of the respective chapter and labeled by section. The text contains both theoretical and applied exercises and examples. SAS supplements are provided to aid in the discussion and solution of applied examples and exercises.

Some sections are labeled "advanced." These sections contain material that is interesting but not essential to the material that follows. They can be omitted if time does not permit their coverage.

The text is intended for a one-semester introductory course in the theory of linear statistical models. It is not an applied regression text, although the basic concepts of this area of statistics are presented in Chapters 3 and 4. The less than full rank model is treated in Chapters 5 and 6; it is illustrated using the one-way, two-way, and randomized blocks designs familiar to most students. Chapter 7 is a brief expository chapter intended to point the student toward additional topics that he or she will encounter in depth in more advanced courses.

We would like to thank the following reviewers: Don Edwards, University of South Carolina; Leon Jay Gleser, Purdue University; Timothy Green, University of Georgia; James E. Holstein, University of Missouri; Ralph Russo, University of Iowa; Harry Schey, Rochester Institute of Technology; and Mary Sue Younger, University of Tennessee.

Thanks go to Mrs. Jo Ann Fisher for typing this manuscript, to Mrs. Linda Seawell for typing the solutions manual, and to Ms. Rosemary Oakes for help in preparing solutions. We wish to express our thanks to Mr. Michael Payne for his encouragement during the writing of the text and to Ms. Helen Walden and Ms. Sally Stickney for the care and skill that they demonstrated during copyediting.

INTRODUCTION

The basic job of an experimenter is to describe what he or she sees, to try to explain what is observed, and to use this knowledge to help answer questions posed in the future. The explanation given often takes the form of a physical model. A *model* is a theoretical explanation of the physical phenomena under study. It is usually expressed verbally at first and then formalized into one or more equations. In this text, we shall be studying both the theory behind and the applications of linear statistical models. Although a more formal definition will be given later, a *linear statistical model* is a model that is expressed mathematically in the form

$$y = \beta_0 + \beta_1 x_1 + \beta_2 x_2 + \cdots \beta_k x_k + \varepsilon$$

where y is a random variable called the response; $x_1, x_2, \ldots, x_k$ are mathematical variables whose values are controlled or at least accurately observed by the experimenter; ε is a random variable that accounts for unexplained random variations in response; and $\beta_0, \beta_1, \ldots, \beta_k$ are constants whose exact values are not known and hence must be estimated from experimental data.

For example, a chemist might theorize that the rate at which a chemical dissolves in water (y) is dependent only on water temperature (x). A linear model expressing this idea formally takes the form

$$y = \beta_0 + \beta_1 x + \varepsilon$$

The first two terms on the right-hand side, $\beta_0 + \beta_1 x$, express the idea that temperature (x) influences the rate at which the chemical dissolves. However, we must admit that even with the temperature held constant, there will be some variation in the rate at which the chemical dissolves. This variation, which cannot be attributed to temperature differences, is due to random or uncontrolled influences. We account for this unavoidable random variation by adding the random variable ε to the model. Thus our final model is

$$\underset{\substack{\uparrow \\ \text{Response}}}{y} = \underset{\substack{\uparrow \\ \text{Explained} \\ \text{variation}}}{\underbrace{\beta_0 + \beta_1 x}} + \underset{\substack{\uparrow \\ \text{Random or unexplained} \\ \text{variation}}}{\varepsilon}$$

The above model assumes that the response, y, is dependent only on one variable x. In most problems of real interest, more than one variable is thought to influence the response significantly. For example, a psychologist might conjecture that the age at which a child begins to speak in sentences depends on IQ (x_1), the number of older siblings (x_2), and the sex of the child (x_3). Here, the linear model is of the form

$$\underset{\substack{\uparrow \\ \text{Response}}}{y} = \underset{\substack{\uparrow \\ \text{Explained} \\ \text{variation}}}{\underbrace{\beta_0 + \beta_1 x_1 + \beta_2 x_2 + \beta_3 x_3}} + \underset{\substack{\uparrow \\ \text{Random} \\ \text{variation}}}{\varepsilon}$$

Again, this is a statistical model. We are admitting that even if we observe two children of the same sex, with the same IQ and the same number of older siblings, they will not necessarily begin to speak at exactly the same age. Random influences such as physiological and environmental differences will come into play to produce variations in response. We add the random variable ε to the basic expression that relates y, x_1, x_2, and x_3 to account for this unavoidable uncertainty.

Data obtained in a typical experiment is a set of observations on the response, y, together with the values of the x's that produced the response. For example, in a study of the age at which a child begins to talk in sentences, perhaps six children are observed and these data are obtained:

Child	y (Age in years at which the child begins to talk)	x_1 (IQ)	x_2 (Number of older siblings)	x_3 (Sex where 1 = female and 0 = male)
1	1.8	130	4	1
2	2.0	110	2	0
3	2.5	100	0	0
4	1.7	120	1	0
5	3.0	90	1	1
6	2.3	100	0	0

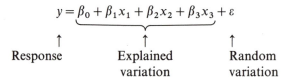

Our model, $y = \beta_0 + \beta_1 x_1 + \beta_2 x_2 + \beta_3 x_3 + \varepsilon$, generates this system of six linear equations:

$$
\begin{aligned}
1.8 &= \beta_0 + \beta_1(130) + \beta_2(4) + \beta_3(1) + \varepsilon_1 \\
2.0 &= \beta_0 + \beta_1(110) + \beta_2(2) + \beta_3(0) + \varepsilon_2 \\
2.5 &= \beta_0 + \beta_1(100) + \beta_2(0) + \beta_3(0) + \varepsilon_3 \\
1.7 &= \beta_0 + \beta_1(120) + \beta_2(1) + \beta_3(0) + \varepsilon_4 \\
3.0 &= \beta_0 + \beta_1(90) \;\; + \beta_2(1) + \beta_3(1) + \varepsilon_5 \\
2.3 &= \beta_0 + \beta_1(100) + \beta_2(0) + \beta_3(0) + \varepsilon_6
\end{aligned}
$$

Notice that even though the x values for child 3 and child 6 are identical with $x_1 = 100$, $x_2 = 0$, and $x_3 = 0$, their responses are different. Child 3 begins to talk at age 2.5 years, whereas child 6 begins to talk a little earlier, at age 2.3 years. This difference in response is accounted for in our model via the random variable ε, the term included in the model to account for random unexplained differences in response. As you shall see later, there are many questions that we must try to answer from the observed data. The most obvious is: What are the true values of the constants $\beta_0, \beta_1, \beta_2$, and β_3? The mathematics needed to answer this and other questions is matrix algebra. For this reason, we begin our study of linear models with a review of the concepts of matrix theory that will be useful later. This review is in no way intended to be a complete development of matrix algebra. For this reason, many results are stated without proof; some of the more easily obtainable results are left to the student as exercises. The proofs of the results given can be found in most linear algebra texts. See, for example, [**1**], [**3**], [**8**], [**11**], and [**15**].

CHAPTER 1

CONCEPTS OF MATRIX ALGEBRA

Techniques of matrix algebra are used to solve problems in many areas of applied mathematics and statistics. In this chapter we present topics that are pertinent to the study of linear statistical models. Although much of this material should be review, some of it might be new to you. You should at least skim this material. If you feel comfortable with the topics covered, you can skip to Chapter 2 to begin your study.

1.1 ELEMENTARY MATRIX OPERATIONS

In this section we define the term *matrix* and introduce the notation that will be used in the remainder of the text. We also review the basic operations of addition, subtraction, and scalar and matrix multiplication and consider their properties.

Definition 1.1.1

A *matrix* is a rectangular array of numbers of the form

$$\begin{bmatrix} x_{11} & x_{12} & \cdots & x_{1k} \\ x_{21} & x_{22} & \cdots & x_{2k} \\ x_{31} & x_{32} & \cdots & x_{3k} \\ \vdots & \vdots & & \vdots \\ x_{n1} & x_{n2} & \cdots & x_{nk} \end{bmatrix}$$

Notice that we are denoting the number in the ith row and the jth column by x_{ij}. That is, the first subscript denotes the row number and the second the column number. We will use uppercase letters to denote matrices in general.

The size of a matrix is specified by giving its *dimensions,* that is, by specifying the number of rows and the number of columns in the matrix. The general matrix referred to in Definition 1.1.1 is of dimension n by k ($n \times k$) since it has n rows and k columns. A matrix of dimension $1 \times k$ has only one row and k columns. Such a matrix is called a *row vector.* A matrix of dimension $n \times 1$ has n rows and one column and is called a *column vector.* If a matrix has the same number of rows as columns, it is called a *square* matrix.

Example 1.1.1 The following are matrices:

$$X = \begin{bmatrix} 2 & 7 & 8 \\ 1 & 1 & 0 \\ 0 & 4 & -5 \end{bmatrix}; \qquad Y = [2 \quad 3 \quad 7 \quad 8]; \qquad Z = \begin{bmatrix} 1 \\ 6 \\ 5 \end{bmatrix}$$

X is a 3×3 square matrix; Y is a 1×4 row vector, and Z is a 3×1 column vector. The entry in the third row and the second column of matrix X, 4, is denoted by x_{32}.

As with other mathematical entities, it is necessary to define the notion of equality. We say that two matrices are *equal* if and only if they agree with one another on an entry to entry basis. For example, in order for

$$X = \begin{bmatrix} 2 & 6 \\ 7 & x_{22} \end{bmatrix}$$

to be equal to

$$Y = \begin{bmatrix} 2 & 6 \\ 7 & -4 \end{bmatrix}$$

we must have $x_{22} = -4$.

There are four important matrix operations: addition, subtraction, scalar multiplication, and matrix multiplication. The first three are defined on a term by term basis. The last is a little more complicated.

Definition 1.1.2

Let X and Y be matrices each of dimension $n \times k$. Let x_{ij} and y_{ij} $i = 1, 2, \ldots, n$ and $j = 1, 2, \ldots, k$ denote the entries in these matrices. Then

1. $X + Y$ is the matrix whose (i, j)th entry is $x_{ij} + y_{ij}$.
2. $X - Y$ is the matrix whose (i, j)th entry is $x_{ij} - y_{ij}$.
3. cX, where c is any real number, is the matrix whose (i, j)th entry is cx_{ij}. Multiplication of a matrix by a real number is called *scalar multiplication.*

To perform the matrix operations of addition and subtraction, the dimensions of X and Y must coincide.

Example 1.1.2 Let

$$X = \begin{bmatrix} 2 & 3 & 1 \\ 8 & 0 & 4 \end{bmatrix}, \qquad Y = \begin{bmatrix} -1 & 0 & 6 \\ 7 & 2 & -3 \end{bmatrix},$$

and let the scalar c be 3. Then

$$X + Y = \begin{bmatrix} 2 & 3 & 1 \\ 8 & 0 & 4 \end{bmatrix} + \begin{bmatrix} -1 & 0 & 6 \\ 7 & 2 & -3 \end{bmatrix} = \begin{bmatrix} 1 & 3 & 7 \\ 15 & 2 & 1 \end{bmatrix}$$

$$X - Y = \begin{bmatrix} 2 & 3 & 1 \\ 8 & 0 & 4 \end{bmatrix} - \begin{bmatrix} -1 & 0 & 6 \\ 7 & 2 & -3 \end{bmatrix} = \begin{bmatrix} 3 & 3 & -5 \\ 1 & -2 & 7 \end{bmatrix}$$

$$cX = 3\begin{bmatrix} 2 & 3 & 1 \\ 8 & 0 & 4 \end{bmatrix} = \begin{bmatrix} 6 & 9 & 3 \\ 24 & 0 & 12 \end{bmatrix}$$

As indicated earlier, matrix multiplication is a little more complicated than addition, subtraction, and scalar multiplication. To be able to multiply two matrices, they must be *conformable.* This means that the number of columns of the first matrix must be the same as the number of rows of the second. If this is true, then matrix multiplication can be defined as follows.

Definition 1.1.3

Let X be a matrix of dimension $n \times k$ and let Y be a matrix of dimension $k \times m$. The *product* XY is defined to be the $n \times m$ matrix whose (i, j)th element is given by

$$\sum_{s=1}^{k} x_{is}y_{sj} = x_{i1}y_{1j} + x_{i2}y_{2j} + x_{i3}y_{3j} + \cdots + x_{ik}y_{kj}$$

This definition is harder to express mathematically than to state verbally. We are saying that to obtain the entry in the ith row and jth column of the product, we multiply term by term the ith row of X by the jth column of Y and add the results. An example should clarify things.

Example 1.1.3 Let

$$X = \begin{bmatrix} 2 & 7 \\ 1 & 2 \\ 3 & 9 \end{bmatrix} \quad \text{and} \quad Y = \begin{bmatrix} 6 & 1 & 6 & 8 & 1 \\ 4 & 7 & 3 & 6 & 0 \end{bmatrix}$$

Notice that X is of dimension 3×2 and Y is of dimension 2×5. Since the number of columns of X is the same as the number of rows of Y, X and Y are conformable. The product will be a matrix of dimension 3×5. To obtain the entry in the first row and the first column of the product, multiply the first row of X by the first column of Y term by term and add the results. Let us call the resulting number z_{11}. It can be seen that

$$z_{11} = 2(6) + 7(4) = 40$$

In a similar manner,

$$\begin{aligned} z_{21} &= 1(6) + 2(4) = 14 \\ z_{31} &= 3(6) + 9(4) = 54 \\ z_{12} &= 2(1) + 7(7) = 51 \end{aligned}$$

The entire product matrix is given by

$$Z = \begin{bmatrix} 40 & 51 & 33 & 58 & 2 \\ 14 & 15 & 12 & 20 & 1 \\ 54 & 66 & 45 & 78 & 3 \end{bmatrix}$$

One unusual situation should be mentioned. Consider a $1 \times n$ row vector and an $n \times 1$ column vector. These special matrices can be multiplied in either order. In one case an $n \times n$ matrix results; in the other, a 1×1 matrix is obtained. In the latter case, the matrix notation can be dropped. The resulting product will be a real number or a function of real variables. For example, let

$$X = [1 \quad 3 \quad 5] \quad \text{and} \quad Y = \begin{bmatrix} y_1 \\ y_2 \\ y_3 \end{bmatrix}$$

Then XY is the 1×1 matrix $[y_1 + 3y_2 + 5y_3]$. We can drop the matrix notation and write $y_1 + 3y_2 + 5y_3$. This expression, when evaluated for various values of y_1, y_2, and y_3, yields a single real number. It is treated as a real number in future matrix manipulations.

It is always interesting to investigate the properties of mathematical operations. We need to know which maneuvers are "legal" when dealing with matrices and which are not. For reference purposes, a list of properties that do hold is given below. You will be asked to verify some of these properties as exercises.

Properties of Matrix Operations

1. If X and Y are both $n \times k$ matrices, then $X + Y = Y + X$. (Matrix addition is commutative.)
2. If X, Y, and Z are all $n \times k$ matrices, then $X + (Y + Z) = (X + Y) + Z$. (Matrix addition is associative.)
3. If X, Y, and Z are conformable, then $X(YZ) = (XY)Z$. (Matrix multiplication is associative.)
4. If X is of dimension $n \times k$ and Y and Z are of dimension $k \times m$, then $X(Y + Z) = XY + XZ$. (Matrix multiplication is left-distributive.)
5. If X is of dimension $n \times k$ and Y and Z are of dimension $m \times n$, then $(Y + Z)X = YX + ZX$. (Matrix multiplication is right-distributive.)
6. If c is a real number, X is of dimension $n \times k$, and Y is of dimension $k \times m$, then $X(cY) = c(XY)$. (Real numbers can be factored from matrices.)
7. If a and b are real numbers and X is an $n \times k$ matrix, then $(a + b)X = aX + bX$. (Matrices distribute over real numbers.)
8. If X and Y are $n \times k$ matrices and c is a real number, then $c(X + Y) = cX + cY$. (Real numbers distribute over matrices.)
9. Let X be an $n \times k$ matrix. There exists a unique matrix 0 called the *zero matrix* such that $X + 0 = 0 + X = X$.
10. Let X be an $n \times k$ matrix. There exists a unique matrix Y, called the *negative* of X, such that $X + Y = 0$.

The distributive properties given in 4 and 5 are stated carefully and must be used carefully. Although these properties closely resemble the corresponding distributive properties for real numbers, there is one important difference. Namely, when multiplying real numbers x and y it is true that $xy = yx$. The product can be written in either order. This is not true in general when *matrices* X and Y are multiplied.

That is, usually $XY \neq YX$. In fact, matrices that are conformable to matrix multiplication in the order XY might not even be conformable if the order is reversed. Even if they are conformable, there is no guarantee that the two products will be the same.

PARTITIONED MATRICES

Occasionally a matrix will need to be partitioned or subdivided into what are called *submatrices* so that multiplication or addition can be performed and displayed in a theoretically convenient format. To illustrate the notation, let

$$X = \begin{bmatrix} 1 & 0 & 1 & 0 \\ 0 & 1 & 3 & -1 \\ 0 & 1 & -1 & 1 \\ 2 & -1 & 0 & 2 \end{bmatrix}$$

and assume that we want to subdivide X as follows:

$$X = \left[\begin{array}{cc|c|c} 1 & 0 & 1 & 0 \\ 0 & 1 & 3 & -1 \\ \hline 0 & 1 & -1 & 1 \\ 2 & -1 & 0 & 2 \end{array}\right]$$

Since each submatrix is a matrix in its own right, it is convenient to denote these matrices using the same labeling convention as that used earlier. That is, X_{ij} denotes the matrix in row i and column j. Here we can write

$$X = \left[\begin{array}{c|c|c} X_{11} & X_{12} & X_{13} \\ \hline X_{21} & X_{22} & X_{23} \end{array}\right]$$

where $\quad X_{11} = \begin{bmatrix} 1 & 0 \\ 0 & 1 \end{bmatrix}, \quad X_{12} = \begin{bmatrix} 1 \\ 3 \end{bmatrix}, \quad \text{and} \quad X_{13} = \begin{bmatrix} 0 \\ -1 \end{bmatrix},$

and $\quad X_{21} = \begin{bmatrix} 0 & 1 \\ 2 & -1 \end{bmatrix}, \quad X_{22} = \begin{bmatrix} -1 \\ 0 \end{bmatrix}, \quad \text{and} \quad X_{23} = \begin{bmatrix} 1 \\ 2 \end{bmatrix}$

Although formal rules for adding, subtracting, and multiplying matrices in partitioned form can be developed, they are not really necessary. These operations simply parallel the definitions given earlier in the chapter. However, before an operation is performed you must take into account the dimensions of the submatrices involved. To add or subtract partitioned matrices, their corresponding submatrices must be of the same dimensions; to multiply, appropriate submatrices must be conformable. Some examples will clarify this point.

Example 1.1.4 Let

$$X = \begin{bmatrix} 1 & 0 & 1 & 0 \\ 0 & 1 & 3 & -1 \\ 0 & 1 & -1 & 1 \end{bmatrix} \quad \text{and} \quad Y = \begin{bmatrix} 2 & 1 & 1 & 1 \\ 1 & 0 & 0 & 1 \\ 0 & 0 & 2 & 3 \end{bmatrix}$$

If we partition X as

$$X = \left[\begin{array}{ccc|c} 1 & 0 & 1 & 0 \\ 0 & 1 & 3 & -1 \\ \hline 0 & 1 & -1 & 1 \end{array}\right] = \left[\begin{array}{c|c} X_{11} & X_{12} \\ \hline X_{21} & X_{22} \end{array}\right]$$

then to add X and Y in partitioned form, we must subdivide Y as

$$Y = \left[\begin{array}{ccc|c} 2 & 1 & 1 & 1 \\ 1 & 0 & 0 & 1 \\ \hline 0 & 0 & 2 & 3 \end{array}\right] = \left[\begin{array}{c|c} Y_{11} & Y_{12} \\ \hline Y_{21} & Y_{22} \end{array}\right]$$

Notice that the partitioning of Y is done in such a way that the dimensions of its submatrices match the corresponding dimensions of the submatrices of X. Hence, we can add X and Y and express the result in partitioned form as

$$X + Y = \left[\begin{array}{c|c} X_{11} + Y_{11} & X_{12} + Y_{12} \\ \hline X_{21} + Y_{21} & X_{22} + Y_{22} \end{array}\right] = \begin{bmatrix} 3 & 1 & 2 & 1 \\ 1 & 1 & 3 & 0 \\ 0 & 1 & 1 & 4 \end{bmatrix}$$

Example 1.1.5 Multiplication is a little trickier. The reason for this is that the partitioning of Y is not unique. We must only be sure that appropriate submatrices are conformable. To illustrate this, let

$$X = \left[\begin{array}{cc|c} 2 & 1 & 0 \\ \hline 3 & 4 & 1 \end{array}\right] = \left[\begin{array}{c|c} X_{11} & X_{12} \\ \hline X_{21} & X_{22} \end{array}\right] \quad \text{and} \quad Y = \left[\begin{array}{cc} 1 & 0 \\ 2 & 4 \\ \hline 3 & -1 \end{array}\right] = \left[\begin{array}{c} Y_{11} \\ \hline Y_{21} \end{array}\right]$$

Since X is 2×3 and Y is 3×2 we know that XY exists. Using the usual definition of matrix multiplication, its value is

$$XY = \begin{bmatrix} 4 & 4 \\ 14 & 15 \end{bmatrix}$$

In partitioned form,

$$XY = \left[\begin{array}{c|c} X_{11} & X_{12} \\ \hline X_{21} & X_{22} \end{array}\right] \cdot \left[\begin{array}{c} Y_{11} \\ \hline Y_{21} \end{array}\right] = \left[\begin{array}{c} X_{11}Y_{11} + X_{12}Y_{21} \\ \hline X_{21}Y_{11} + X_{22}Y_{21} \end{array}\right]$$

Now

$$\begin{aligned} X_{11}Y_{11} + X_{12}Y_{21} &= [2 \quad 1]\begin{bmatrix} 1 & 0 \\ 2 & 4 \end{bmatrix} + [0][3 \quad -1] \\ &= [4 \quad 4] + [0 \quad 0] = [4 \quad 4] \\ X_{21}Y_{11} + X_{22}Y_{21} &= [3 \quad 4]\begin{bmatrix} 1 & 0 \\ 2 & 4 \end{bmatrix} + [1][3 \quad -1] \\ &= [11 \quad 16] + [3 \quad -1] = [14 \quad 15] \end{aligned}$$

Hence

$$XY = \begin{bmatrix} 4 & 4 \\ 14 & 15 \end{bmatrix}$$

as before.

Consider a second partitioning of Y:

$$Y = \left[\begin{array}{c|c} 1 & 0 \\ 2 & 4 \\ \hline 3 & -1 \end{array}\right] = \left[\begin{array}{c|c} Y_{11} & Y_{12} \\ \hline Y_{21} & Y_{22} \end{array}\right]$$

In this form,

$$XY = \left[\begin{array}{c|c} X_{11} & X_{12} \\ \hline X_{21} & X_{22} \end{array}\right] \left[\begin{array}{c|c} Y_{11} & Y_{12} \\ \hline Y_{21} & Y_{22} \end{array}\right] = \left[\begin{array}{c|c} X_{11}Y_{11} + X_{12}Y_{21} & X_{11}Y_{12} + X_{12}Y_{22} \\ \hline X_{21}Y_{11} + X_{22}Y_{21} & X_{21}Y_{12} + X_{22}Y_{22} \end{array}\right]$$

Considering each of these submatrices individually,

$$\begin{aligned} X_{11}Y_{11} + X_{12}Y_{21} &= [2 \quad 1]\begin{bmatrix} 1 \\ 2 \end{bmatrix} + [0][3] = [4] \\ X_{21}Y_{11} + X_{22}Y_{21} &= [3 \quad 4]\begin{bmatrix} 1 \\ 2 \end{bmatrix} + [1][3] = [14] \\ X_{11}Y_{12} + X_{12}Y_{22} &= [2 \quad 1]\begin{bmatrix} 0 \\ 4 \end{bmatrix} + [0][-1] = [4] \\ X_{21}Y_{12} + X_{22}Y_{22} &= [3 \quad 4]\begin{bmatrix} 0 \\ 4 \end{bmatrix} + [1][-1] = [15] \end{aligned}$$

Thus

$$XY = \left[\begin{array}{c|c} 4 & 4 \\ \hline 14 & 15 \end{array}\right]$$

as before.

As you can see, there is no particular advantage to performing simple matrix multiplication or addition via partitioned matrices. The purpose of these examples is to show you the idea of partitioning. We will use partitioned matrices in a theoretical setting later in this book. These examples should help you visualize what is being done.

1.2

TRANSPOSE AND VECTOR NOTATION

In this section we consider the *transpose* of a matrix and investigate its properties. The notion of a *symmetric* matrix arises naturally in the course of this discussion. We also reconsider those matrices that consist of either exactly one row or exactly one column. Remember that such matrices are called row and column vectors, respectively. We also introduce some special vector notation that will appear throughout the remainder of the text.

Definition 1.2.1

Let X be an $n \times k$ matrix. The *transpose* of X, denoted by X', is the $k \times n$ matrix obtained by interchanging the rows and columns of X.

Example 1.2.1 Let

$$X = \begin{bmatrix} 2 & 0 \\ 1 & 1 \\ 7 & 2 \end{bmatrix}$$

Since X is of dimension 3×2, X' is of dimension 2×3. It is formed by writing the columns of X as the rows of X'. Thus

$$X' = \begin{bmatrix} 2 & 1 & 7 \\ 0 & 1 & 2 \end{bmatrix}$$

It can be seen from this example that usually $X \neq X'$. In fact, if X is not square, then X cannot equal its transpose because the dimensions of the two matrices will differ. However, if X is square then it is possible that $X = X'$. In this case we say that X is *symmetric*. For example, the matrix

$$X = \begin{bmatrix} 1 & 3 & 2 \\ 3 & 4 & -1 \\ 2 & -1 & 6 \end{bmatrix}$$

is symmetric. If we interchange its rows and columns the resulting matrix is equal to X.

These properties of transposes will be useful in future work.

Properties of the Transpose

1. Let X be an $n \times k$ matrix and c a real number. Then $(cX)' = cX'$.
2. Let X and Y be $n \times k$ matrices. Then $(X \pm Y)' = X' \pm Y'$.
3. Let X be an $n \times k$ matrix. Then $(X')' = X$.
4. Let X be an $n \times k$ matrix and Y a $k \times m$ matrix. Then $(XY)' = Y'X'$.

Properties 1–3 are rather obvious. Property 4 is a little harder to see. Verbally it expresses the idea that the transpose of a product is the product of the transposes in reverse order. You are asked to verify this result numerically in Exercises 11 and 12.

In the work to come, matrices of the form $X'X$ play an important role. The following theorem, which is easy to prove, gives you a look at the nature of such a matrix.

Theorem 1.2.1 Let

$$X = \begin{bmatrix} x_{11} & x_{12} & \cdots & x_{1k} \\ x_{21} & x_{22} & \cdots & x_{2k} \\ x_{31} & x_{32} & \cdots & x_{3k} \\ \vdots & \vdots & & \vdots \\ x_{n1} & x_{n2} & \cdots & x_{nk} \end{bmatrix}$$

Then $X'X$ is a $k \times k$ symmetric matrix of sums of squares and sums of cross products. It assumes the form

$$X'X = \begin{bmatrix} \sum_{i=1}^{n} x_{i1}^2 & \sum_{i=1}^{n} x_{i1}x_{i2} & \cdots & \sum_{i=1}^{n} x_{i1}x_{ik} \\ \sum_{i=1}^{n} x_{i2}x_{i1} & \sum_{i=1}^{n} x_{i2}^2 & & \sum_{i=1}^{n} x_{i2}x_{ik} \\ \vdots & \vdots & \ddots & \\ \sum_{i=1}^{n} x_{ik}x_{i1} & \sum_{i=1}^{n} x_{ik}x_{i2} & & \sum_{i=1}^{n} x_{ik}^2 \end{bmatrix}$$

■

Notice that the entries appearing on what is called the *main diagonal* of $X'X$, the diagonal that starts at the upper-left corner of the matrix and ends in the lower-right corner, are sums of squares of the terms in the columns of X. The off-diagonal entries are sums of cross products. They are found by multiplying the columns of X on a term by term basis and adding the results. An example will demonstrate how easy it is to form $X'X$ by inspection.

Example 1.2.2 Let

$$X = \begin{bmatrix} 1 & -1 & 0 \\ 0 & 2 & 4 \end{bmatrix}$$

By Theorem 1.2.1, $X'X$ is a 3×3 symmetric matrix of sums of squares and sums of cross products. The main diagonal contains the sums of squares of the terms of the columns of X. Thus the first entry on the main diagonal is $1^2 + 0^2 = 1$, the second is $(-1)^2 + 2^2 = 5$, and the third is $0^2 + 4^2 = 16$. At this point $X'X$ takes the form

$$X'X = \begin{bmatrix} 1 & & ? \\ & 5 & \\ ? & & 16 \end{bmatrix}$$

The off-diagonal elements are sums of cross products. For example, the entry in column 1 row 2 is found by multiplying columns 1 and 2 of X on a term by term basis and adding the results. This entry is $1(-1) + 0(2) = -1$; the entry in column 1 row 3 is found by multiplying columns 1 and 3 of X on a term by term basis and adding the results. This entry is $1(0) + 0(4) = 0$. Continuing in a similar manner and using symmetry, the resulting matrix is

$$X'X = \begin{bmatrix} 1 & -1 & 0 \\ -1 & 5 & 8 \\ 0 & 8 & 16 \end{bmatrix}$$

This can be verified by computing $X'X$ directly.

Even though a column vector is a matrix, it is convenient to emphasize its special nature by using a different notation. Instead of using an uppercase letter as we have done up to this point, we will begin using a lowercase boldface letter. Thus, a column vector

$$\begin{bmatrix} x_1 \\ x_2 \\ \vdots \\ x_n \end{bmatrix}$$

where $x_1, x_2, \ldots, x_n$ are real numbers is denoted by $\mathbf{x}$. A row vector can be viewed as being the transpose of a column vector. Thus we denote row vectors in general by $\mathbf{x}'$.

In the statistical applications to come, you must be adept at manipulating matrices and vectors in general form. These results, whose proofs are left as exercises, come into play often.

Theorem 1.2.2 Let

$$\mathbf{x} = \begin{bmatrix} x_1 \\ x_2 \\ \vdots \\ x_n \end{bmatrix}$$

Then

$$\mathbf{x}'\mathbf{x} = \sum_{i=1}^{n} x_i^2$$

■

Because $\mathbf{x}'\mathbf{x}$ is a 1×1 matrix, matrix notation is not needed. Notice that the resulting function is a sum of squares; it is the sum of the squares of the entries in $\mathbf{x}$. When $x_1, x_2, \ldots, x_n$ assume numerical values, $\mathbf{x}'\mathbf{x}$ is a real number.

Theorem 1.2.3

Let

$$\mathbf{x} = \begin{bmatrix} x_1 \\ x_2 \\ \vdots \\ x_n \end{bmatrix}$$

Then $\mathbf{x}\mathbf{x}'$ is an $n \times n$ symmetric matrix of squares and cross products. ■

The exercises for Section 1.2 will provide practice in matrix and vector manipulation.

1.3

INVERSES OF MATRICES AND ORTHOGONALITY

Those of you who are more theoretically oriented probably recognized the matrix that we called the "zero" matrix in Section 1.1 as being the identity for addition; similarly, the "negative" matrix plays the role of the additive inverse. Although these matrices are interesting, they do not play a major role in statistical theory.

However, their multiplicative counterparts are vital in the work to come. Thus, we begin by considering the idea of an identity for multiplication.

Definition 1.3.1

Let I_p denote the $p \times p$ diagonal matrix

$$\begin{bmatrix} 1 & 0 & 0 & \cdots & 0 \\ 0 & 1 & 0 & \cdots & 0 \\ \vdots & & & \ddots & \\ 0 & 0 & 0 & & 1 \end{bmatrix}$$

1. The $k \times k$ matrix I_k is called the *right identity* for the set of all $n \times k$ matrices.
2. The $n \times n$ matrix I_n is called the *left identity* for the set of all $n \times k$ matrices.
3. If $n = k$ then $I_n = I_k = I$ is called the *identity* for the set of all $n \times n$ matrices.

Note that in case 1, $XI_k = X$; in case 2, $I_n X = X$; and in case 3, $IX = XI = X$. Thus the matrices I_k, I_n, and I are multiplicative identities in that they preserve the identity of the original matrix X under the operation of matrix multiplication.

Once the idea of an identity for multiplication is available, we can define a multiplicative inverse. We shall restrict ourselves to square matrices.

Definition 1.3.2

Let X be a $k \times k$ matrix. The *inverse* of X, denoted by X^{-1}, is the $k \times k$ matrix such that

$$XX^{-1} = X^{-1}X = I$$

If such a matrix exists, then X is said to be *invertible* or *nonsingular*; otherwise, X is said to be *noninvertible* or *singular*.

Example 1.3.1 Let

$$X = \begin{bmatrix} 2 & 0 & 0 \\ 0 & 3 & 0 \\ 0 & 0 & 5 \end{bmatrix}$$

Since X is diagonal, it is not hard to guess its inverse. All we have to do is find a matrix that when multiplied by X will yield a diagonal matrix of ones. Such a matrix is

$$X^{-1} = \begin{bmatrix} \frac{1}{2} & 0 & 0 \\ 0 & \frac{1}{3} & 0 \\ 0 & 0 & \frac{1}{5} \end{bmatrix}$$

You can check for yourself to see that $XX^{-1} = X^{-1}X = I$.

Example 1.3.2 Let

$$X = \begin{bmatrix} 2 & 0 \\ 4 & -1 \end{bmatrix}$$

By direct multiplication, you can verify that the inverse for X is

$$X^{-1} = \begin{bmatrix} \frac{1}{2} & 0 \\ 2 & -1 \end{bmatrix}$$

This, in turn, implies that X is nonsingular.

As you can see, inverting a diagonal matrix is easy. However, you are probably wondering how we found the inverse for the matrix in Example 1.3.2. There is no obvious pattern to the numbers given. To find X^{-1} we used one of the many computing algorithms available for inverting a nonsingular matrix. One such algorithm is given in Exercise 28. Fortunately, in the work to come we will not need to invert matrices by hand so there is no need to present these algorithms here. If you are curious about them, they can be found in any text on matrix or linear algebra. In our statistical applications we will allow the computer to generate any inverses that are needed in our computations. However, it is necessary to understand the rules governing inverses. These rules are summarized below.

Properties of Inverse

1. If X is nonsingular, then X^{-1} is nonsingular and $(X^{-1})^{-1} = X$.
2. If X and Y are both $k \times k$ nonsingular matrices, then XY is nonsingular and $(XY)^{-1} = Y^{-1}X^{-1}$.
3. If X is nonsingular, then X' is nonsingular and $(X')^{-1} = (X^{-1})'$.

Note that property 2 parallels property 4 of transposes. Note also that property 3 allows us to interchange the inverse and transpose operations whenever necessary.

ORTHOGONAL MATRICES

In statistical work, we often deal with a special type of nonsingular matrix called an *orthogonal* matrix. Such a matrix is defined as follows.

Definition 1.3.3

Let X be a $k \times k$ matrix such that $X'X = I$. Then X is said to be orthogonal.

This definition has several implications. First, every orthogonal matrix is square. Second, since $X'X = I$, we are saying that $X^{-1} = X'$. That is, to find the inverse of an orthogonal matrix, we need only form its transpose.

Example 1.3.3 Let

$$X = \begin{bmatrix} \frac{2}{3} & -\frac{2}{3} & \frac{1}{3} \\ \frac{2}{3} & \frac{1}{3} & -\frac{2}{3} \\ \frac{1}{3} & \frac{2}{3} & \frac{2}{3} \end{bmatrix}$$

Using the computational shortcut demonstrated in Section 1.2, it is easy to see that $X'X = I$. Hence X is an orthogonal matrix.

An important theorem concerning orthogonal matrices is useful in the statistical setting. This theorem allows us to recognize an orthogonal matrix X without actually having to compute $X'X$ as was done in Example 1.3.3. To understand the theorem, some new terminology must be introduced.

Definition 1.3.4

Let $\mathbf{x}$ and $\mathbf{y}$ be $n \times 1$ vectors. If

$$\mathbf{x}'\mathbf{y} = \sum_{i=1}^{n} x_i y_i = 0$$

then $\mathbf{x}$ and $\mathbf{y}$ are said to be *orthogonal.*

As an example, consider $\mathbf{x}_1$ and $\mathbf{x}_2$, the first two columns of the matrix X of Example 1.3.3. Since $\mathbf{x}_1'\mathbf{x}_2 = 0$, these vectors are orthogonal.

In two-dimensional space, orthogonal vectors can be interpreted geometrically.

Consider the two vectors

$$\mathbf{x} = \begin{bmatrix} x_1 \\ x_2 \end{bmatrix} \quad \text{and} \quad \mathbf{y} = \begin{bmatrix} y_1 \\ y_2 \end{bmatrix}$$

These vectors can be pictured as line segments drawn from the origin to the points (x_1, x_2) and (y_1, y_2), respectively, as shown in Figure 1.1. It is known that these line segments are perpendicular if and only if the slope of one is the negative reciprocal of the slope of the other. In this case, we must have

$$\frac{x_2 - 0}{x_1 - 0} = -\frac{y_1 - 0}{y_2 - 0}$$

or

$$x_1 y_1 + x_2 y_2 = 0$$

This is exactly the condition needed for **x** and **y** to be orthogonal. In other words, two vectors in two-dimensional space are orthogonal if and only if they are geometrically perpendicular. Although orthogonality is harder to picture in higher dimensions, the terminology used in the two-dimensional case persists. Thus, the terms *orthogonal* and *perpendicular* are used interchangeably.

FIGURE 1.1 **Orthogonal vectors in two-dimensional space are equivalent to line segments that are perpendicular**

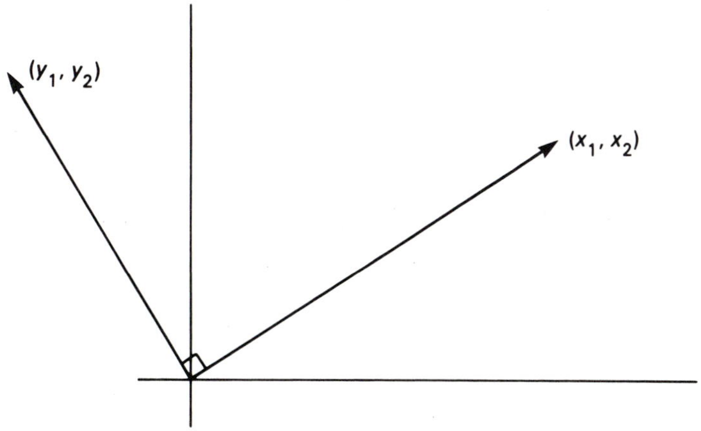

Consider a 2 × 1 vector

$$\mathbf{x} = \begin{bmatrix} x_1 \\ x_2 \end{bmatrix}$$

We now ask, What do we mean by the "length" of this vector? To answer this question, again imagine drawing a line segment from the origin (0, 0) to the point

(x_1, x_2) in two-dimensional space. It is natural to define the length of $\mathbf{x}$ to be the length of this line segment. Using the usual distance formula from plane geometry, this length is given by $\sqrt{x_1^2 + x_2^2}$.

The next definition extends this idea of length to an $n \times 1$ vector $\mathbf{x}$.

Definition 1.3.5

Let $\mathbf{x}$ be an $n \times 1$ vector. The *length* of $\mathbf{x}$, denoted by $|\mathbf{x}|$, is given by

$$|\mathbf{x}| = \sqrt{\mathbf{x}'\mathbf{x}} = \sqrt{x_1^2 + x_2^2 + \cdots + x_n^2}$$

Notice that each column of the matrix X of Example 1.3.3 is a 3×1 vector of length 1.

The last definition we need gives a special name to sets of orthogonal vectors of unit length.

Definition 1.3.6

Let $\{\mathbf{x}_1, \mathbf{x}_2, \ldots, \mathbf{x}_k\}$ be a set of $n \times 1$ orthogonal vectors. If each is of unit length, then these vectors form what is called an *orthonormal* set.

The next theorem gives necessary and sufficient conditions for a matrix X to be orthogonal. Its proof is outlined in Exercise 38.

Theorem 1.3.1 Let X be a $k \times k$ matrix. X is orthogonal if and only if its columns form an orthonormal set. ■

It is easy to see that the columns of the matrix of Example 1.3.3 form an orthonormal set. As we verified earlier, X is orthogonal.

1.4

EIGENVALUES AND RANK

In this section we introduce some concepts that play a major role in both the theory and applications of linear statistical models. We begin by defining an eigenvalue, or a characteristic root of a matrix.

Definition 1.4.1

Let A be a $k \times k$ matrix and $\mathbf{x}$ a $k \times 1$ nonzero vector. An *eigenvalue*, or *characteristic root*, of A is a number λ such that

$$A\mathbf{x} = \lambda\mathbf{x}$$

A vector $\mathbf{x}$ satisfying this equation is called an *eigenvector* associated with λ.

This definition is theoretical, but it is helpful to rephrase it in such a way that the technique used to find eigenvalues is evident. Notice that if

$$A\mathbf{x} = \lambda\mathbf{x}$$

then we can conclude that

$$A\mathbf{x} - \lambda\mathbf{x} = \mathbf{0}$$

or that

$$(A - \lambda I)\mathbf{x} = \mathbf{0}$$

In this form, we are dealing with a system of linear equations each of which has a right-hand-side zero. It can be shown that such a system has a solution other than the zero vector if and only if the determinant of the matrix $A - \lambda I$ is 0 [**11**] [**1**]. Thus, to obtain a nonzero solution we must find the values of λ for which the determinant of $A - \lambda I$ is 0. That is, the eigenvalues are found by solving the equation

$$|A - \lambda I| = 0$$

where $|A - \lambda I|$ denotes the determinant of the matrix $A - \lambda I$. A full discussion of the definition and properties of determinants can be found in any linear algebra text. Here they are used primarily in a theoretical setting. It will not be necessary to compute them by hand in any but the simplest cases. In Example 1.4.1 a method for finding the determinant of a 2×2 matrix is demonstrated. Some useful properties of determinants are pointed out in Exercise 45.

Example 1.4.1 Let

$$A = \begin{bmatrix} 1 & 1 \\ -2 & 4 \end{bmatrix}$$

To find the eigenvalues of A, solve the equation

$$|A - \lambda I| = 0$$

Here

$$A - \lambda I = \begin{bmatrix} 1 & 1 \\ -2 & 4 \end{bmatrix} - \begin{bmatrix} \lambda & 0 \\ 0 & \lambda \end{bmatrix} = \begin{bmatrix} 1-\lambda & 1 \\ -2 & 4-\lambda \end{bmatrix}$$

The determinant of a 2×2 matrix is easy to find. Simply multiply the elements along the main (upper-left to lower-right) diagonal and then subtract the product of the elements along the secondary (lower-left to upper-right) diagonal. In this case,

$$\begin{aligned} |A - \lambda I| &= (1-\lambda)(4-\lambda) - (-2) \\ &= \lambda^2 - 5\lambda + 6 \end{aligned}$$

Solving for λ, we obtain two distinct eigenvalues, namely $\lambda_1 = 2$ and $\lambda_2 = 3$. To find the eigenvectors associated with these eigenvalues, substitute back into the original definition. Thus to find the eigenvectors associated with the eigenvalue $\lambda_1 = 2$, solve

$$A\mathbf{x} = 2\mathbf{x}$$

Here we have

$$\begin{bmatrix} 1 & 1 \\ -2 & 4 \end{bmatrix} \begin{bmatrix} x_1 \\ x_2 \end{bmatrix} = 2 \begin{bmatrix} x_1 \\ x_2 \end{bmatrix}$$

or

$$\begin{aligned} x_1 + x_2 &= 2x_1 \\ -2x_1 + 4x_2 &= 2x_2 \end{aligned}$$

Solving, we see that we must have $x_1 = x_2$. Thus any vector whose entries are identical will be an eigenvector associated with the eigenvalue $\lambda_1 = 2$. One such eigenvector is

$$\mathbf{x} = \begin{bmatrix} 1 \\ 1 \end{bmatrix}$$

Check for yourself to see that for this vector, $A\mathbf{x} = 2\mathbf{x}$.

As in the case with inverses, we will not often be called upon to compute eigenvalues and eigenvectors by hand, so you need not become an expert on determinants. A computer will be used to do such calculations whenever necessary. However, you do need to know what is meant by the terms *eigenvalue* and *eigenvector* because they are used to characterize the matrices that arise in the theoretical study of linear models and are also used to help assess the worth of a model in a practical sense.

The solutions to $|A - \lambda I| = 0$ are not always real numbers; they might be imaginary. Even if they are all real, they might not be distinct as they were in our example. Property 1 below gives a condition that guarantees the existence of real eigenvalues.

Properties of Eigenvalues

1. If A is a $k \times k$ *symmetric* matrix, then the eigenvalues of A are all real.
2. If A is a $k \times k$ matrix and C is a $k \times k$ orthogonal matrix, then the eigenvalues of $C'AC$ are the same as the eigenvalues of A. (See Exercise 45.)
3. If A is a symmetric $k \times k$ matrix, then the eigenvectors associated with distinct eigenvalues of A are orthogonal. (See Exercise 46.)

The next result is of major importance in the work to come. Although we do not attempt a proof, we label it as a theorem to emphasize its importance.

Theorem 1.4.1 Let A be a symmetric $k \times k$ matrix. Then an orthogonal matrix P exists such that

$$P'AP = \begin{bmatrix} \lambda_1 & 0 & 0 & 0 \\ 0 & \lambda_2 & 0 & 0 \\ 0 & 0 & \lambda_3 & 0 \\ \vdots & \vdots & \vdots & \vdots \\ 0 & 0 & 0 & \lambda_k \end{bmatrix}$$

where λ_i $i = 1, 2, 3, \ldots, k$ are eigenvalues of A. ■

There are several things to notice here. Since $P'AP$ is a diagonal matrix, we say that P has "diagonalized" A. The diagonalization is special in that the resulting matrix displays the eigenvalues of A along the main diagonal. As you have probably guessed, not just any orthogonal matrix P will accomplish this diagonalization. The form of P is very special. It can be shown that the columns of P must form an orthonormal set and that they must also be eigenvectors of A. That is, the first column of P is an eigenvector of length 1 associated with λ_1, the second is an eigenvector of length 1 associated with λ_2, and so forth. Exercises 47 and 48 illustrate this idea.

RANK OF A MATRIX

The next major topic is rank. However, before you can really understand this concept, we must take a brief look at the notions of linear dependence and independence. We begin with the basic definition.

Definition 1.4.2

Let $\{\mathbf{x}_1, \mathbf{x}_2, \ldots, \mathbf{x}_k\}$ be a set of k column vectors. If real numbers $a_1, a_2, \ldots, a_k$ not all zero such that

$$a_1\mathbf{x}_1 + a_2\mathbf{x}_2 + \cdots + a_k\mathbf{x}_k = \mathbf{0}$$

exist, then the vectors $\mathbf{x}_1, \mathbf{x}_2, \ldots, \mathbf{x}_k$ are said to be *linearly dependent*. Otherwise, they are said to be *linearly independent*.

This definition basically states that if one or more vectors can be written as a linear combination of the others in the set, then they are not linearly independent. An example will help.

Example 1.4.2 Consider the vectors

$$\mathbf{x}_1 = \begin{bmatrix} 1 \\ 0 \\ 0 \end{bmatrix}, \quad \mathbf{x}_2 = \begin{bmatrix} 0 \\ 1 \\ 0 \end{bmatrix}, \quad \text{and} \quad \mathbf{x}_3 = \begin{bmatrix} 0 \\ 0 \\ 1 \end{bmatrix}$$

You can probably see that none of these vectors can be expressed in terms of the other two. That is, these vectors are linearly independent. To check, try to find real numbers a_1, a_2, and a_3 such that

$$a_1\mathbf{x}_1 + a_2\mathbf{x}_2 + a_3\mathbf{x}_3 = \mathbf{0}$$

Here we want

$$a_1 \begin{bmatrix} 1 \\ 0 \\ 0 \end{bmatrix} + a_2 \begin{bmatrix} 0 \\ 1 \\ 0 \end{bmatrix} + a_3 \begin{bmatrix} 0 \\ 0 \\ 1 \end{bmatrix} = \begin{bmatrix} 0 \\ 0 \\ 0 \end{bmatrix}$$

or

$$\begin{bmatrix} a_1 \\ a_2 \\ a_3 \end{bmatrix} = \begin{bmatrix} 0 \\ 0 \\ 0 \end{bmatrix}$$

This can only occur if $a_1 = a_2 = a_3 = 0$, implying that $\mathbf{x}_1$, $\mathbf{x}_2$, and $\mathbf{x}_3$ are linearly independent as suspected.

Example 1.4.3 Let

$$\mathbf{x}_1 = \begin{bmatrix} 2 \\ 3 \\ 1 \end{bmatrix}, \quad \mathbf{x}_2 = \begin{bmatrix} 1 \\ 0 \\ 2 \end{bmatrix}, \quad \text{and} \quad \mathbf{x}_3 = \begin{bmatrix} 3 \\ 3 \\ 3 \end{bmatrix}$$

It is easy to see that $\mathbf{x}_3 = \mathbf{x}_1 + \mathbf{x}_2$ or that $\mathbf{x}_1 + \mathbf{x}_2 - \mathbf{x}_3 = \mathbf{0}$. Hence there are constants not all zero, namely $a_1 = 1$, $a_2 = 1$, and $a_3 = -1$, such that $a_1\mathbf{x}_1 + a_2\mathbf{x}_2 + a_3\mathbf{x}_3 = \mathbf{0}$. This means that $\mathbf{x}_1$, $\mathbf{x}_2$, and $\mathbf{x}_3$ are linearly dependent. Are $\mathbf{x}_1$ and $\mathbf{x}_2$ linearly independent? To check we note that

$$a_1\mathbf{x}_1 + a_2\mathbf{x}_2 = a_1 \begin{bmatrix} 2 \\ 3 \\ 1 \end{bmatrix} + a_2 \begin{bmatrix} 1 \\ 0 \\ 2 \end{bmatrix} = \begin{bmatrix} 2a_1 + a_2 \\ 3a_1 \\ a_1 + 2a_2 \end{bmatrix}$$

Equating this to the zero vector, we see that

$$\begin{bmatrix} 2a_1 + a_2 \\ 3a_1 \\ a_1 + 2a_2 \end{bmatrix} = \begin{bmatrix} 0 \\ 0 \\ 0 \end{bmatrix}$$

Solving for a_1 and a_2 we must have

$$\begin{aligned} 2a_2 + a_2 &= 0 \\ 3a_1 &= 0 \\ a_1 + 2a_2 &= 0 \end{aligned}$$

This can only occur if $a_1 = a_2 = 0$. Hence even though the set $\{\mathbf{x}_1, \mathbf{x}_2, \mathbf{x}_3\}$ is not a set of linearly independent vectors, there is a subset, namely $\{\mathbf{x}_1, \mathbf{x}_2\}$, whose elements are linearly independent.

With the idea of linear independence in mind, we can now define the rank of a matrix. To do so, first recall that we have been writing a matrix X in the following general form:

$$X = \begin{bmatrix} x_{11} & x_{12} & \cdots & x_{1k} \\ x_{21} & x_{22} & & x_{2k} \\ x_{31} & x_{32} & & x_{3k} \\ \vdots & \vdots & & \vdots \\ x_{n1} & x_{n2} & & x_{nk} \end{bmatrix}$$

Each column of X is a column vector in its own right. Denoting the jth column of X by $\mathbf{x}_j$, we can rewrite X in column vector form as

$$X = [\mathbf{x}_1 \quad \mathbf{x}_2 \quad \mathbf{x}_3 \ldots \mathbf{x}_k]$$

The *rank* of X, denoted by $r(X)$, is defined to be the greatest number of linearly independent vectors in the set $\{\mathbf{x}_1, \mathbf{x}_2, \mathbf{x}_3, \ldots, \mathbf{x}_k\}$. As you can see, the rank of an $n \times k$ matrix is at most k.

Example 1.4.4 Consider

$$X = \begin{bmatrix} 1 & 0 & \frac{1}{2} \\ 0 & 1 & \frac{1}{2} \\ 1 & -1 & 0 \\ 1 & 1 & 1 \end{bmatrix} = [\mathbf{x}_1 \quad \mathbf{x}_2 \quad \mathbf{x}_3]$$

This matrix is of dimension 4×3 and hence its rank is *at most* 3. However, by inspection, it is easy to see that $\mathbf{x}_3 = \frac{1}{2}\mathbf{x}_1 + \frac{1}{2}\mathbf{x}_2$. It is also easy to see that $\mathbf{x}_1$ and $\mathbf{x}_2$ are linearly independent. Thus the actual rank of X is 2, and we write $r(X) = 2$.

In the statistical setting, we often deal with $n \times k$ matrices of rank k where $n \geqslant k$. Such matrices are said to be of *full rank*.

Some properties of rank are listed below.

Properties of Rank

1. Let X be an $n \times k$ matrix of rank k where $n \geqslant k$. That is, let X be full rank. Then $r(X) = r(X') = r(X'X) = k$.
2. Let X be a $k \times k$ matrix. Then X is nonsingular if and only if $r(X) = k$.
3. Let X be an $n \times k$ matrix, P be an $n \times n$ nonsingular matrix, and Q a $k \times k$ nonsingular matrix. Then $r(X) = r(PX) = r(XQ)$.
4. The rank of a diagonal matrix is equal to the number of nonzero columns in the matrix.
5. The rank of XY is less than or equal to the rank of X and is less than or equal to the rank of Y.

Property 3 can be remembered easily as simply stating that multiplying a matrix by a nonsingular matrix does not affect the rank of the product. Other properties concerning rank will be presented as needed.

1.5

IDEMPOTENT MATRICES AND THE TRACE

In statistical work, we often deal with square matrices A that have the property that $A^2 = A$. That is, they are unchanged when squared. Such matrices are said to be *idempotent*. A rather trivial example of an idempotent matrix is the identity matrix I. It is evident that for any identity matrix, $I^2 = I$. Example 1.5.1 points out a nontrivial idempotent matrix that is especially useful in the development of the theory of linear statistical models.

Example 1.5.1 Let X be an $n \times k$ matrix of full rank. The $n \times n$ matrix

$$H = X(X'X)^{-1}X'$$

is idempotent. First note that since X is of full rank, $r(X) = k$. Since $r(X) = r(X'X)$, $r(X'X) = k$. Furthermore, note that $X'X$ is a $k \times k$ matrix. We saw in Section 1.4 that any $k \times k$ matrix of rank k is nonsingular. Hence $(X'X)^{-1}$ does exist. To see that H is idempotent, we form its square. Here

$$H^2 = [X(X'X)^{-1}X'][X(X'X)^{-1}X']$$

Applying the associative property for matrix multiplication,

$$H^2 = X(X'X)^{-1}(X'X)(X'X)^{-1}X'$$

Since $(X'X)^{-1}(X'X) = I$, we can simplify to see that

$$H^2 = X(X'X)^{-1}X' = H$$

Thus, H is idempotent as claimed.

THE TRACE OF A MATRIX

Before stating some useful properties of idempotent matrices, we need to define the trace of a square matrix.

Definition 1.5.1

The trace of a $k \times k$ matrix X, denoted by $\text{tr}(X)$, is defined to be the sum of the entries along its main diagonal. That is,

$$\text{tr}(X) = \sum_{i=1}^{k} x_{ii}$$

Example 1.5.2 Let

$$X = \begin{bmatrix} 2 & 0 & 1 \\ 1 & 1 & 0 \\ 3 & 2 & -1 \end{bmatrix} \quad \text{and} \quad Y = \begin{bmatrix} -1 & 1 & 0 \\ 2 & 4 & 0 \\ 0 & 0 & 3 \end{bmatrix}$$

Then $\text{tr}(X) = 2 + 1 + (-1) = 2$ and $\text{tr}(Y) = (-1) + 4 + 3 = 6$.

Three useful properties of traces are given below. The first two are obvious; the derivation of the third is left as an exercise (Exercise 65).

Properties of the Trace

1. Let c be a real number. Then $\text{tr}(cX) = c\,\text{tr}(X)$.
2. $\text{tr}(X \pm Y) = \text{tr}(X) \pm \text{tr}(Y)$.
3. If X is $n \times p$ and Y is $p \times n$, then $\text{tr}(XY) = \text{tr}(YX)$.

Example 1.5.3 Let us verify property 3.

$$XY = \begin{bmatrix} 2 & 0 \\ 1 & 1 \\ 3 & 2 \end{bmatrix} \cdot \begin{bmatrix} -1 & 1 & 0 \\ 2 & 4 & 0 \end{bmatrix} = \begin{bmatrix} -2 & 2 & 0 \\ 1 & 5 & 0 \\ 1 & 11 & 0 \end{bmatrix}$$

and $\text{tr}(XY) = (-2) + 5 + (0) = 3$.

$$YX = \begin{bmatrix} -1 & 1 & 0 \\ 2 & 4 & 0 \end{bmatrix} \cdot \begin{bmatrix} 2 & 0 \\ 1 & 1 \\ 3 & 2 \end{bmatrix} = \begin{bmatrix} -1 & 1 \\ 8 & 4 \end{bmatrix}$$

and $\text{tr}(YX) = (-1) + 4 = 3$.

Even though $XY \neq YX$, their traces are equal, as claimed in property 3.

As mentioned earlier, idempotent matrices play a vital role in the theory of linear statistical models. Rather than simply listing some important results concerning such matrices as we have done earlier in this chapter, we will begin considering them one at a time. We will prove the given theorems whenever it is possible to do so from material presented in this text. Many of these results are probably new to you, so you should try to justify the steps in the proof.

Theorem 1.5.1 The eigenvalues of idempotent matrices are always either zeros or ones.

Proof

Let A be an idempotent matrix and let λ be an eigenvalue of A. By definition, there exists a nonzero vector $\mathbf{x}$ such that

$$A\mathbf{x} = \lambda\mathbf{x}$$

Multiplying each side of this equation by A and simplifying, we see that

$$\begin{aligned} A^2\mathbf{x} &= A(\lambda\mathbf{x}) \\ &= \lambda(A\mathbf{x}) \end{aligned}$$

Since A is idempotent, $A^2 = A$ and hence by the previous equation

$$\begin{aligned} A^2\mathbf{x} &= A\mathbf{x} \\ &= \lambda(A\mathbf{x}) \end{aligned}$$

By assumption, $A\mathbf{x} = \lambda\mathbf{x}$ so that substitution now yields

$$\begin{aligned} \lambda\mathbf{x} &= \lambda(\lambda\mathbf{x}) \\ &= \lambda^2\mathbf{x} \end{aligned}$$

By subtracting $\lambda\mathbf{x}$ from each side of the equation, we obtain

$$\lambda^2\mathbf{x} - \lambda\mathbf{x} = \mathbf{0}$$

or

$$(\lambda^2 - \lambda)\mathbf{x} = \mathbf{0}$$

Since $\mathbf{x} \neq \mathbf{0}$, we can conclude that $\lambda^2 - \lambda = 0$. Solving for λ, it is easy to see that $\lambda = 0$ or $\lambda = 1$ as claimed. ■

Theorem 1.5.2 Let A be a $k \times k$ symmetric and idempotent matrix of rank r. Then the rank of A is equal to its trace. That is, $r(A) = \text{tr}(A)$.

Proof

Let A be symmetric and idempotent of rank r. Theorem 1.4.1 guarantees that there exists an orthogonal matrix P such that

$$P'AP = \begin{bmatrix} \lambda_1 & 0 & 0 \\ 0 & \lambda_2 & 0 \\ \vdots & \vdots & \vdots \\ 0 & 0 & \lambda_k \end{bmatrix}$$

where $\lambda_1, \lambda_2, \ldots, \lambda_k$ are the eigenvalues of A. Since P is orthogonal, it is also nonsingular, as is P'. Property 3 of ranks guarantees that $r(P'A) = r(A)$ and that $r(P'AP) = r(P'A) = r(A)$. Thus to find the rank of A, we must find the rank of $P'AP$. This is easy to do because $P'AP$ is diagonal. Property 4 of ranks ensures that $r(P'AP)$ is the number of nonzero columns; that is, it is the number of nonzero eigenvalues. Since A is idempotent, we know from Theorem 1.5.1 that the eigenvalues are all either 0 or 1. Thus to count the number of nonzero eigenvalues, we need only sum the eigenvalues. Since the eigenvalues appear along the main diagonal, this is equivalent to finding its trace. We now know that

$$r(A) = r(P'AP) = \text{tr}(P'AP)$$

Applying property 3 of traces, we see that

$$\text{tr}(P'AP) = \text{tr}(PP'A)$$

Since P is orthogonal, $P' = P^{-1}$ and $PP' = I$. Substituting, we can conclude that $\text{tr}(PP'A) = \text{tr}(A)$ and hence that $r(A) = \text{tr}(P'AP) = \text{tr}(PP'A) = \text{tr}(A)$ as claimed. ■

The next two theorems, which we will not attempt to prove, will allow us to justify a result that will figure extensively in the work to come.

Theorem 1.5.3 Let $A_1, A_2, \ldots, A_m$ be a collection of $k \times k$, symmetric matrices. A necessary and sufficient condition for the existence of a common orthogonal matrix P such that $P'A_iP$ is diagonal for $i = 1, 2, 3, \ldots, m$ is that $A_iA_j = A_jA_i$ for every pair (i, j). ■

Since each of the matrices $A_1, A_2, \ldots, A_m$ is symmetric, we already know that there exist orthogonal matrices $P_1, P_2, \ldots, P_m$ that might differ from one another that diagonalize $A_1, A_2, \ldots, A_m$, respectively (see Theorem 1.4.1). Theorem 1.4.1 states that if all pairs in the collection commute then there exists a *single* common matrix P that diagonalizes each of the matrices in the collection simultaneously.

Theorem 1.5.4 Let $A_1, A_2, \ldots, A_m$ be a collection of symmetric $k \times k$ matrices. Then any two of the following implies the third:

1. All A_i $i = 1, 2, \ldots, m$ are idempotent.
2. $\Sigma_{i=1}^{m} A_i$ is idempotent.
3. $A_iA_j = 0$ for $i \neq j$. ■

The next theorem, whose proof follows from Theorems 1.5.2–1.5.4, is the main result of this section in terms of applications in the statistical setting. It will be used extensively in Chapter 2.

Theorem 1.5.5 Let $A_1, A_2, \ldots, A_m$ be a collection of symmetric $k \times k$ matrices. Let r denote the rank of $\Sigma_{i=1}^{m} A_i$ and let r_i denote the rank of A_i $i = 1, 2, \ldots, m$. If any two of the following are true, then $r = \Sigma_{i=1}^{m} r_i$. That is, any two of the following guarantee that the rank of the sum is equal to the sum of the ranks.

1. All A_i $i = 1, 2, \ldots, m$ are idempotent.
2. $\Sigma_{i=1}^{m} A_i$ is idempotent.
3. $A_iA_j = 0$ for $i \neq j$.

Proof

Assume that any two of the three conditions hold. By Theorem 1.5.4, the other condition also holds. In particular, we know that $A_1, A_2, \ldots, A_m$ is a collection of $k \times k$ symmetric and idempotent matrices and that $A_iA_j = 0$ for $i \neq j$. This in turn implies that $A_iA_j = A_jA_i$ for $i \neq j$ since each of these products is the zero matrix. We can apply Theorem 1.5.3 to conclude that there exists a single orthogonal matrix P such that for $i = 1, 2, \ldots, m$, $P'A_iP$ is diagonal. Now let

$$S = P'A_1P + P'A_2P + \cdots + P'A_mP = P'\sum_{i=1}^{m} A_iP$$

Note that

$$S' = \left(P' \sum_{i=1}^{m} A_i P\right)' = P'\left(\sum_{i=1}^{m} A_i\right)'(P')' = P'\left(\sum_{i=1}^{m} A_i\right)P$$

Since the sum of symmetric matrices is symmetric, $(\Sigma_{i=1}^{m} A_i)' = \Sigma_{i=1}^{m} A_i$. Hence $S' = P'\Sigma_{i=1}^{m} A_i P = S$, and we have shown that S is symmetric. To show that S is idempotent, consider its square.

$$\begin{aligned} S^2 &= \left[P' \sum_{i=1}^{m} A_i P\right]\left[P' \sum_{i=1}^{m} A_i P\right] \\ &= P'\left(\sum_{i=1}^{m} A_i\right)PP'\left(\sum_{i=1}^{m} A_i\right)P \qquad \text{(Associativity)} \\ &= P'\left(\sum_{i=1}^{m} A_i\right)^2 P \qquad (P \text{ is orthogonal.}) \end{aligned}$$

By condition 2, $\Sigma_{i=1}^{m} A_i$ is idempotent and hence we have

$$S^2 = P' \sum_{i=1}^{m} A_i P = S$$

showing that S is idempotent as claimed. We now know that S is a $k \times k$ symmetric and idempotent matrix and can therefore apply Theorem 1.5.2 to conclude that the rank of S is equal to its trace. To see the implications of this statement, note that

$$\begin{aligned} r\left(P' \sum_{i=1}^{m} A_i P\right) &= \text{tr}\left(P' \sum_{i=1}^{m} A_i P\right) \\ &= \text{tr}\left(PP' \sum_{i=1}^{m} A_i\right) \\ &= \text{tr}\left(\sum_{i=1}^{m} A_i\right) \qquad (P \text{ is orthogonal.}) \\ &= \sum_{i=1}^{m} \text{tr}(A_i) \end{aligned}$$

Since each of the A_i $i = 1, 2, \ldots, m$ is symmetric and idempotent, we know from Theorem 1.5.2 that $\text{tr}(A_i) = r(A_i) = r_i$. Substituting,

$$r\left(P' \sum_{i=1}^{m} A_i P\right) = \sum_{i=1}^{m} r_i$$

However, since P is orthogonal, it as well as its transpose is nonsingular. Since multiplication by nonsingular matrices does not affect rank, we also know that

$$r\left(P' \sum_{i=1}^{m} A_i P\right) = r\left(\sum_{i=1}^{m} A_i\right)$$

Equating the two expressions for $r(P'\Sigma_{i=1}^{m} A_i P)$, we can conclude that

$$r = r\left(\sum_{i=1}^{m} A_i\right) = \sum_{i=1}^{m} r_i$$

as claimed. ■

EXERCISES

Section 1.1

Let

$$X = \begin{bmatrix} 2 & 3 \\ -1 & 4 \end{bmatrix}; \quad Y = \begin{bmatrix} 2 & 0 & 1 \\ 1 & -2 & 3 \end{bmatrix}; \quad Z = \begin{bmatrix} 1 & 1 \\ -1 & 1 \\ 0 & 2 \end{bmatrix};$$

$$W = \begin{bmatrix} 1 & 0 \\ 8 & 3 \end{bmatrix}; \quad T = \begin{bmatrix} 1 & 0 \\ 0 & 1 \end{bmatrix}$$

Use these matrices for Exercises 1–5.

1. If possible, perform each of these operations. If the indicated operation cannot be performed, explain why.

a. $X + Y$	d. XY	g. XT	j. $Z(T + W)$
b. $X + W$	e. YX	h. TX	k. $Y(T + W)$
c. $X - T$	f. $3X$	i. $X + (Y + Z)$	

2. Find $(W + T)Y$, WY, and TY. Show that $(W + T)Y = WY + TY$, thus illustrating property 5 for matrix operations.

3. Find the zero matrix associated with X; with Y. Can you generalize to describe the zero matrix for any $n \times k$ matrix?

4. Find the negative of X; of Y. Can you generalize to describe the form of the negative matrix for any $n \times k$ matrix?

5. Find $2Y$, $X(2Y)$, XY, and then $2(XY)$. Show that $X(2Y) = 2(XY)$, thus illustrating property 6 for matrix operations.

6. Let

$$X = \begin{bmatrix} x_{11} & x_{12} \\ x_{21} & x_{22} \\ x_{31} & x_{32} \end{bmatrix} \quad \text{and} \quad Y = \begin{bmatrix} y_{11} & y_{12} & y_{13} & y_{14} \\ y_{21} & y_{22} & y_{23} & y_{24} \end{bmatrix}$$

What are the dimensions of XY? Using Definition 1.1.3, find the entry in row 1 and column 1 of the product XY. Also find the entry in row 2 and column 3 of the product.

7. Let $\mathbf{x}' = [x_1 \quad x_2 \quad x_3]$ and $\mathbf{y}' = [y_1 \quad y_2 \quad y_3]$. What are the dimensions of $\mathbf{x}'\mathbf{y}$? of $\mathbf{y}\mathbf{x}'$? Find the expression for $\mathbf{x}'\mathbf{y}$ and evaluate this expression when $x_1 = 3$, $x_2 = -1$, and $x_3 = 6$.

8. If we define X^2 to mean X multiplied by itself, what must be true of X in order for this matrix to exist?

9. Let

$$X = \begin{bmatrix} 1 & 3 & 0 & -2 \\ 0 & -1 & 1 & 3 \end{bmatrix} \quad \text{and} \quad Y = \begin{bmatrix} 2 & 1 & 1 \\ -1 & 0 & 0 \\ 0 & 1 & 0 \\ 1 & 1 & 3 \end{bmatrix}$$

a. Find XY.

b. Let X and Y be partitioned as shown:

$$X = \left[\begin{array}{cc|cc} 1 & 3 & 0 & -2 \\ 0 & -1 & 1 & 3 \end{array}\right] \quad \text{and} \quad Y = \left[\begin{array}{ccc} 2 & 1 & 1 \\ -1 & 0 & 0 \\ \hline 0 & 1 & 0 \\ 1 & 1 & 3 \end{array}\right]$$

Express X, Y, and XY in submatrix notation. Find XY by multiplying in partitioned form and verify your answer via your answer to part a.

c. Let X and Y be partitioned as shown:

$$X = \left[\begin{array}{c|ccc} 1 & 3 & 0 & -2 \\ 0 & -1 & 1 & 3 \end{array}\right] \quad Y = \left[\begin{array}{ccc} 2 & 1 & 1 \\ \hline -1 & 0 & 0 \\ 0 & 1 & 0 \\ 1 & 1 & 3 \end{array}\right]$$

Follow the instructions in part b.

d. Let X and Y be partitioned as shown:

$$X = \left[\begin{array}{cc|cc} 1 & 3 & 0 & -2 \\ \hline 0 & -1 & 1 & 3 \end{array}\right] \quad \text{and} \quad Y = \left[\begin{array}{c|cc} 2 & 1 & 1 \\ -1 & 0 & 0 \\ \hline 0 & 1 & 0 \\ 1 & 1 & 3 \end{array}\right]$$

Follow the instructions in part b.

e. Let

$$X = \left[\begin{array}{cc|cc} 1 & 3 & 0 & -2 \\ 0 & -1 & 1 & 3 \end{array}\right] = [X_{11} \mid X_{12}]$$

Show that

$$X' = \left[\begin{array}{c} X'_{11} \\ \hline X'_{12} \end{array}\right]$$

Now let $\mathbf{y}$ denote any 2×1 vector. Show that

$$X'\mathbf{y} = \left[\begin{array}{c} X'_{11}\ \mathbf{y} \\ \hline X'_{12}\ \mathbf{y} \end{array}\right]$$

Section 1.2

10. Let

$$X = \begin{bmatrix} 2 & 1 & 1 \\ 1 & -1 & 3 \\ 0 & 1 & 2 \end{bmatrix} \quad \text{and} \quad Y = \begin{bmatrix} 1 & 1 & 0 \\ 0 & 2 & 1 \\ -1 & 2 & 3 \end{bmatrix}$$

Find X', Y', $X' + Y'$, $X + Y$, and $(X + Y)'$. Verify that $(X + Y)' = X' + Y'$.

11. Let

$$X = \begin{bmatrix} 1 & 1 \\ 1 & 1 \end{bmatrix} \quad \text{and} \quad Y = \begin{bmatrix} 1 & 3 & 0 \\ 2 & 5 & -1 \end{bmatrix}$$

Find X', Y', $Y'X'$, XY, and $(XY)'$. Verify that $(XY)' = Y'X'$.

12. Let X, Y, and Z be conformable. Prove that $(XYZ)' = Z'Y'X'$.

13. Let

$$X = \begin{bmatrix} 3 & 0 & 8 & -2 \\ 1 & 2 & 5 & 0 \end{bmatrix}$$

What are the dimensions of X'? Find X'. Is X symmetric?

14. Let

$$X = \begin{bmatrix} 2 & 1 & 4 \\ 1 & -3 & x_{23} \\ 4 & 2 & 0 \end{bmatrix}$$

What must x_{23} equal in order for X to be symmetric?

15. Let

$$\mathbf{x} = \begin{bmatrix} 0 \\ 1 \\ -1 \end{bmatrix}$$

Find $\mathbf{x}'\mathbf{x}$ and verify the statement of Theorem 1.2.2.

16. Let

$$\mathbf{x} = \begin{bmatrix} 2 \\ 4 \\ 1 \end{bmatrix}$$

Find $\mathbf{x}\mathbf{x}'$ and verify the statement of Theorem 1.2.3.

17. Prove Theorem 1.2.2.

18. Prove Theorem 1.2.3.

19. Let

$$X = \begin{bmatrix} 2 & 0 & 1 \\ 1 & 3 & -1 \end{bmatrix}$$

What are the dimensions of $X'X$? Find $X'X$ by inspection. Verify your answer by direct multiplication. Is $X'X$ symmetric as expected?

20. Let

$$X = \begin{bmatrix} 2 & 8 \\ 1 & 0 \\ 4 & -3 \end{bmatrix}$$

Find $X'X$ by inspection and verify your answer by direct multiplication.

21. Let

$$X = \begin{bmatrix} 1 & 1 & 1 \\ 1 & 1 & -1 \\ 1 & -1 & 1 \\ 1 & -1 & -1 \end{bmatrix}$$

Find $X'X$ by inspection and verify your answer by direct multiplication.

22. Prove Theorem 1.2.1.

23. Suppose that X is an $n \times k$ matrix whose first column is a column of ones. What will be the first main diagonal entry of $X'X$?

24. Use the properties of transposes to prove that $X'X$ is symmetric. That is, prove that $(X'X)' = X'X$.

25. Show that if X and Y are both $n \times k$ symmetric matrices, then $X + Y$ and $X - Y$ are also symmetric.

Section 1.3

26. Let

$$X = \begin{bmatrix} 2 & 4 & 8 \\ 1 & 3 & 2 \end{bmatrix}$$

Show that

$$I_3 = \begin{bmatrix} 1 & 0 & 0 \\ 0 & 1 & 0 \\ 0 & 0 & 1 \end{bmatrix}$$

is the right identity for X. What are the dimensions of its left identity? Does it have an overall identity?

27. Find the inverse of

$$X = \begin{bmatrix} 3 & 0 & 0 \\ 0 & 7 & 0 \\ 0 & 0 & -2 \end{bmatrix}$$

and verify your answer.

28. a. Let

$$X = \begin{bmatrix} 3 & 0 \\ 6 & 1 \end{bmatrix}$$

Show that

$$X^{-1} = \begin{bmatrix} \frac{1}{3} & 0 \\ -2 & 1 \end{bmatrix}$$

*b. Let

$$A = \begin{bmatrix} a & b \\ c & d \end{bmatrix}$$

be any nonsingular 2×2 matrix. By direct multiplication show that

$$A^{-1} = \begin{bmatrix} \dfrac{d}{ad - bc} & \dfrac{-b}{ad - bc} \\ \dfrac{-c}{ad - bc} & \dfrac{a}{ad - bc} \end{bmatrix}$$

c. Let

$$A = \begin{bmatrix} 2 & 4 \\ 1 & -3 \end{bmatrix}$$

Use the algorithm of part b to find A^{-1}.

29. Prove that $(XY)^{-1} = Y^{-1}X^{-1}$.

***30.** Prove that if X, Y, and Z are conformable and nonsingular then $(XYZ)^{-1} = Z^{-1}Y^{-1}X^{-1}$.

***31.** Prove that $(X')^{-1} = (X^{-1})'$. (*Hint*: Show that $X'(X^{-1})' = I$.)

32. Use the computational shortcut of Section 1.2 to show that

$$X = \begin{bmatrix} \dfrac{1}{\sqrt{2}} & \dfrac{1}{\sqrt{2}} \\ \dfrac{-1}{\sqrt{2}} & \dfrac{1}{\sqrt{2}} \end{bmatrix}$$

is orthogonal.

33. Is

$$X = \begin{bmatrix} -1 & 1 \\ 1 & 1 \end{bmatrix}$$

orthogonal? If not, what value of c makes the matrix cX orthogonal?

34. Is

$$X = \begin{bmatrix} 1 & 1 & -1 & -1 \\ 1 & -1 & 1 & -1 \\ 1 & -1 & -1 & 1 \\ 1 & 1 & 1 & 1 \end{bmatrix}$$

orthogonal? If not, what value of c makes the matrix cX orthogonal?

***35.** Prove that if X is orthogonal, X^{-1} and X' are orthogonal.

36. Let

$$\mathbf{x} = \begin{bmatrix} 2 \\ 1 \\ 3 \end{bmatrix} \quad \text{and} \quad \mathbf{y} = \begin{bmatrix} 1 \\ 2 \\ 0 \end{bmatrix}$$

a. Are $\mathbf{x}$ and $\mathbf{y}$ orthogonal?
b. What is the length of $\mathbf{x}$? of $\mathbf{y}$?

37. Let

$$\mathbf{x}_1 = \begin{bmatrix} 1 \\ 1 \\ -1 \\ -1 \end{bmatrix}, \quad \mathbf{x}_2 = \begin{bmatrix} 1 \\ -1 \\ 1 \\ -1 \end{bmatrix}, \quad \text{and} \quad \mathbf{x}_3 = \begin{bmatrix} 1 \\ 1 \\ 1 \\ 1 \end{bmatrix}$$

Is $\{\mathbf{x}_1, \mathbf{x}_2, \mathbf{x}_3\}$ an orthonormal set? If not, find a constant c such that $\{c\mathbf{x}_1, c\mathbf{x}_2, c\mathbf{x}_3\}$ is an orthonormal set. Is the choice of c unique?

38. Prove Theorem 1.3.1. (*Hint*: Consider Theorem 1.2.1.)

39. Prove that if X is orthogonal then its rows also form an orthonormal set.

Section 1.4

40. Find the eigenvectors associated with the eigenvalue λ_2 of Example 1.4.1.

41. Consider the matrix

$$A = \begin{bmatrix} 1 & -1 \\ 2 & 4 \end{bmatrix}$$

Is there any guarantee that the eigenvalues of this matrix are real? If they are real, find the eigenvalues and eigenvectors associated with A.

42. Consider the matrix

$$A = \begin{bmatrix} 1 & 1 \\ -1 & 2 \end{bmatrix}$$

Is there any guarantee that the eigenvalues of this matrix are real? If they are real, find the eigenvalues and eigenvectors associated with A.

43. Consider the matrix

$$A = \begin{bmatrix} 1 & 1 \\ 1 & 1 \end{bmatrix}$$

Is there any guarantee that the eigenvalues of this matrix are real? If they are real, find the eigenvalues and eigenvectors associated with A.

44. Show that

$$C = \begin{bmatrix} \frac{1}{\sqrt{2}} & \frac{1}{\sqrt{2}} \\ \frac{-1}{\sqrt{2}} & \frac{1}{\sqrt{2}} \end{bmatrix}$$

is orthogonal. Let A be as defined in Exercise 41 and find $C'AC$. Find the eigenvalues of $C'AC$ and show that they are the same as the eigenvalues of A, thus illustrating property 2 of eigenvalues.

***45.** Use these properties of determinants to prove parts a and b:

1. $|X| = |X'|$ (The determinant of a matrix is equal to the determinant of its transpose.)
2. If X and Y are conformable, then $|XY| = |X||Y|$ (The determinant of a product is equal to the product of the determinants.)

a. Let C be orthogonal. Show that $|C| = \pm 1$.

b. Prove property 2 of eigenvalues. (*Hint*: Note that

$$\begin{aligned} C'AC - \lambda I &= C'AC - \lambda C'C \\ &= C'(A - \lambda I)C \end{aligned}$$

and apply the rules for determinants.)

46. Prove property 3 of eigenvalues. (*Hint*: Assume that $\mathbf{x}_1$ and $\mathbf{x}_2$ are the eigenvectors associated with λ_1 and λ_2, respectively, and that $\lambda_1 \neq \lambda_2$. Use Definition 1.4.1 and the properties of transposes to show that $\lambda_1 \mathbf{x}_1' \mathbf{x}_2 = \lambda_2 \mathbf{x}_1' \mathbf{x}_2$ and then argue that $\mathbf{x}_1' \mathbf{x}_2 = 0$ as claimed.)

47. Let

$$A = \begin{bmatrix} 2 & 2 \\ 2 & 2 \end{bmatrix}$$

a. Show that $\lambda_1 = 0$ and $\lambda_2 = 4$ are eigenvalues of A.

b. Show that

$$\begin{bmatrix} \frac{1}{\sqrt{2}} \\ \frac{-1}{\sqrt{2}} \end{bmatrix} \quad \text{and} \quad \begin{bmatrix} \frac{1}{\sqrt{2}} \\ \frac{1}{\sqrt{2}} \end{bmatrix}$$

form an orthonormal set.

c. Show that

$$\begin{bmatrix} \frac{1}{\sqrt{2}} \\ \frac{-1}{\sqrt{2}} \end{bmatrix}$$

is an eigenvector associated with $\lambda_1 = 0$.

d. Show that

$$\begin{bmatrix} \frac{1}{\sqrt{2}} \\ \frac{1}{\sqrt{2}} \end{bmatrix}$$

is an eigenvector associated with $\lambda_2 = 4$.

e. Let

$$P = \begin{bmatrix} \frac{1}{\sqrt{2}} & \frac{1}{\sqrt{2}} \\ \frac{-1}{\sqrt{2}} & \frac{1}{\sqrt{2}} \end{bmatrix}$$

and show that

$$P'AP = \begin{bmatrix} \lambda_1 & 0 \\ 0 & \lambda_2 \end{bmatrix}$$

thus illustrating Theorem 1.4.1.

48. Let

$$A = \begin{bmatrix} 2 & 1 \\ 1 & 2 \end{bmatrix}$$

a. Show that the eigenvalues of A are $\lambda_1 = 3$ and $\lambda_2 = 1$.
b. Show that all eigenvectors associated with λ_1 are such that $x_1 = x_2$. Find an eigenvector $\mathbf{x}_1$ associated with λ_1 of length 1.
c. Show that eigenvectors associated with λ_2 are such that $x_1 = -x_2$. Find an eigenvector $\mathbf{x}_2$ associated with λ_2 of length 1.
d. Show that $\mathbf{x}_1$ and $\mathbf{x}_2$ are orthogonal as claimed in property 3 for eigenvalues.
e. Form the orthogonal matrix P that diagonalizes A and show that

$$P'AP = \begin{bmatrix} \lambda_1 & 0 \\ 0 & \lambda_2 \end{bmatrix}$$

49. Let

$$\mathbf{x}_1 = \begin{bmatrix} 1 \\ 4 \\ 1 \end{bmatrix}, \quad \mathbf{x}_2 = \begin{bmatrix} 1 \\ 0 \\ 1 \end{bmatrix}, \quad \text{and} \quad \mathbf{x}_3 = \begin{bmatrix} 2 \\ 0 \\ 0 \end{bmatrix}$$

Are these vectors linearly independent?

50. What is the rank of the matrix

$$X = \begin{bmatrix} 1 & 1 & 2 \\ 4 & 0 & 0 \\ 1 & 1 & 0 \end{bmatrix}$$

Is X of full rank?

51. Let

$$\mathbf{x}_1 = \begin{bmatrix} 1 \\ 1 \\ 2 \end{bmatrix}, \quad \mathbf{x}_2 = \begin{bmatrix} 2 \\ -3 \\ -1 \end{bmatrix}, \quad \text{and} \quad \mathbf{x}_3 = \begin{bmatrix} 3 \\ 4 \\ 7 \end{bmatrix}$$

Are these vectors linearly independent?

52. What is the rank of the matrix

$$Y = \begin{bmatrix} 1 & 2 & 3 \\ 1 & -3 & 4 \\ 2 & -1 & 7 \end{bmatrix}$$

Is Y of full rank?

53. Let X be a $k \times k$ orthogonal matrix. Prove that X is of full rank.

54. Suppose that X is $k \times k$ of rank less than k. Is X nonsingular? Explain.

55. Consider the matrix X of Exercise 50. Is X nonsingular? Explain.

56. Consider the matrices X and Y of Exercises 50 and 52, respectively. What is the rank of the matrix XY? of the matrix YX? Explain.

57. Let

$$X = \begin{bmatrix} 1 & 1 & 0 \\ 0 & 1 & 1 \\ 0 & 0 & -1 \\ 0 & 1 & 0 \\ 1 & 0 & 0 \end{bmatrix}$$

Is X of full rank? Is X nonsingular?

58. Consider the following statements. If a statement is false, provide a counterexample to show that it is false.

a. Every orthogonal matrix is nonsingular.
b. Every nonsingular matrix is orthogonal.
c. Every matrix of full rank is square.
d. Every square matrix is of full rank.
e. Every nonsingular matrix is of full rank.
f. Every square matrix of full rank is nonsingular.
g. Every orthogonal matrix is of full rank.
h. Every matrix of full rank is orthogonal.

59. Consider the matrix

$$X = \begin{bmatrix} 2 & 0 & 0 & 0 \\ 0 & 4 & 0 & 0 \\ 0 & 0 & 0 & 0 \\ 0 & 0 & 0 & 1 \end{bmatrix}$$

What is the rank of X? Since $r(X) \neq 4$, the columns of X are not linearly independent. Demonstrate this by finding constants a_1, a_2, a_3, a_4 not *all* zero such that $a_1\mathbf{x}_1 + a_2\mathbf{x}_2 + a_3\mathbf{x}_3 + a_4\mathbf{x}_4 = \mathbf{0}$. Are these values unique? That is, could you find a second set of values that would also satisfy the criterion for linear dependence?

Section 1.5

60. Consider the matrix

$$X = \begin{bmatrix} \frac{1}{2} & \frac{1}{2} \\ \frac{1}{2} & \frac{1}{2} \end{bmatrix}$$

Show that X is idempotent.

61. Is it possible for an idempotent matrix to have a negative value on its main diagonal? If so, give an example of such a matrix.

62. Is it possible for an idempotent matrix to have a negative value off the main diagonal? If so, give an example of such a matrix.

63. Let I be the $n \times n$ identity matrix and let X be an $n \times k$ matrix of full rank. Show that $I - X(X'X)^{-1}X'$ is idempotent.

64. Let

$$X = \begin{bmatrix} 1 & 2 & 0 \\ 0 & 1 & 0 \\ 1 & 3 & 4 \end{bmatrix}$$

and let

$$Y = \begin{bmatrix} 0 & 1 & 0 \\ 1 & -1 & 2 \\ 3 & 2 & 0 \end{bmatrix}$$

Find tr(X), tr(Y), tr(XY), and tr(YX). Verify that $\text{tr}(XY) = \text{tr}(YX)$.

65. Prove that $\text{tr}(XY) = \text{tr}(YX)$. (*Hint*: Apply the theoretical definition of matrix multiplication and use the commutative property for multiplication of real numbers.)

66. Consider the matrix of Exercise 60. What are the possible values for the eigenvalues of this matrix? Verify your answer by finding the eigenvalues.

67. What is the rank of the matrix of Exercise 60?

68. Consider a 4×4 diagonal idempotent matrix A of rank 2. How many columns of A are nonzero? Give an example of such a matrix. Is the example unique? That is, is your example the only one possible?

***69.** Let X be a 10×5 matrix of full rank. Consider the matrix $H = X(X'X)^{-1}X'$. Find the numerical value for the trace of H. In general, if X is $n \times k$ of full rank, what is the value of the trace of H?

70. Consider the following matrices:

$$A = \begin{bmatrix} 1 & 0 \\ 0 & 1 \end{bmatrix} \quad B = \begin{bmatrix} \frac{1}{3} & \frac{1}{3} & \frac{1}{3} \\ \frac{1}{3} & \frac{1}{3} & \frac{1}{3} \\ \frac{1}{3} & \frac{1}{3} & \frac{1}{3} \end{bmatrix} \quad C = \begin{bmatrix} -1 & 0 \\ 0 & 1 \end{bmatrix}$$

$$D = \begin{bmatrix} 2 & 0 \\ 0 & 1 \end{bmatrix} \quad E = \begin{bmatrix} \frac{1}{2} & -\frac{1}{2} \\ -\frac{1}{2} & \frac{1}{2} \end{bmatrix} \quad F = \begin{bmatrix} 1 & 0 \\ -1 & 0 \end{bmatrix}$$

$$G = \begin{bmatrix} 1 & 0 & -1 \\ 0 & 1 & 1 \\ 1 & 0 & -1 \\ 0 & 1 & 1 \end{bmatrix}$$

We have considered these special types of matrices: diagonal, symmetric, nonsingular, square, orthogonal, idempotent, and full rank. For each of the above matrices, list those properties possessed by the matrix. For which matrices can we conclude that its rank is equal to its trace based on Theorem 1.5.2?

***71.** Let X be $n \times p$, Y be $p \times m$, and Z be $m \times n$.

a. Show that XYZ, ZXY, and YZX all exist. Are each of these products square matrices?

b. Show that $\text{tr}(XYZ) = \text{tr}(ZXY) = \text{tr}(YZX)$. That is, show that the trace of the product of conformable matrices is invariant under any cyclic permutation of the matrices.

c. Let

$$X = \begin{bmatrix} 2 & 1 & 0 \\ 3 & 7 & 0 \end{bmatrix}, \quad Y = \begin{bmatrix} 2 \\ 7 \\ -1 \end{bmatrix}, \quad \text{and} \quad Z = [8 \quad 4]$$

Use these matrices to verify the invariance property stated in part b.

CHAPTER 2

QUADRATIC FORMS AND THEIR DISTRIBUTIONS

In this chapter we begin to link the theory of matrices to the theory of statistics. To do so, we introduce the notion of quadratic forms, a topic that is not in itself statistical. However, we then make the assumption that the vectors involved are random vectors, which immediately injects a statistical flavor into the discussion. We will be particularly interested in investigating the probability distribution of these quadratic forms. You will see how nicely the matrix theory presented in Chapter 1 comes into play. It will be assumed that you are familiar with the basic concepts of statistical inference. In particular, you should review the Student's t, X^2, F, and normal distributions as well as the notions of expected value, variance, covariance, correlation, and independence. These topics are discussed in [**13**], [**18**], and [**14**].

2.1

QUADRATIC FORMS

We begin by considering the general definition of the quadratic form.

Definition 2.1.1

Let A be a $k \times k$ matrix and

$$\mathbf{y} = \begin{bmatrix} y_1 \\ y_2 \\ \vdots \\ y_k \end{bmatrix}$$

be a $k \times 1$ column vector of real variables. Then

$$q = \mathbf{y}' A \mathbf{y}$$

is called a quadratic form in $\mathbf{y}$ and A is called the matrix of the quadratic form.

At this point, the most important element of a quadratic form is its dimensions. Since $\mathbf{y}$ is a $k \times 1$ column vector, $\mathbf{y}'$ is a $1 \times k$ row vector. Hence $q = \mathbf{y}'A\mathbf{y}$ is a 1×1 matrix whose only entry is a function of the variables $y_1, y_2, \ldots, y_k$. Thus when $y_1, y_2, \ldots, y_k$ assume numeric values, q is just a very special scalar. Usually we work with these functions in matrix form, but occasionally we will need to express them in expanded form as a sum of squares and cross products of the y's. It is not hard to show that

$$q = \sum_{i=1}^{k} \sum_{j=1}^{k} a_{ij} y_i y_j$$

(See Exercise 1.)

An example will clarify the notation.

Example 2.1.1 Let

$$\mathbf{y} = \begin{bmatrix} y_1 \\ y_2 \\ y_3 \end{bmatrix} \quad \text{and} \quad A = \begin{bmatrix} 2 & 3 & 1 \\ 1 & 2 & 0 \\ 4 & 6 & 3 \end{bmatrix}$$

In expanded form, $\mathbf{y}'A\mathbf{y}$ is given by

$$\begin{aligned}\mathbf{y}'A\mathbf{y} = a_{11}y_1y_1 + a_{21}y_1y_2 + a_{31}y_1y_3 + a_{12}y_1y_2 + a_{22}y_2y_2 \\ + a_{32}y_2y_3 + a_{13}y_1y_3 + a_{23}y_2y_3 + a_{33}y_3y_3\end{aligned}$$

Substituting the appropriate values from A,

$$\mathbf{y}'A\mathbf{y} = 2y_1^2 + y_1y_2 + 4y_1y_3 + 3y_1y_2 + 2y_2^2 + 6y_2y_3 + y_1y_3 + 3y_3^2$$

Combining terms and rearranging,

$$\mathbf{y}'A\mathbf{y} = 2y_1^2 + 2y_2^2 + 3y_3^2 + 4y_1y_2 + 5y_1y_3 + 6y_2y_3$$

Finding $\mathbf{y}'A\mathbf{y}$ using matrix multiplication,

$$\begin{aligned}\mathbf{y}'A\mathbf{y} &= [y_1 \quad y_2 \quad y_3]\begin{bmatrix} 2 & 3 & 1 \\ 1 & 2 & 0 \\ 4 & 6 & 3 \end{bmatrix}\begin{bmatrix} y_1 \\ y_2 \\ y_3 \end{bmatrix} \\ &= [2y_1 + y_2 + 4y_3 \quad 3y_1 + 2y_2 + 6y_3 \quad y_1 + 3y_3]\begin{bmatrix} y_1 \\ y_2 \\ y_3 \end{bmatrix} \\ &= 2y_1^2 + y_1y_2 + 4y_1y_3 + 3y_1y_2 + 2y_2^2 + 6y_2y_3 + y_1y_3 + 3y_3^2 \\ &= 2y_1^2 + 2y_2^2 + 3y_3^2 + 4y_1y_2 + 5y_1y_3 + 6y_2y_3\end{aligned}$$

as before. As you can see, this quadratic form is a sum of squares and cross products as claimed.

Quadratic forms fall into several different categories. In particular, quadratic forms that are positive definite must be distinguished from those that are positive semidefinite. These terms are defined below.

Definition 2.1.2

The quadratic form $\mathbf{y}'A\mathbf{y}$ is said to be *positive definite* if $\mathbf{y}'A\mathbf{y} > 0$ for all $\mathbf{y} \neq \mathbf{0}$; it is said to be *positive semidefinite* if $\mathbf{y}'A\mathbf{y} \geqslant 0$ for all $\mathbf{y}$ and $\mathbf{y}'A\mathbf{y} = \mathbf{0}$ for some $\mathbf{y} \neq \mathbf{0}$.

These terms are also used to refer to matrices. That is, a matrix is positive definite if its corresponding quadratic form is positive definite; if its quadratic form is positive semidefinite, then so is the matrix.

Example 2.1.2 Let

$$A = \begin{bmatrix} 2 & -1 \\ -1 & 2 \end{bmatrix}$$

Then

$$\mathbf{y}'A\mathbf{y} = [y_1 \quad y_2] \begin{bmatrix} 2 & -1 \\ -1 & 2 \end{bmatrix} \begin{bmatrix} y_1 \\ y_2 \end{bmatrix} = 2y_1^2 + 2y_2^2 - 2y_1 y_2$$

For the moment, let us use only the definition and elementary algebra to argue that A is positive definite. To do so, we must show that $2y_1^2 + 2y_2^2 - 2y_1 y_2 > 0$ for all $\mathbf{y} \neq \mathbf{0}$. First assume that $0 \leqslant y_1 \leqslant y_2$. Multiplying by y_2, we can conclude that $y_1 y_2 \leqslant y_2^2$ and also that $y_1^2 + y_2^2 \geqslant y_1 y_2$. This in turn implies that $2y_1^2 + 2y_2^2 - 2y_1 y_2 \geqslant 0$. A similar argument will show that this inequality holds if $y_1 \leqslant y_2 \leqslant 0$. It obviously holds if y_1 and y_2 differ in sign. To show that the right-hand side is not zero for $\mathbf{y} \neq \mathbf{0}$, we try to solve

$$y_1^2 - y_1 y_2 + y_2^2 = 0$$

treating this as a quadratic equation in y_1. The discriminant, $(b^2 - 4ac)$, is $(-y_2)^2 - 4y_2^2 = -3y_2^2$. The only value of y_2 for which this discriminant is nonnegative is $y_2 = 0$. In this case, y_1 is also 0. Thus, we have shown that the only real solution to the equation $y_1^2 - y_1 y_2 + y_2^2 = 0$ is $\mathbf{y} = \mathbf{0}$. This guarantees that for all $\mathbf{y} \neq \mathbf{0}$

$$2y_1^2 + 2y_2^2 - 2y_1 y_2 = \mathbf{y}'A\mathbf{y} > 0$$

That is, A is positive definite as claimed.

Arguing that a matrix is positive definite or positive semidefinite on the basis of the definition alone is tricky business, especially if the matrix is larger than 2×2. For this reason, we need some theorems that will allow us to identify the type of matrix involved without having to resort to the definition itself. Theorems 2.1.1 and 2.1.2 serve this purpose. Exercise 10 partially outlines the proof of Theorem 2.1.1.

Theorem 2.1.1 A symmetric matrix A is positive definite if and only if all its eigenvalues are positive. ■

Theorem 2.1.2 A symmetric matrix A is positive semidefinite if and only if its eigenvalues are all nonnegative and at least one eigenvalue is zero. ■

Notice that even though the definitions of the terms *quadratic form, positive definite*, and *positive semidefinite* do not require that the matrix involved be symmetric, these theorems do have that restriction. This is not a problem in the statistical setting since most of the quadratic forms encountered there have symmetric associated matrices.

Example 2.1.3 Let us reconsider Example 2.1.2 in light of these two theorems. Recall that to find the eigenvalues of a matrix A, we must find the solutions to the equation $A - \lambda I = 0$. Here

$$A = \begin{bmatrix} 2 & -1 \\ -1 & 2 \end{bmatrix}$$

and $|A - \lambda I| = (2 - \lambda)^2 - 1$. We must solve the equation

$$\lambda^2 - 4\lambda + 3 = 0$$

It is easy to see that the solutions are $\lambda = 3$ and $\lambda = 1$. Since A is symmetric and each of these eigenvalues is positive, we can apply Theorem 2.1.1 to conclude that A is positive definite as expected.

INVERSE OF POSITIVE DEFINITE SYMMETRIC MATRICES

Before closing this section, we present one more theoretical result that plays an important role in the development of statistical tests. We derive a method for writing the inverse of a submatrix of a symmetric positive definite matrix in partitioned form. The proof of the required theorem is based in part on a result from linear algebra concerning the principal minors of a matrix. A *principal minor* is a submatrix obtained by deleting a specified set of rows and like-numbered columns from the original matrix. For example, principal minors can be obtained by striking the first three rows and the first three columns or the first and last rows and the first and last columns. The algebraic result that is needed is the fact that the principal minors of a positive definite matrix are positive definite and their determinants are positive [**8**]. This in turn implies that the principal minors are each nonsingular matrices [**11**].

Theorem 2.1.3 Let A be a positive definite symmetric matrix written in partitioned form as

$$A = \left[\begin{array}{c|c} A_{11} & A_{12} \\ \hline A_{21} & A_{22} \end{array}\right]$$

where A_{11} and A_{22} are square matrices. Let $B = A^{-1}$ where

$$B = \left[\begin{array}{c|c} B_{11} & B_{12} \\ \hline B_{21} & B_{22} \end{array}\right]$$

and the dimensions of B_{11} and B_{22} are the same as those of A_{11} and A_{22}, respectively. Then

$$A_{11}^{-1} = B_{11} - B_{12}B_{22}^{-1}B_{21}$$

Proof

Since A and B are inverses, $AB = I$. In partitioned form

$$\left[\begin{array}{c|c} A_{11} & A_{12} \\ \hline A_{21} & A_{22} \end{array}\right]\left[\begin{array}{c|c} B_{11} & B_{12} \\ \hline B_{21} & B_{22} \end{array}\right] = \left[\begin{array}{c|c} A_{11}B_{11} + A_{12}B_{21} & A_{11}B_{12} + A_{12}B_{22} \\ \hline A_{21}B_{11} + A_{22}B_{21} & A_{21}B_{12} + A_{22}B_{22} \end{array}\right] = I$$

From this, it can be seen that

$$A_{11}B_{11} + A_{12}B_{21} = I \quad \text{and} \quad A_{11}B_{12} + A_{12}B_{22} = 0$$

Since A_{11} is a principal minor of the positive definite matrix A, A_{11} is nonsingular. Since $B = A^{-1}$, B is also positive definite [7] and B_{22}, a principal minor, is nonsingular. This information can be used to find A_{11}^{-1}. First solve the equation

$$A_{11}B_{12} + A_{12}B_{22} = 0$$

for A_{12} to obtain

$$A_{12} = -A_{11}B_{12}B_{22}^{-1}$$

Substituting into the equation

$$A_{11}B_{11} + A_{12}B_{21} = I$$

and solving for A_{11}^{-1}, we obtain

$$A_{11}^{-1} = A_{11}^{-1}A_{11}B_{11} + A_{11}^{-1}(-A_{11}B_{12}B_{22}^{-1})B_{21}$$

or

$$A_{11}^{-1} = B_{11} - B_{12}B_{22}^{-1}B_{21}$$

as claimed. ■

Example 2.1.4 Consider the matrix A partitioned as here:

$$A = \left[\begin{array}{cc|c} 2 & -1 & 2 \\ -1 & 2 & 1 \\ \hline 2 & 1 & 4 \end{array}\right]$$

You can verify by direct multiplication that

$$B = A^{-1} = -\frac{1}{2}\left[\begin{array}{cc|c} 7 & 6 & -5 \\ 6 & 4 & -4 \\ \hline -5 & -4 & 3 \end{array}\right]$$

and that

$$A_{11}^{-1} = \begin{bmatrix} \frac{2}{3} & \frac{1}{3} \\ \frac{1}{3} & \frac{2}{3} \end{bmatrix}$$

Direct multiplication will show that, as claimed in Theorem 2.1.3,

$$A_{11}^{-1} = B_{11} - B_{12}B_{22}^{-1}B_{21}$$

2.2

DIFFERENTIATION OF QUADRATIC FORMS AND EXPECTATION AND VARIANCE OF VECTORS AND MATRICES

As indicated earlier, this chapter provides a link between matrix theory and statistical theory. In this section we develop some rules for differentiating vector expressions and introduce the basic terminology used in the study of the probability distribution of certain quadratic forms. Most of what is done here is rather mechanical, but you must become adept at using these rules in order to understand the statistical arguments to come.

We begin with a notational convention. Let us assume that the scalar z can be written as a function of the k variables $y_1, y_2, \ldots, y_k$. We will express this idea by writing

$$z = f(y_1, y_2, \ldots, y_k) = f(\mathbf{y})$$

As you know, we can take k partial derivatives of such a function, one with respect to each of the y variables. By $\partial z/\partial \mathbf{y}$ we shall mean the column vector whose ith entry is $\partial z/\partial y_i$. That is,

$$\frac{\partial z}{\partial \mathbf{y}} = \begin{bmatrix} \partial z/\partial y_1 \\ \partial z/\partial y_2 \\ \vdots \\ \partial z/\partial y_k \end{bmatrix}$$

Example 2.2.1 illustrates the use of this notation.

Example 2.2.1 Let

$$A = \begin{bmatrix} 1 & -1 & 2 \\ -1 & 1 & 3 \\ 2 & 3 & 2 \end{bmatrix} \quad \text{and} \quad \mathbf{y} = \begin{bmatrix} y_1 \\ y_2 \\ y_3 \end{bmatrix}$$

Consider the quadratic form $z = \mathbf{y}'A\mathbf{y}$. Here

$$z = y_1^2 + y_2^2 + 2y_3^2 - 2y_1y_2 + 4y_1y_3 + 6y_2y_3$$

Taking partial derivatives,

$$\begin{aligned} \partial z/\partial y_1 &= 2y_1 - 2y_2 + 4y_3 \\ \partial z/\partial y_2 &= 2y_2 - 2y_1 + 6y_3 \\ \partial z/\partial y_3 &= 4y_3 + 4y_1 + 6y_2 \end{aligned}$$

and hence

$$\partial z/\partial \mathbf{y} = \begin{bmatrix} 2y_1 - 2y_2 + 4y_3 \\ 2y_2 - 2y_1 + 6y_3 \\ 4y_3 + 4y_1 + 6y_2 \end{bmatrix}$$

Using this notation, three useful rules for differentiating vector expressions can be derived.

Rules for Differentiation

1. Let $z = \mathbf{a}'\mathbf{y}$ where $\mathbf{a}$ is a vector of scalars. Then $\partial z/\partial \mathbf{y} = \mathbf{a}$.
2. Let $z = \mathbf{y}'\mathbf{y}$. Then $\partial z/\partial \mathbf{y} = 2\mathbf{y}$.
3. Let $z = \mathbf{y}'A\mathbf{y}$ where A is a $k \times k$ matrix. Then $\partial z/\partial \mathbf{y} = A\mathbf{y} + A'\mathbf{y}$.

The first two rules are obvious. (See Exercises 11 and 12.) Rule 3 can be verified by writing z as a sum of squares and cross products as was done in Exercise 1, Section 2.1, differentiating with respect to y_i and showing that the resulting term is in fact the ith entry in the vector $A\mathbf{y} + A'\mathbf{y}$. A small example should convince you that rule 3 works.

Example 2.2.2 Let

$$A = \begin{bmatrix} 2 & 0 & 2 \\ 1 & -1 & -2 \\ 3 & 4 & 5 \end{bmatrix}, \quad y = \begin{bmatrix} y_1 \\ y_2 \\ y_3 \end{bmatrix}, \quad \text{and} \quad z = \mathbf{y}'A\mathbf{y}$$

Then

$$z = 2y_1^2 - y_2^2 + 5y_3^2 + y_1 y_2 + 5y_1 y_3 + 2y_2 y_3$$

By definition,

$$\partial z/\partial \mathbf{y} = \begin{bmatrix} 4y_1 + y_2 + 5y_3 \\ -2y_2 + y_1 + 2y_3 \\ 10y_3 + 5y_1 + 2y_2 \end{bmatrix}$$

Now

$$A\mathbf{y} = \begin{bmatrix} 2 & 0 & 2 \\ 1 & -1 & -2 \\ 3 & 4 & 5 \end{bmatrix} \begin{bmatrix} y_1 \\ y_2 \\ y_3 \end{bmatrix} = \begin{bmatrix} 2y_1 + 2y_3 \\ y_1 - y_2 - 2y_3 \\ 3y_1 + 4y_2 + 5y_3 \end{bmatrix}$$

and

$$A'\mathbf{y} = \begin{bmatrix} 2 & 1 & 3 \\ 0 & -1 & 4 \\ 2 & -2 & 5 \end{bmatrix} \begin{bmatrix} y_1 \\ y_2 \\ y_3 \end{bmatrix} = \begin{bmatrix} 2y_1 + y_2 + 3y_3 \\ -y_2 + 4y_3 \\ 2y_1 - 2y_2 + 5y_3 \end{bmatrix}$$

Adding,

$$A\mathbf{y} + A'\mathbf{y} = \begin{bmatrix} 2y_1 + 2y_3 \\ y_1 - y_2 - 2y_3 \\ 3y_1 + 4y_2 + 5y_3 \end{bmatrix} + \begin{bmatrix} 2y_1 + y_2 + 3y_3 \\ -y_2 + 4y_3 \\ 2y_1 - 2y_2 + 5y_3 \end{bmatrix}$$

$$= \begin{bmatrix} 4y_1 + y_2 + 5y_3 \\ y_1 - 2y_2 + 2y_3 \\ 5y_1 + 2y_2 + 10y_3 \end{bmatrix} = \partial z/\partial \mathbf{y}$$

as claimed in rule 3.

EXPECTED VALUE OF RANDOM VECTORS

Recall that the expected value of a random variable Y is its theoretical average value. We denote this expectation by $E[Y]$ or μ. We now need to be able to talk about the expected value of a *random vector*, a vector whose entries are random variables rather than real variables or scalars. This expectation is defined as you should expect. Namely, the expected value of a random vector is the vector of associated expectations. Definition 2.2.1 formalizes this idea.

Definition 2.2.1

Let

$$\mathbf{Y} = \begin{bmatrix} Y_1 \\ Y_2 \\ \vdots \\ Y_k \end{bmatrix}$$

be a vector of random variables and let $E[Y_i] = \mu_i \; i = 1, 2, \ldots, k$. By $E[\mathbf{Y}]$ we mean the vector $\boldsymbol{\mu}$ given by

$$\boldsymbol{\mu} = \begin{bmatrix} E[Y_1] \\ E[Y_2] \\ \vdots \\ E[Y_k] \end{bmatrix} = \begin{bmatrix} \mu_1 \\ \mu_2 \\ \vdots \\ \mu_k \end{bmatrix}$$

A few rules govern the behavior of expectations of random vectors. These rules should come as no surprise since they parallel the rules for expectation that govern the behavior of one-dimensional random variables.

Rules for Expectation

1. Let $\mathbf{a}$ be a vector of real numbers. Then $E[\mathbf{a}] = \mathbf{a}$.
2. Let $\mathbf{a}$ be a $k \times 1$ vector of scalars and $\mathbf{Y}$ a $k \times 1$ random vector with expectation $\boldsymbol{\mu}$. Then

$$E[\mathbf{a}'\mathbf{Y}] = \mathbf{a}'E[\mathbf{Y}] = \mathbf{a}'\boldsymbol{\mu}$$

3. Let A be an $n \times k$ matrix and $\mathbf{Y}$ a $k \times 1$ random vector with expectation $\boldsymbol{\mu}$. Then

$$E[A\mathbf{Y}] = AE[\mathbf{Y}] = A\boldsymbol{\mu}$$

Example 2.2.3 illustrates the use of this notation and the rules just given.

Example 2.2.3 Let

$$A = \begin{bmatrix} 2 & 3 \\ 1 & 4 \end{bmatrix} \quad \text{and} \quad \mathbf{Y} = \begin{bmatrix} Y_1 \\ Y_2 \end{bmatrix}$$

Assume that $E[Y_1] = 10$ and $E[Y_2] = 20$ so that

$$\boldsymbol{\mu} = \begin{bmatrix} E[Y_1] \\ E[Y_2] \end{bmatrix} = \begin{bmatrix} 10 \\ 20 \end{bmatrix}$$

Now

$$AY = \begin{bmatrix} 2 & 3 \\ 1 & 4 \end{bmatrix} \cdot \begin{bmatrix} Y_1 \\ Y_2 \end{bmatrix} = \begin{bmatrix} 2Y_1 + 3Y_2 \\ Y_1 + 4Y_2 \end{bmatrix}$$

and by definition,

$$E[AY] = \begin{bmatrix} E[2Y_1 + 3Y_2] \\ E[Y_1 + 4Y_2] \end{bmatrix} = \begin{bmatrix} 80 \\ 90 \end{bmatrix}$$

To illustrate rule 3, find $AE[\mathbf{Y}] = A\boldsymbol{\mu}$. Multiplying,

$$A\boldsymbol{\mu} = \begin{bmatrix} 2 & 3 \\ 1 & 4 \end{bmatrix} \begin{bmatrix} 10 \\ 20 \end{bmatrix} = \begin{bmatrix} 80 \\ 90 \end{bmatrix} = E[A\mathbf{Y}]$$

as expected.

VARIANCE OF RANDOM VECTORS

Recall that the variance of an individual random variable Y is a special expectation that measures the variability of Y about its mean, μ. *Variance* is denoted by σ^2 and is defined by $\sigma^2 = E[(Y - \mu)^2]$. The notion of variability is harder to carry over to random vectors than is that of expectation. The reason for this is simple. When dealing with more than one random variable, a second important measure of variability comes into play, namely, the covariance between pairs of variables. Recall that given two random variables Y_i and Y_j with means μ_i and μ_j, respectively, the *covariance* is given by $\text{cov}(Y_i, Y_j) = E[(Y_i - \mu_i)(Y_j - \mu_j)]$. Hence we will *not* define the variance of a random vector to be the vector of associated variances. Rather we shall define it in such a way that both variances and covariances are accounted for and displayed simultaneously. To do so, Definition 2.2.1 must be extended. In particular, we define the expected value of a matrix A of random variables to be the matrix whose entries are term by term the expected values of the elements of A. This definition, of course, includes Definition 2.2.1 as a special case. Rules governing expectations of random matrices are given in Exercise 21.

Definition 2.2.2

Let

$$\mathbf{Y} = \begin{bmatrix} Y_1 \\ Y_2 \\ \vdots \\ Y_k \end{bmatrix}$$

be a random vector with var $Y_i = \sigma_{ii} = \sigma_i^2$ $i = 1, 2, \ldots, k$; $\text{cov}(Y_i, Y_j) = \sigma_{ij}$ $i \neq j$; and $E[\mathbf{Y}] = \boldsymbol{\mu}$. The variance of $\mathbf{Y}$, denoted by var $\mathbf{Y}$ or V, is the $k \times k$ matrix given by

$$\text{var } \mathbf{Y} = V = E[(\mathbf{Y} - \boldsymbol{\mu})(\mathbf{Y} - \boldsymbol{\mu})']$$

Direct multiplication should convince you that V does what is claimed. The entries on the main diagonal, v_{ii}, are the variances of the random variables; that is, $v_{ii} = \sigma_{ii} = \sigma_i^2$ $i = 1, 2, \ldots, k$. The off-diagonal entries are covariances with $v_{ij} = \sigma_{ij}$ for $i \neq j$. (See Exercise 20.) For this reason, V is usually called the *variance-covariance* matrix for the random vector $\mathbf{Y}$.

Throughout the remainder of the text, we will encounter questions concerning the distribution of random vectors that arise in the study of linear statistical models. The variance-covariance structure of such vectors is always of prime concern. These rules for variance will help simplify complex expressions so that their variances can be determined more easily.

Rules for Variance

1. Let $\mathbf{Y}$ be a random vector with var $\mathbf{Y} = V$. Let $\mathbf{Z} = \mathbf{a}'\mathbf{Y}$ where $\mathbf{a}$ is a vector of real numbers. Then

$$\text{var } \mathbf{a}'\mathbf{Y} = \mathbf{a}'V\mathbf{a}$$

2. Let $\mathbf{Y}$ be a random vector with var $\mathbf{Y} = V$. Let A be a $k \times k$ matrix. If $\mathbf{Z} = A\mathbf{Y}$, then

$$\text{var } \mathbf{Z} = AVA'$$

These rules follow directly from Definition 2.2.1 and the rules for expectation. Their derivations are left as exercises. (See Exercises 23 and 24.)

Thus far, we have carefully denoted random vectors with uppercase letters to be consistent with the usual labeling convention for random variables. In the statistical work to come, this convention will become awkward since we will be denoting matrices by capital letters. For this reason, in the statistical setting we shall not distinguish notationally between vectors containing only scalars and random vectors. The context of the problem will be the guide that distinguishes the two.

Example 2.2.4 illustrates rule 2 for variance.

Example 2.2.4 Let

$$\mathbf{y} = \begin{bmatrix} y_1 \\ y_2 \\ y_3 \end{bmatrix}$$

be a random vector such that var $y_i = \sigma^2$ $i = 1, 2, 3$. That is, the random variables y_1, y_2, and y_3 have common variance σ^2. Assume that y_1, y_2, and y_3 are independent and recall that this implies that $\sigma_{12} = \sigma_{13} = \sigma_{23} = 0$. That is, all covariances are zero. The variance-covariance matrix for $\mathbf{y}$ is

$$V = \begin{bmatrix} \sigma^2 & 0 & 0 \\ 0 & \sigma^2 & 0 \\ 0 & 0 & \sigma^2 \end{bmatrix} = \sigma^2 I$$

Assume that X is an $n \times k$ matrix of full rank and recall that this implies that the $k \times k$ matrix $X'X$ is nonsingular. Let us find the variance of $\mathbf{z}$ when $\mathbf{z} = (X'X)^{-1}X'\mathbf{y}$. Applying rule 2 with $A = (X'X)^{-1}X'$, we find that

$$\operatorname{var} \mathbf{z} = AVA' = [(X'X)^{-1}X']\sigma^2 I[(X'X)^{-1}X']'$$

Simplifying using the rules for transposes and inverses,

$$\begin{aligned}\operatorname{var} \mathbf{z} &= (X'X)^{-1}X'(X')'[(X'X)^{-1}]'\sigma^2 \\ &= (X'X)^{-1}(X'X)[(X'X)']^{-1}\sigma^2 \\ &= (X'X)^{-1}\sigma^2\end{aligned}$$

Example 2.2.4 was not just an idle exercise in matrix manipulation. You will see that its generalization plays a vital role in the work to come.

Before closing, we consider some special "quadratic forms." The term *quadratic forms* is in quotes because those considered here are a bit different from those defined in Section 2.1. Earlier, the vector $\mathbf{y}$ was assumed to be a vector of real variables; that is, $y_1, y_2, \ldots, y_k$ denoted real numbers and $\mathbf{y}'A\mathbf{y}$ was a function of real variables. Here, the vector $\mathbf{y}$ is assumed to be random; $y_1, y_2, \ldots, y_k$ denote random variables; and $\mathbf{y}'A\mathbf{y}$ is a *random variable* that can be expressed as a sum of squares and cross products of the y's. We should call these special forms *random quadratic forms*. However, because it will be clear from the context of the problem whether or not a quadratic form is random, we shall not make this distinction. Any expression of the form $\mathbf{y}'A\mathbf{y}$ will be referred to simply as a quadratic form.

Since $\mathbf{y}'A\mathbf{y}$ is a random variable, it makes sense to ask for its expected value. The next theorem speaks to this problem. In the proof we use an alternative form for covariance. You are asked to derive this form in Exercise 27. We also use an expanded form of the trace of the product of two matrices. This form is illustrated in Exercise 28. It would be helpful if you would try these exercises before continuing.

Theorem 2.2.1 Let $\mathbf{y}$ be a $k \times 1$ random vector with $E[\mathbf{y}] = \boldsymbol{\mu}$ and $\operatorname{var} \mathbf{y} = V$. Let A be a $k \times k$ matrix of real numbers. Then

$$E[\mathbf{y}'A\mathbf{y}] = \operatorname{tr}(AV) + \boldsymbol{\mu}'A\boldsymbol{\mu}$$

Proof

We know that for $i \neq j$,

$$\sigma_{ij} = E[y_i y_j] - \mu_i \mu_j$$

and that for $i = j$,

$$\sigma_{ij} = \sigma_{ii} = \sigma_i^2 = E[y_i^2] - \mu_i^2$$

Writing the quadratic form $\mathbf{y}'A\mathbf{y}$ in expanded form as was done in Section 2.1,

$$\begin{aligned} E[\mathbf{y}'A\mathbf{y}] &= E\left[\sum_{i=1}^{k}\sum_{j=1}^{k} a_{ij}y_iy_j\right] \\ &= \sum_{i=1}^{k}\sum_{j=1}^{k} E[a_{ij}y_iy_j] \end{aligned}$$

We know that $\sigma_{ij} = E[y_iy_j] - \mu_i\mu_j$ and hence $E[y_iy_j] = \sigma_{ij} + \mu_i\mu_j$. Substituting,

$$\begin{aligned} E[\mathbf{y}'A\mathbf{y}] &= \sum_{i=1}^{k}\sum_{j=1}^{k} a_{ij}(\sigma_{ij} + \mu_i\mu_j) \\ &= \sum_{i=1}^{k}\sum_{j=1}^{k} a_{ij}\sigma_{ij} + \sum_{i=1}^{k}\sum_{j=1}^{k} a_{ij}\mu_i\mu_j \end{aligned}$$

Noting that the terms σ_{ij} are the entries of the variance-covariance matrix and that $\sigma_{ij} = \sigma_{ji}$, we use Exercise 28 to claim that

$$\sum_{i=1}^{k}\sum_{j=1}^{k} a_{ij}\sigma_{ij} = \operatorname{tr}(AV)$$

The term $\Sigma_{i=1}^{k}\Sigma_{j=1}^{k} a_{ij}\mu_i\mu_j$ is the expanded form of $\boldsymbol{\mu}'A\boldsymbol{\mu}$ developed in Exercise 1, Section 2.1. Substituting, it is easy to see that

$$E[\mathbf{y}'A\mathbf{y}] = \operatorname{tr}(AV) + \boldsymbol{\mu}'A\boldsymbol{\mu}$$

as claimed. ■

The next example illustrates Theorem 2.2.1 in a familiar setting.

Example 2.2.5 Let $\mathbf{y}$ be a 2×1 random vector with

$$\boldsymbol{\mu} = \begin{bmatrix} 1 \\ 3 \end{bmatrix} \quad \text{and} \quad V = \begin{bmatrix} 2 & 1 \\ 1 & 5 \end{bmatrix}$$

Let

$$A = \begin{bmatrix} 4 & 1 \\ 1 & 2 \end{bmatrix}$$

Let us first find $E[\mathbf{y}'A\mathbf{y}]$ using only the definitions of variance and covariance from elementary statistics. To do so, note that

$$\mathbf{y}'A\mathbf{y} = 4y_1^2 + 2y_1y_2 + 2y_2^2$$

and

$$E[\mathbf{y}'A\mathbf{y}] = 4E[y_1^2] + 2E[y_1y_2] + E[y_2^2]$$

From the variance-covariance matrix and $\boldsymbol{\mu}$, it can be seen that

$$\begin{aligned} 2 = \text{var}(y_1) &= E[y_1^2] - E[y_1]^2 \\ &= E[y_1^2] - 1 \end{aligned}$$

Solving for $E[y_1^2]$, we see that $E[y_1^2] = 3$. In a similar manner, it can be seen that $E[y_2^2] = 14$. Applying the formula for covariance given in Exercise 27,

$$\begin{aligned} 1 = \text{cov}(y_1, y_2) &= E[y_1 y_2] - E[y_1]E[y_2] \\ &= E[y_1 y_2] - 1(3) \end{aligned}$$

Solving for $E[y_1 y_2]$, we see that $E[y_1 y_2] = 4$. Substitution now yields

$$\begin{aligned} E[\mathbf{y}'A\mathbf{y}] &= 4(3) + 2(4) + 2(14) \\ &= 48 \end{aligned}$$

To find this expectation by means of Theorem 2.2.1, we need only note that

$$AV = \begin{bmatrix} 4 & 1 \\ 1 & 2 \end{bmatrix} \begin{bmatrix} 2 & 1 \\ 1 & 5 \end{bmatrix} = \begin{bmatrix} 9 & 9 \\ 4 & 11 \end{bmatrix}$$

and $\text{tr}(AV) = 20$. Matrix multiplication yields $\boldsymbol{\mu}'A\boldsymbol{\mu} = 28$. By Theorem 2.2.1,

$$\begin{aligned} E[\mathbf{y}'A\mathbf{y}] &= \text{tr}(AV) + \boldsymbol{\mu}'A\boldsymbol{\mu} \\ &= 48 \end{aligned}$$

as expected.

2.3

DISTRIBUTION OF SOME SPECIAL QUADRATIC FORMS

When a statistician uses the term *distribution*, he or she means several things. Primarily, distribution means the group or family to which the random variable under study belongs. We want to know if we are dealing with the normal, t, chi-squared (X^2), F, binomial, Poisson, or perhaps some less frequently encountered type of random variable. Once the family to which the variable belongs is identified, we must determine the value(s) of the parameter(s) that characterize the distribution. For example, in the case of the normal family, we need to find the mean and variance to completely describe the distribution; for the t, X^2, and F families, we need to determine degrees of freedom. In this section we introduce a distribution that underlies much of the theory of linear statistical models. This distribution, the noncentral chi-squared distribution, is probably new to you. As you shall see, the usual chi-squared distribution is a special important member of the family of noncentral chi-squared random variables.

Definition 2.3.1

Let $\mathbf{y}$ be a $k \times 1$ normally distributed random vector with mean $\boldsymbol{\mu}$ and variance I. Then $\mathbf{y}'\mathbf{y}$ follows the *noncentral chi-squared distribution* with k degrees of freedom and noncentrality parameter $\lambda = (\frac{1}{2})\boldsymbol{\mu}'\boldsymbol{\mu}$. We shall denote such a random variable by $X^2_{k,\lambda}$.

This definition has many facets. First, when we say that $\mathbf{y}$ is normally distributed with mean $\boldsymbol{\mu}$, we are implying that the random variables $y_1, y_2, \ldots, y_k$ that constitute the random vector $\mathbf{y}$ are normally distributed with means $\mu_1, \mu_2, \ldots, \mu_k$, respectively. These means are not necessarily equal to one another. Second, to say that var $\mathbf{y} = I$ means that the variance-covariance matrix of $\mathbf{y}$ is the identity matrix. More important, it means that the variance of each of the random variables $y_1, y_2, \ldots, y_k$ is 1 and that their covariances are all 0. Recall that in the case of normal random variables, zero covariance implies independence. So we are dealing with a collection of k independent normally distributed random variables with means $\mu_1, \mu_2, \ldots \mu_k$ and common variance 1. Third, the random variable $\mathbf{y}'\mathbf{y}$ is a sum of squares. That is, $\mathbf{y}'\mathbf{y} = \Sigma_{i=1}^k y_i^2$. The definition thus says that the sum of squares of k independent normally distributed random variables each with variance 1 follows what will be called a *noncentral chi-squared* distribution. This distribution will be characterized by two parameters: k, called degrees of freedom, and λ, called the noncentrality parameter.

Example 2.3.1 Let y_1, y_2, and y_3 be independent normally distributed random variables with common variance 1 and means 4, 2, and -2, respectively. Here

$$\boldsymbol{\mu} = \begin{bmatrix} 4 \\ 2 \\ -2 \end{bmatrix} \quad \text{and} \quad \text{var } \mathbf{y} = I = \begin{bmatrix} 1 & 0 & 0 \\ 0 & 1 & 0 \\ 0 & 0 & 1 \end{bmatrix}$$

By Definition 2.3.1, $\mathbf{y}'\mathbf{y} = \Sigma_{i=1}^k y_i^2$ follows a noncentral chi-squared distribution with $k = 3$ degrees of freedom and noncentrality parameter

$$\lambda = (\tfrac{1}{2})\boldsymbol{\mu}'\boldsymbol{\mu} = (\tfrac{1}{2})[4 \quad 2 \quad -2] \begin{bmatrix} 4 \\ 2 \\ -2 \end{bmatrix} = 12$$

We denote this random variable by $X^2_{3,12}$.

One way that authors introduce the ordinary chi-squared distribution is to define a chi-squared random variable with k degrees of freedom to be a random variable that can be expressed as the sum of squares of k independent standard normal random variables, $z_1, z_2, \ldots, z_k$. Since the term *standard normal* implies mean 0 and variance 1, $\mathbf{z}$ is normally distributed with $\boldsymbol{\mu} = \mathbf{0}$ and $\operatorname{var} \mathbf{z} = I$. By Definition 2.3.1,

$$X^2_{k,0} = \sum_{i=1}^{k} z_i^2 = \mathbf{z}'\mathbf{z}$$

follows a noncentral chi-squared distribution with k degrees of freedom and noncentrality parameter $\lambda = (\frac{1}{2})\boldsymbol{\mu}'\boldsymbol{\mu} = 0$. That is, an *ordinary chi-squared random variable is simply a noncentral chi-squared random variable with noncentrality parameter 0*. Notationally, we write X^2_k with the understanding that $\lambda = 0$. In the future, this random variable will be referred to simply as a chi-squared random variable.

One very important property of noncentral chi-squared random variables is of great help in the development of the theory of linear statistical models. This property, called the *additivity property*, essentially states that a sum of independent noncentral chi-squared random variables is itself a noncentral chi-squared random variable and that both degrees of freedom and noncentrality parameters can be added. This should not be too surprising since a similar result holds for ordinary chi-squared random variables. The proof of this result uses moment-generating function techniques and is found in more advanced linear models texts [7] [17]. The theorem is stated here without proof.

Theorem 2.3.1 Let $X^2_{k_1,\lambda_1}, X^2_{k_2,\lambda_2}, \ldots, X^2_{k_n,\lambda_n}$ be a collection of n independent noncentral chi-squared random variables with $k_1, k_2, \ldots, k_n$ degrees of freedom and noncentrality parameters $\lambda_1, \lambda_2, \ldots, \lambda_n$, respectively. Then

$$\sum_{i=1}^{n} X^2_{k_i,\lambda_i}$$

follows a noncentral chi-squared distribution with $k = k_1 + k_2 + \cdots + k_n$ degrees of freedom and noncentrality parameter $\lambda = \lambda_1 + \lambda_2 + \cdots \lambda_n$. That is,

$$\sum_{i=1}^{n} X^2_{k_i,\lambda_i} = X^2_{k,\lambda}$$

■

The next series of theorems and corollaries concerns the distribution of quadratic forms under various assumptions on the form of the variance-covariance matrix of the random vector $\mathbf{y}$. The first proof given entails the use of partitioned matrices and relies on some simple manipulations of these matrices that have not been pointed out. In order not to interrupt the flow of the proof, these ideas are outlined in Exercise 35. You should try this exercise before continuing.

Theorem 2.3.2 Let $\mathbf{y}$ be an $n \times 1$ normally distributed random vector with mean $\boldsymbol{\mu}$ and variance I. Let A be an $n \times n$ symmetric matrix. Then $\mathbf{y}'A\mathbf{y}$ follows a noncentral chi-squared distribution with k degrees of freedom and noncentrality parameter $\lambda = (\frac{1}{2})\boldsymbol{\mu}'A\boldsymbol{\mu}$ if and only if A is idempotent of rank k.

Proof

Assume that A is symmetric and idempotent of rank k. Theorem 1.4.1 guarantees that there exists an orthogonal matrix P that diagonalizes A and displays its eigenvalues along the main diagonal. That is, there exists an orthogonal matrix P such that

$$P'AP = \begin{bmatrix} \lambda_1 & 0 & \cdots & 0 \\ 0 & \lambda_2 & & \vdots \\ \vdots & & \ddots & \\ 0 & & & \lambda_n \end{bmatrix}$$

Since A is idempotent, these eigenvalues are all either 0 or 1 (Theorem 1.5.1). Since the rank of a symmetric idempotent matrix is equal to its trace (Theorem 1.5.2), $r(A) = \text{tr}(A) = k$. However, $\text{tr}(P'AP) = \text{tr}(PP'A) = \text{tr}(A)$, and we can conclude that $\text{tr}(P'AP) = \Sigma_{i=1}^{n} \lambda_i = k$. This implies that k of these eigenvalues must be ones and the remaining $n - k$ are zeros. We can therefore rearrange the columns of $P'AP$ and partition the resulting matrix as

$$P'AP = \left[\begin{array}{c|c} I & 0 \\ \hline 0 & 0 \end{array}\right]$$

where I is a $k \times k$ identity matrix. Now define a random vector $\mathbf{z}$ by $\mathbf{z} = P'\mathbf{y}$. We can write $\mathbf{z}$ in partitioned form as

$$\mathbf{z} = \left[\begin{array}{c} \mathbf{z}_1 \\ \hline \mathbf{z}_2 \end{array}\right]$$

where $\mathbf{z}_1$ is a $k \times 1$ vector and $\mathbf{z}_2$ is an $(n-k) \times 1$ vector. Similarly, P is partitioned so that

$$P = [P_1 \mid P_2]$$

where P_1 is an $n \times k$ matrix and P_2 is an $n \times (n-k)$ matrix. Substituting,

$$\left[\begin{array}{c} \mathbf{z}_1 \\ \hline \mathbf{z}_2 \end{array}\right] = \mathbf{z} = P'\mathbf{y} = \left[\begin{array}{c} P_1' \\ \hline P_2' \end{array}\right]\mathbf{y} = \left[\begin{array}{c} P_1'\mathbf{y} \\ \hline P_2'\mathbf{y} \end{array}\right]$$

where $P_1'\mathbf{y}$ is a $k \times 1$ vector and $P_2'\mathbf{y}$ is an $(n-k) \times 1$ vector. In particular, we can conclude that $\mathbf{z}_1 = P_1'\mathbf{y}$. Notice that the entries in the vector $\mathbf{z}_1$ are linear combinations of the independent normally distributed random variables $y_1, y_2, \ldots, y_n$ and hence each entry is itself a normally distributed random variable. Now consider the quadratic form $\mathbf{y}'A\mathbf{y}$. Since $\mathbf{z} = P'\mathbf{y}$, we know that $P\mathbf{z} = PP'\mathbf{y}$. However, P is orthogonal and so $PP' = I$, and we can conclude that $\mathbf{y} = P\mathbf{z}$. Substituting,

$$\begin{aligned}
\mathbf{y}'A\mathbf{y} &= (P\mathbf{z})'A(P\mathbf{z}) \\
&= \mathbf{z}'P'AP\mathbf{z} \\
&= [\mathbf{z}_1' \mid \mathbf{z}_2'] \left[\begin{array}{c|c} I & 0 \\ \hline 0 & 0 \end{array}\right] \left[\begin{array}{c} \mathbf{z}_1 \\ \hline \mathbf{z}_2 \end{array}\right] \\
&= \mathbf{z}_1'\mathbf{z}_1
\end{aligned}$$

To determine the distribution of $\mathbf{y}'A\mathbf{y}$, we need to consider the distribution of $\mathbf{z}_1'\mathbf{z}_1$. We have already noted that the entries in $\mathbf{z}_1$ are all normally distributed random variables. Hence $\mathbf{z}_1'\mathbf{z}_1$ is a sum of squares of normally distributed random variables. To find $\boldsymbol{\mu}$, use the rules for expectation (Section 2.2). Since $\mathbf{z}_1 = P_1'\mathbf{y}$, $E[\mathbf{z}_1] = E[P_1'\mathbf{y}] = P_1'E[\mathbf{y}] = P_1'\boldsymbol{\mu}$. The rules for variance (Section 2.2) are used to see that $\text{var}\ \mathbf{z}_1 = \text{var}\ P_1'\mathbf{y} = P_1'VP_1$ where V is the variance-covariance matrix of $\mathbf{y}$. In this case, $V = I$ and we have $\text{var}\ \mathbf{z}_1 = P_1'P_1$. It can be shown that $P_1'P_1 = I$ (see Exercise 36). Hence we have shown that $\mathbf{z}_1$ is a $k \times 1$ normally distributed random vector with mean $P_1'\boldsymbol{\mu}$ and variance I. By definition, $\mathbf{z}_1'\mathbf{z}_1$ follows a noncentral chi-squared distribution with k degrees of freedom and noncentrality parameter

$$\lambda = (\tfrac{1}{2})(P_1'\boldsymbol{\mu})'(P_1'\boldsymbol{\mu}) = (\tfrac{1}{2})\boldsymbol{\mu}'P_1P_1'\boldsymbol{\mu}$$

By Exercise 36, $P_1P_1' = A$ and hence $\lambda = (\frac{1}{2})\boldsymbol{\mu}'A\boldsymbol{\mu}$ as claimed. Although the proof of the converse is beyond the scope of this text, it can be found in [3]. ∎

Two corollaries follow easily from this theorem. Their proofs are left as exercises (Exercises 37–38).

Corollary 2.3.1 Let $\mathbf{y}$ be an $n \times 1$ normally distributed random vector with mean $\mathbf{0}$ and variance I. Let A be an $n \times n$ symmetric matrix. Then $\mathbf{y}'A\mathbf{y}$ follows a chi-squared distribution with k degrees of freedom if and only if A is idempotent of rank k. ∎

Corollary 2.3.2 Let $\mathbf{y}$ be an $n \times 1$ normally distributed random vector with mean $\boldsymbol{\mu}$ and variance $\sigma^2 I$ where $\sigma^2 > 0$. Let A be an $n \times n$ symmetric matrix. Then $(1/\sigma^2)\mathbf{y}'A\mathbf{y}$ follows a noncentral chi-squared distribution with k degrees of freedom and $\lambda = (1/2\sigma^2)\boldsymbol{\mu}'A\boldsymbol{\mu}$ if and only if A is idempotent of rank k. ∎

MULTIVARIATE NORMAL DISTRIBUTION

Thus far we have assumed that the variance-covariance matrix for $\mathbf{y}$ is at least diagonal. This means, of course, that the covariances are all zero and that the normally distributed random variables are independent. Suppose that this is not true. Can we say anything about the distribution of $\mathbf{y}'A\mathbf{y}$ in this case? Can we prove a theorem similar to Theorem 2.3.2, which drops the assumption that var $\mathbf{y} = I$? The answer is yes, but to do so we must define the multivariate normal distribution.

Definition 2.3.2

Let $\mathbf{y}$ be an $n \times 1$ normally distributed random vector with mean $\boldsymbol{\mu}$ and variance I. Let C be an $n \times n$ nonsingular matrix and define the random vector $\mathbf{z}$ by

$$\mathbf{z} = C'\mathbf{y}$$

Then $\mathbf{z}$ is said to follow the *multivariate normal distribution.* We call $\mathbf{z}$ a multivariate normal random variable.

This definition has several important implications. First, each entry in $\mathbf{z}$ is a linear combination of the independent normally distributed random variables $y_1, y_2, \ldots, y_n$, and hence each entry is itself a normally distributed random variable. Second, the rules for expectation and variance given in Section 2.2 can be applied to show that $E[\mathbf{z}] = C'\boldsymbol{\mu}$ and that var $\mathbf{z} = C'IC = C'C$. That is, the variance-covariance matrix of a multivariate normal random variable can always be expressed in the form $C'C$ for some nonsingular matrix C. Third, the normally distributed random vectors $\mathbf{y}$ that we studied earlier are special multivariate normal random variables. To see this, note that the identity matrix I is nonsingular and that $\mathbf{y} = I'\mathbf{y}$. With these points made, we can extend Theorem 2.3.2 to the general multivariate normal case. You will be asked to verify some of the statements in the proof as exercises.

Theorem 2.3.3 Let $\mathbf{y}$ be an $n \times 1$ multivariate normal random variable with mean $\boldsymbol{\mu}$ and variance V. Let A be an $n \times n$ symmetric matrix. Then $\mathbf{y}'A\mathbf{y}$ follows a noncentral chi-squared distribution with k degrees of freedom and noncentrality parameter $\lambda = (\frac{1}{2})\boldsymbol{\mu}'A\boldsymbol{\mu}$ if and only if AV is idempotent of rank k.

Proof

Since $\mathbf{y}$ is a multivariate normal random variable, there exists a nonsingular matrix C such that $V = C'C$. Now let us define a random vector $\mathbf{z}$ by

$$\mathbf{z} = (C')^{-1}(\mathbf{y} - \boldsymbol{\mu})$$

The rules for expectation and variance can be used to show that $E[\mathbf{z}] = \mathbf{0}$ and var $\mathbf{z} = I$ (see Exercise 40). Notice that $\mathbf{y}$ can be written in terms of $\mathbf{z}$ by multiplying $\mathbf{z}$ by C' and solving for $\mathbf{y}$ to obtain

$$\mathbf{y} = C'\mathbf{z} + \boldsymbol{\mu}$$

Substituting, the quadratic form $\mathbf{y}'A\mathbf{y}$ is written as

$$\mathbf{y}'A\mathbf{y} = (C'\mathbf{z} + \boldsymbol{\mu})'A(C'\mathbf{z} + \boldsymbol{\mu})$$

It can be shown that

$$(C'\mathbf{z} + \boldsymbol{\mu})'A(C'\mathbf{z} + \boldsymbol{\mu}) = \mathbf{u}'B\mathbf{u}$$

where $\mathbf{u} = \mathbf{z} + (C')^{-1}\boldsymbol{\mu}$ and $B = CAC'$ (see Exercise 41). Note that $\mathbf{u}$ is normally distributed with mean $(C')^{-1}\boldsymbol{\mu}$ and variance I (see Exercise 42). We now know that

$$\mathbf{y}'A\mathbf{y} = \mathbf{u}'B\mathbf{u}$$

and that the random vector $\mathbf{u}$ satisfies the conditions of Theorem 2.3.2. Applying this theorem, we can conclude that $\mathbf{y}'A\mathbf{y} = \mathbf{u}'B\mathbf{u}$ follows a noncentral chi-squared distribution with k degrees of freedom and noncentrality parameter $\lambda = (\frac{1}{2}) \times [(C')^{-1}\boldsymbol{\mu}]'B[(C')^{-1}\boldsymbol{\mu}]$ if and only if $B = CAC'$ is idempotent of rank k. It is easy to show that $\lambda = (\frac{1}{2})\boldsymbol{\mu}'A\boldsymbol{\mu}$ (see Exercise 43). Hence the proof will be complete when we show that B is idempotent of rank k if and only if AV is idempotent of rank k. For B to be idempotent, we must have $B^2 = B$. Substituting and remembering that $C'C = V$, we must have

$$(CAC')(CAC') = CAC'$$
$$CA(C'C)AC' = CAC'$$
$$CAVAC' = CAC'$$

This equality will hold if and only if

$$C^{-1}CAVAC'C = C^{-1}CAC'C$$
$$(AV)(AV) = AV$$

That is, B is idempotent if and only if AV is idempotent. To complete the proof we need only note that we have a noncentral chi-squared distribution if and only if $r(B) = k$. But $r(B) = r(CAC') = r(AC'C) = r(AV)$. Hence we must have $r(AV) = k$ as claimed. ■

Corollaries 2.3.3 and 2.3.4 follow easily from Theorem 2.3.3.

Corollary 2.3.3 Let $\mathbf{y}$ be an $n \times 1$ multivariate normal random variable with mean $\mathbf{0}$ and variance V. Let A be an $n \times n$ symmetric matrix. Then $\mathbf{y}'A\mathbf{y}$ follows a chi-squared distribution with k degrees of freedom if and only if AV is idempotent of rank k. ■

Corollary 2.3.4 Let $\mathbf{y}$ be an $n \times 1$ multivariate normal random variable with mean $\boldsymbol{\mu}$ and variance V. Then $\mathbf{y}'V^{-1}\mathbf{y}$ follows a noncentral chi-squared distribution with n degrees of freedom and noncentrality parameter $\lambda = (\frac{1}{2})\boldsymbol{\mu}'V^{-1}\boldsymbol{\mu}$. ■

The importance of these results to the theory of linear statistical models will become evident in Chapter 3.

2.4

INDEPENDENCE OF QUADRATIC FORMS

We have used the term *independent random variables* assuming that you are familiar with its definition. Recall that if $\mathbf{y}$ is an $n \times 1$ random vector then the quadratic forms $\mathbf{y}'A\mathbf{y}$ and $\mathbf{y}'B\mathbf{y}$ are random variables in their own right. Each is a sum of squares and cross products of the random variables $y_1, y_2, \ldots, y_n$. In this section we investigate conditions under which these quadratic forms are independent. We also consider what is meant by independence between a quadratic form $\mathbf{y}'A\mathbf{y}$ and a random vector $B\mathbf{y}$.

Our first theorem gives sufficient conditions for quadratic forms of the type encountered in statistical models to be independent. Its proof depends on Theorem 1.5.3. This theorem is restated here as a lemma for easy reference.

Lemma 2.4.1 Let $A_1, A_2, \ldots, A_m$ be a collection of $k \times k$ symmetric matrices. A necessary and sufficient condition for the existence of a common orthogonal matrix P such that $P'A_iP$ is diagonal is that $A_iA_j = A_jA_i$ for every pair (i, j). ■

Theorem 2.4.1 Let $\mathbf{y}$ be an $n \times 1$ multivariate normal random variable with mean $\boldsymbol{\mu}$ and variance V. Let A and B be symmetric $n \times n$ matrices of ranks r_1 and r_2, respectively. If $AVB = 0$, then $\mathbf{y}'A\mathbf{y}$ and $\mathbf{y}'B\mathbf{y}$ are independent. Furthermore, if $\mathbf{y}'A\mathbf{y}$ and $\mathbf{y}'B\mathbf{y}$ are independent, then $AVB = 0$.

Proof

(Sufficiency) Let $\mathbf{y}$ be an $n \times 1$ multivariate normal random variable with mean $\boldsymbol{\mu}$ and variance V. Let A and B be symmetric $n \times n$ matrices of ranks r_1 and r_2, respectively. Assume that $AVB = 0$. Since $\mathbf{y}$ is a multivariate normal random variable, by definition there exists an $n \times n$ nonsingular matrix C such that $V = C'C$. Substituting, we are assuming that $AC'CB = 0$. Multiply on the left by C and on the right by C' to conclude that $(CAC')(CBC') = 0$. Let $R = CAC'$ and

$S = CBC'$. We know that $RS = 0$ and hence that $(RS)' = 0$. However,

$$\begin{aligned}(RS)' &= [(CAC')(CBC')]' \\ &= (CBC')'(CAC') \\ &= (CB'C')(CA'C')\end{aligned}$$

Since A and B are assumed to be symmetric, $A' = A$ and $B' = B$. Thus

$$(RS)' = (CBC')(CAC') = SR = 0$$

From this, we can conclude that $RS = SR$. It is easy to see that R and S are each symmetric and hence we can apply Lemma 2.4.1 to conclude that there exists a common orthogonal matrix P that diagonalizes both R and S simultaneously. That is, there exists an orthogonal matrix P such that

$$P'RP = \left[\begin{array}{c|c} D_1 & 0 \\ \hline 0 & 0 \end{array}\right] \quad \text{and} \quad P'SP = \left[\begin{array}{c|c} 0 & 0 \\ \hline 0 & D_2 \end{array}\right]$$

where D_1 is an $r_1 \times r_1$ diagonal matrix and D_2 is an $(n - r_1) \times (n - r_1)$ diagonal matrix. Define the random vector $\mathbf{z}$ by

$$\mathbf{z} = P'(C')^{-1}\mathbf{y}$$

It is not hard to show that $\mathbf{z}$ is a multivariable normal random variable with mean $P'(C')^{-1}\boldsymbol{\mu}$ and variance I (see Exercise 51). Hence $z_1, z_2, \ldots, z_n$, the elements of $\mathbf{z}$, are independent. Let us now partition $\mathbf{z}$ as follows:

$$\mathbf{z} = \left[\begin{array}{c} \mathbf{z}_1 \\ \hline \mathbf{z}_2 \end{array}\right]$$

Here $\mathbf{z}_1$ is an $r_1 \times 1$ vector and $\mathbf{z}_2$ is an $(n - r_1) \times 1$ vector. Let us now rewrite the quadratic forms $\mathbf{y}'A\mathbf{y}$ and $\mathbf{y}'B\mathbf{y}$ and consider their relationship to one another. To do so note that

$$\begin{aligned}\mathbf{y} &= C'P\mathbf{z} \\ A &= C^{-1}R(C')^{-1} \\ B &= C^{-1}S(C')^{-1}\end{aligned}$$

(See Exercise 52.) Substituting,

$$\begin{aligned}\mathbf{y}'A\mathbf{y} &= \mathbf{z}'P'CC^{-1}R(C')^{-1}C'P\mathbf{z} \\ &= \mathbf{z}'P'RP\mathbf{z} \\ &= [\mathbf{z}_1' \mid \mathbf{z}_2'] \left[\begin{array}{c|c} D_1 & 0 \\ \hline 0 & 0 \end{array}\right] \left[\begin{array}{c} \mathbf{z}_1 \\ \hline \mathbf{z}_2 \end{array}\right]\end{aligned}$$

Note that $\mathbf{y}'A\mathbf{y}$ is a function only of the first r_1 elements of $\mathbf{z}$. Now

$$\begin{aligned}\mathbf{y}'B\mathbf{y} &= \mathbf{z}'P'CC^{-1}S(C')^{-1}C'P\mathbf{z}\\ &= \mathbf{z}'P'SP\mathbf{z}\\ &= [\mathbf{z}_1' \mid \mathbf{z}_2'] \left[\begin{array}{c|c} 0 & 0 \\ \hline 0 & D_2 \end{array}\right]\end{aligned}$$

From this it is clear that $\mathbf{y}'B\mathbf{y}$ depends only on the last $n-r_1$ elements of $\mathbf{z}$. Since the elements of $\mathbf{z}$ are independent of one another, $\mathbf{y}'A\mathbf{y}$ and $\mathbf{y}'B\mathbf{y}$ are independent as claimed. The proof of the converse depends upon results not presented in this text. It can be found in [7]. ■

In the study of linear statistical models, it is often assumed that $y_1, y_2, \ldots, y_n$ are independent normally distributed random variables with common variance σ^2. In this case the variance-covariance matrix is the diagonal matrix $\sigma^2 I$. The next corollary gives conditions under which $\mathbf{y}'A\mathbf{y}$ and $\mathbf{y}'B\mathbf{y}$ are independent in this special setting.

Corollary 2.4.1 Let $\mathbf{y}$ be an $n \times 1$ multivariate normal random variable with mean $\boldsymbol{\mu}$ and variance $\sigma^2 I$. Let A and B be symmetric $n \times n$ matrices of ranks r_1 and r_2, respectively. If $AB = 0$, then $\mathbf{y}'A\mathbf{y}$ and $\mathbf{y}'B\mathbf{y}$ are independent. Furthermore, if $\mathbf{y}'A\mathbf{y}$ and $\mathbf{y}'B\mathbf{y}$ are independent, then $AB = 0$. ■

Thus far we have considered independence between two random variables that are expressed as quadratic forms in $\mathbf{y}$. Suppose we are interested in a quadratic form $\mathbf{y}'A\mathbf{y}$ and a random vector $B\mathbf{y}$. What do we mean when we say that these two entities are independent? We answer this question in a logical way. In particular, we will consider $\mathbf{y}'A\mathbf{y}$ and $B\mathbf{y}$ to be *independent* if and only if $\mathbf{y}'A\mathbf{y}$ is independent of each entry of $B\mathbf{y}$. The next theorem gives conditions under which independence is guaranteed. Its proof is found in [7].

Theorem 2.4.2 Let $\mathbf{y}$ be an $n \times 1$ multivariate normal random variable with mean $\boldsymbol{\mu}$ and variance V. Let A be an $n \times n$ symmetric matrix and let B be an $m \times n$ matrix. If $BVA = 0$, then $\mathbf{y}'A\mathbf{y}$ and $B\mathbf{y}$ are independent. Furthermore, if $\mathbf{y}'A\mathbf{y}$ and $B\mathbf{y}$ are independent, then $BVA = 0$. ■

In testing various statistical hypotheses later, we will need to consider collections of quadratic forms in $\mathbf{y}$. We will want to answer two important questions:

1. What is the distribution of each of the quadratic forms in the collection?
2. Are these quadratic forms independent?

The next theorem allows us to answer these questions based on knowledge of the characteristics of the matrices of the quadratic forms.

Theorem 2.4.3 Let $\mathbf{y}$ be an $n \times 1$ multivariate normal random variable with mean $\boldsymbol{\mu}$ and variance I. Let $\mathbf{y}'A_1\mathbf{y}$, $\mathbf{y}'A_2\mathbf{y}$, ..., $\mathbf{y}'A_m\mathbf{y}$ be a collection of m quadratic forms where for each $i = 1, 2, \ldots, m$, A_i is symmetric of rank r_i. If any two of the following three statements are true, then for each i, $\mathbf{y}'A_i\mathbf{y}$ follows a noncentral chi-squared distribution with r_i degrees of freedom and noncentrality parameter $\lambda_i = (\frac{1}{2})\boldsymbol{\mu}'A_i\boldsymbol{\mu}$. Furthermore, $\mathbf{y}'A_i\mathbf{y}$ and $\mathbf{y}'A_j\mathbf{y}$ are independent for $i \neq j$ and $\Sigma_{i=1}^{m} r_i = r$ where r denotes the rank of $\Sigma_{i=1}^{m} A_i$.

1. All A_i are idempotent.
2. $\Sigma_{i=1}^{m} A_i$ is idempotent.
3. $A_iA_j = 0$; $i \neq j$.

Proof

By Theorem 1.5.4, the truth of any two of the conditions guarantees the truth of the third. Hence if any two are true, then A_i is idempotent of rank r_i. By Theorem 2.3.2, $\mathbf{y}'A_i\mathbf{y}$ follows a noncentral chi-squared distribution with r_i degrees of freedom and noncentrality parameter $\lambda = (\frac{1}{2})\boldsymbol{\mu}'A_i\boldsymbol{\mu}$ as claimed. We know from condition 3 that $A_iA_j = 0$ for $i \neq j$. Applying Corollary 2.4.1 with $\sigma^2 = 1$, we can conclude that $\mathbf{y}'A_i\mathbf{y}$ and $\mathbf{y}'A_j\mathbf{y}$ are independent. By Theorem 1.5.5, $\Sigma_{i=1}^{m} r_i = r$, and the proof is complete. ■

This theorem is very important in the work to come. Many of the tests developed in Chapters 4 and 6 depend on a general technique called *analysis of variance.* The basic idea behind this technique is to partition a sum of squares into a sum of meaningful quadratic forms. That is, the task of the statistician is to express $\mathbf{y}'\mathbf{y}$ as

$$\mathbf{y}'\mathbf{y} = \mathbf{y}'A_1\mathbf{y} + \mathbf{y}'A_2\mathbf{y} + \cdots + \mathbf{y}'A_m\mathbf{y}$$

Since $\Sigma_{i=1}^{m} A_i = I$ is idempotent, under appropriate conditions Theorem 2.4.3 can be used to establish the distribution of the quadratic forms $\mathbf{y}'A_i\mathbf{y}$ by showing simply that each of the matrices $A_1, A_2, \ldots, A_m$ is idempotent.

In the statistical work to come, we will encounter various ratios of random variables. For these ratios to follow some of the classic distributions such as Student's t or F, it is necessary that the numerator be independent of the denominator. The theorems presented in this section will be helpful in showing independence in this setting.

EXERCISES

Section 2.1

1. Show that $\mathbf{y}'A\mathbf{y} = \Sigma_{i=1}^{k}\Sigma_{j=1}^{k} a_{ij}y_i y_j$ by writing $\mathbf{y}'$ as $[y_1 \quad y_2 \quad \cdots \quad y_k]$ and writing A in the form

$$A = \begin{bmatrix} a_{11} & a_{12} & \cdots & a_{1k} \\ a_{21} & a_{22} & \cdots & a_{2k} \\ \vdots & \vdots & & \vdots \\ a_{k1} & a_{k2} & & a_{kk} \end{bmatrix}$$

and multiplying.

2. Let

$$\mathbf{y} = \begin{bmatrix} y_1 \\ y_2 \end{bmatrix} \quad \text{and} \quad A = \begin{bmatrix} 2 & 4 \\ 1 & 6 \end{bmatrix}$$

Find $\mathbf{y}'A\mathbf{y}$ via the addition formula of Exercise 1 and by direct matrix multiplication.

3. Let

$$\mathbf{y} = \begin{bmatrix} y_1 \\ y_2 \\ y_3 \end{bmatrix} \quad \text{and} \quad A = \begin{bmatrix} 1 & 2 & 0 \\ 0 & 1 & -1 \\ 1 & 3 & 6 \end{bmatrix}$$

Find $\mathbf{y}'A\mathbf{y}$ via the addition formula of Exercise 1 and by direct matrix multiplication. Do you notice any pattern to the coefficients that would allow you to form $\mathbf{y}'A\mathbf{y}$ by inspection alone?

4. Let

$$\mathbf{y} = \begin{bmatrix} y_1 \\ y_2 \\ y_3 \\ y_4 \end{bmatrix} \quad \text{and} \quad A = \begin{bmatrix} 1 & 4 & 8 & 0 \\ 0 & 6 & 1 & 1 \\ 3 & 1 & 2 & 4 \\ 1 & -1 & 2 & 3 \end{bmatrix}$$

Without doing any computations, what is the coefficient of the term y_3^2 in the expression for $\mathbf{y}'A\mathbf{y}$? What is the coefficient for $y_1 y_3$ in this expression? For $y_2 y_3$?

5. Consider the matrix

$$A = \begin{bmatrix} 2 & -1 \\ -1 & 2 \end{bmatrix}$$

of Example 2.1.2. Complete the argument that A is positive definite by showing that if $y_1 \leqslant y_2 \leqslant 0$, then $2y_1^2 + 2y_2^2 - 2y_1 y_2 \geqslant 0$.

6. Let

$$A = \begin{bmatrix} 2 & 1 \\ 1 & 2 \end{bmatrix}$$

Show that A is positive definite via Definition 2.1.2; verify your answer by applying Theorem 2.1.1 to this matrix.

7. Let

$$A = \begin{bmatrix} 1 & 1 \\ 1 & 2 \end{bmatrix}$$

Is A positive definite, positive semidefinite, or neither?

8. Let

$$A = \begin{bmatrix} 2 & 2 \\ 2 & 2 \end{bmatrix}$$

Is A positive definite, positive semidefinite, or neither?

9. Let

$$A = \begin{bmatrix} 0 & -1 \\ -1 & 0 \end{bmatrix}$$

Is A positive definite, positive semidefinite, or neither?

***10.** Assume that A is a symmetric matrix whose eigenvalues are all positive. Show that A is positive definite, thus partially proving Theorem 2.1.1. (*Hint*: Let $\mathbf{y}$ denote a nonzero vector. Write $\mathbf{y}'A\mathbf{y}$ as $\mathbf{y}'P(P'AP)P'\mathbf{y} = \mathbf{z}'(P'AP)\mathbf{z}$ where P is the orthogonal matrix that diagonalizes A guaranteed in Theorem 1.4.1. Argue that if all eigenvalues are positive then $\mathbf{z}'(P'AP)\mathbf{z} > 0$ by expanding the quadratic form with $P'AP$ in diagonal form.)

Section 2.2

11. Let

$$\mathbf{a} = \begin{bmatrix} 3 \\ 8 \\ 2 \end{bmatrix} \quad \text{and} \quad \mathbf{y} = \begin{bmatrix} y_1 \\ y_2 \\ y_3 \end{bmatrix}$$

a. Write $\mathbf{z} = \mathbf{a}'\mathbf{y}$ in expanded form and find $\partial z/\partial y_1$, $\partial z/\partial y_2$, $\partial z/\partial y_3$.
b. Verify that $\partial z/\partial \mathbf{y} = \mathbf{a}$ as claimed in rule 1 for derivatives.

12. Let $\mathbf{y}$ be a 5×1 column vector and let $z = \mathbf{y}'\mathbf{y}$.
a. Express z in expanded form and find $\partial z/\partial y_i$ for $i = 1, 2, 3, 4, 5$.
b. Verify that $\partial z/\partial \mathbf{y} = 2\mathbf{y}$ as claimed in rule 2 for derivatives.

13. Let

$$\mathbf{y} = \begin{bmatrix} y_1 \\ y_2 \\ y_3 \end{bmatrix}$$

and let $z = \mathbf{y}'\mathbf{y}$. Suppose that when $\partial z/\partial \mathbf{y}$ is evaluated at the point $\mathbf{y}_0$, we obtain

$$\partial z/\partial \mathbf{y} = \begin{bmatrix} 6 \\ -4 \\ 10 \end{bmatrix}$$

Find $\mathbf{y}_0$.

14. Let

$$A = \begin{bmatrix} 2 & 3 \\ 1 & 6 \end{bmatrix}$$

and let $z = \mathbf{y}'A\mathbf{y}$. Find $\partial z/\partial \mathbf{y}$ using rule 3 for derivatives, and check your answer by direct differentiation.

15. Let

$$A = \begin{bmatrix} 2 & 6 \\ 4 & -3 \end{bmatrix}$$

and let $z = \mathbf{y}'A\mathbf{y}$. Find $\partial z/\partial \mathbf{y}$.

16. Let

$$A = \begin{bmatrix} 3 & 1 & 8 \\ 1 & 0 & 1 \\ 2 & 1 & -4 \end{bmatrix}$$

and let $z = \mathbf{y}'A\mathbf{y}$. Find $\partial z/\partial \mathbf{y}$.

17. Let

$$A = \begin{bmatrix} 2 & 1 & 4 \\ 1 & 5 & 3 \\ 4 & 3 & -1 \end{bmatrix}$$

and let $z = \mathbf{y}'A\mathbf{y}$. Find $\partial z/\partial \mathbf{y}$. Do you notice anything unusual about the answer in this case? What property of A causes this unusual result?

18. Show that if A is symmetric and $z = \mathbf{y}'A\mathbf{y}$, then $\partial z/\partial \mathbf{y} = 2A\mathbf{y}$.

19. Let

$$\mathbf{Y} = \begin{bmatrix} Y_1 \\ Y_2 \\ Y_3 \end{bmatrix}, \quad \boldsymbol{\mu} = \begin{bmatrix} 3 \\ 7 \\ 1 \end{bmatrix}, \quad \text{and} \quad \mathbf{a} = \begin{bmatrix} -1 \\ 2 \\ 1 \end{bmatrix}$$

Find $E[\mathbf{a}'\mathbf{Y}]$.

20. Let

$$\mathbf{Y} = \begin{bmatrix} Y_1 \\ Y_2 \\ Y_3 \end{bmatrix}$$

with $E[\mathbf{Y}] = \boldsymbol{\mu}$. Use Definition 2.2.2 and the rules for expectation to show that V assumes the form

$$V = \begin{bmatrix} \sigma_1^2 & \sigma_{12} & \sigma_{13} \\ \sigma_{21} & \sigma_2^2 & \sigma_{23} \\ \sigma_{31} & \sigma_{32} & \sigma_3^2 \end{bmatrix}$$

as claimed.

21. (Additional rules for expectation) The three basic rules for expectation stated in the text are not sufficient to justify the rules for variance completely. Additional rules that pertain to general matrices of random variables can be developed. In particular, these five results will be useful:

1. Let $\mathbf{a}$ be a $k \times 1$ vector of real numbers and $\mathbf{Y}$ a $k \times 1$ vector of random variables. Then
$$E[\mathbf{aY}'] = \mathbf{a}E[\mathbf{Y}']$$
2. Let X be an $n \times k$ matrix of random variables. Then
$$E[X'] = (E[X])'$$
3. Let $\mathbf{a}$ be a $k \times 1$ vector of real numbers and $\mathbf{Y}$ a $k \times 1$ vector of random variables. Then
$$E[\mathbf{Ya}'] = E[\mathbf{Y}]\mathbf{a}'$$
4. Let $\mathbf{a}$ be a $k \times 1$ vector of real numbers and X a $k \times k$ matrix of random variables. Then
$$E[\mathbf{a}'X\mathbf{a}] = \mathbf{a}'E[X]\mathbf{a}$$
5. Let A be a $k \times k$ matrix of real numbers and X a $k \times k$ matrix of random variables. Then
$$E[AXA'] = AE[X]A'$$

a. Prove rule 1.
b. Prove rule 2.
c. Prove rule 3.
d. Verify rule 4 for $k = 3$.
e. Verify rule 5 with $k = 3$ and
$$A = \begin{bmatrix} 2 & 1 \\ 3 & -4 \end{bmatrix}$$

***22.** Let $\mathbf{Y}$ be a random vector with $E[\mathbf{Y}] = \boldsymbol{\mu}$. Use the rules for expectation to show that
$$\text{var } \mathbf{Y} = E[\mathbf{YY}'] - \boldsymbol{\mu\mu}'$$

***23.** Let $\mathbf{Y}$ be a random vector with variance-covariance matrix V and let $\mathbf{a}$ be a vector of real numbers. Use the rules of expectation and Exercise 22 to show that
$$\text{var } \mathbf{a}'\mathbf{Y} = \mathbf{a}'V\mathbf{a}$$

***24.** Let $\mathbf{Y}$ be a random vector with variance-covariance matrix V and let A be a $k \times k$ matrix of scalars. Use the rules of expectation and Exercise 22 to show that
$$\text{var } A\mathbf{Y} = AVA'$$

25. Let
$$\mathbf{Y} = \begin{bmatrix} Y_1 \\ Y_2 \\ Y_3 \end{bmatrix}$$
be a random vector with variance-covariance matrix
$$V = \begin{bmatrix} 3 & 3 & -1 \\ 3 & 2 & 0 \\ -1 & 0 & 5 \end{bmatrix}$$

a. What is the variance of Y_1? of Y_2? of Y_3?
b. What is the covariance between Y_1 and Y_2? between Y_1 and Y_3? between Y_2 and Y_3?
c. Are Y_1 and Y_2 independent? Explain.
d. Are Y_1 and Y_3 independent? Explain.
e. Are Y_2 and Y_3 uncorrelated? Explain.

26. The technical definition of a random sample of size n is "a collection of n independent and identically distributed random variables". Suppose that the random vector

$$\begin{bmatrix} Y_1 \\ Y_2 \\ Y_3 \end{bmatrix}$$

represents a random sample of size 3 from a distribution with variance 6. What is the variance-covariance matrix for $\mathbf{Y}$?

27. Let Y_i and Y_j be random variables with means μ_i and μ_j, respectively. Then $\text{cov}(Y_i, Y_j) = E[Y_iY_j] - E[Y_i]E[Y_j]$. That is, $\sigma_{ij} = E[Y_iY_j] - \mu_i\mu_j$. Verify this result. (*Hint*: Recall that $\text{cov}(Y_i, Y_j) = E[(Y_i - \mu_i)(Y_j - \mu_j)]$. Multiply $(Y_i - \mu_i)$ by $(Y_j - \mu_j)$ and simplify the resulting expression.)

28. a. Let

$$A = \begin{bmatrix} a_{11} & a_{12} & a_{13} \\ a_{21} & a_{22} & a_{23} \\ a_{31} & a_{32} & a_{33} \end{bmatrix} \quad \text{and} \quad B = \begin{bmatrix} b_{11} & b_{12} & b_{13} \\ b_{21} & b_{22} & b_{23} \\ b_{31} & b_{32} & b_{33} \end{bmatrix}$$

Show that $\text{tr}(AB) = \Sigma_{i=1}^{3} \Sigma_{j=1}^{3} a_{ij}b_{ji}$.

b. Let A and B be $k \times k$ matrices. Show that $\text{tr}(AB) = \Sigma_{i=1}^{k} \Sigma_{j=1}^{k} a_{ij}b_{ji}$.

29. Let

$$\mathbf{y} = \begin{bmatrix} y_1 \\ y_2 \\ y_3 \end{bmatrix}$$

be a random vector with

$$\boldsymbol{\mu} = \begin{bmatrix} 1 \\ 3 \\ 2 \end{bmatrix}$$

Assume that $\sigma_{ij} = 0$, $i \neq j$, and that $\sigma_i^2 = 4$, $i = 1, 2, 3$.

a. Find var $\mathbf{y}$.
b. Let

$$A = \begin{bmatrix} 2 & -3 & 1 \\ 1 & 2 & 0 \\ -1 & 6 & 1 \end{bmatrix}$$

and find $E[\mathbf{y}'A\mathbf{y}]$.

30. Let

$$\mathbf{y} = \begin{bmatrix} y_1 \\ y_2 \end{bmatrix}$$

be a random vector with

$$\boldsymbol{\mu} = \begin{bmatrix} 3 \\ -2 \end{bmatrix}$$

Assume that $\sigma_{21} = 3$, $\sigma_1^2 = 5$, and $\sigma_2^2 = 4$.

a. Find var **y**.
b. Let

$$A = \begin{bmatrix} 2 & 4 \\ 1 & 6 \end{bmatrix}$$

and find $E[\mathbf{y}'A\mathbf{y}]$.
c. Verify your answer to part b by expanding $\mathbf{y}'A\mathbf{y}$ and taking the expectation of this quadratic form as demonstrated in Example 2.2.5.

31. Let y_1, y_2, y_3 be a random sample from a distribution that has mean 3 and variance 2. Let

$$A = \begin{bmatrix} 1 & -1 & 0 \\ 1 & 2 & 0 \\ 3 & 0 & 1 \end{bmatrix}$$

Find $E[\mathbf{y}'A\mathbf{y}]$.

Section 2.3

32. Let y_1, y_2, and y_3 be independent normally distributed random variables with common variance 1 and means 0, 1, and 5, respectively. Show that $\mathbf{y}'\mathbf{y}$ follows a noncentral chi-squared distribution and find the numerical values of k and λ.

33. Let $X^2_{2,5}$, $X^2_{3,7}$, and $X^2_{10,4}$ be independent noncentral chi-squared random variables. What is the distribution of the sum of these three random variables?

34. Let **y** be a $k \times 1$ normally distributed random vector with mean $\boldsymbol{\mu}$ and variance D where D is a diagonal matrix. In this exercise, we shall show that $\mathbf{y}'D^{-1}\mathbf{y}$ follows a noncentral chi-squared distribution with k degrees of freedom and $\lambda = (\frac{1}{2})\boldsymbol{\mu}'D^{-1}\boldsymbol{\mu}$.

a. Write the matrix D in general form.
b. Argue that the main diagonal entries of D are all positive so that D^{-1} exists. Write D^{-1} in general form.
c. Write $\mathbf{y}'D^{-1}\mathbf{y}$ in summation form.
d. Let $z_i = y_i/\sigma_i$. What is the distribution of z_i?
e. Show that $\mathbf{y}'D^{-1}\mathbf{y} = \mathbf{z}'\mathbf{z}$ where $\mathbf{z}' = [z_1 \quad z_2 \quad \ldots \quad z_k]$.
f. What is the mean and variance of the random vector **z**?
g. Use Definition 2.3.1 to argue that $\mathbf{y}'D^{-1}\mathbf{y}$ follows a noncentral chi-squared distribution with k degrees of freedom and $\lambda = (\frac{1}{2})\boldsymbol{\mu}'D^{-1}\boldsymbol{\mu}$.

35. Let P be an $n \times n$ matrix partitioned as $P = [P_1 \mid P_2]$ where P_1 is an $n \times k$ matrix and P_2 is an $n \times (n-k)$ matrix.

a. What are the dimensions of P'? of P_1'? of P_2'? Show that

$$P' = \begin{bmatrix} P_1' \\ \hline P_2' \end{bmatrix}$$

b. Let $\mathbf{y}$ be an $n \times 1$ vector. Show that

$$P'\mathbf{y} = \begin{bmatrix} P_1'\mathbf{y} \\ \hline P_2'\mathbf{y} \end{bmatrix}$$

What are the dimensions of $P_1'\mathbf{y}$ and $P_2'\mathbf{y}$?

c. Let A be an $n \times n$ matrix. Argue that

$$P'AP = \left[\begin{array}{c|c} P_1'AP_1 & P_1'AP_2 \\ \hline P_2'AP_1 & P_2'AP_2 \end{array}\right]$$

What are the dimensions of each of these submatrices?

36. These two results are used in the proof of Theorem 2.3.2, and the matrices refer to those defined there.

a. Show that $A = P_1 P_1'$. (*Hint*: Write

$$P'AP = \left[\begin{array}{c|c} I & 0 \\ \hline 0 & 0 \end{array}\right]$$

Premultiply by the orthogonal matrix P and postmultiply by P' to get an expression for A. Write P and P' in partitioned form and multiply.)

b. Show that $P_1'P_1 = I$.

37. Prove Corollary 2.3.1.

38. Prove Corollary 2.3.2. (*Hint*: Let $\mathbf{z} = (1/\sigma)\mathbf{y}$. Find the distribution of $\mathbf{z}$ and of $\mathbf{z}'A\mathbf{z}$. Show that $\mathbf{z}'A\mathbf{z} = (1/\sigma^2)\mathbf{y}'A\mathbf{y}$, thus completing the proof.)

39. Let $\mathbf{y}$ be an $n \times 1$ multivariate normal random variable with variance-covariance matrix V. Show that $r(V) = n$.

40. Consider the random vector $\mathbf{z} = (C')^{-1}(\mathbf{y} - \boldsymbol{\mu})$ defined in the proof of Theorem 2.3.3. Use the rules for expectation and variance given in Section 2.2 to show that $E[\mathbf{z}] = \mathbf{0}$ and $\text{var}\ \mathbf{z} = I$. (*Hint*: In showing that $\text{var}\ \mathbf{z} = I$, remember that transposes and inverses are interchangeable. That is, in general $(A')^{-1} = (A^{-1})'$.)

41. Consider the quadratic form $(C'\mathbf{z} + \boldsymbol{\mu})'A(C'\mathbf{z} + \boldsymbol{\mu})$ given in the proof of Theorem 2.3.3. Show that this quadratic form can be written as $\mathbf{u}'B\mathbf{u}$ where $\mathbf{u} = \mathbf{z} + (C')^{-1}\boldsymbol{\mu}$ and $B = CAC'$. (*Hint*: As a first step, insert two identity matrices into the expression and note that

$$(C'\mathbf{z} + \boldsymbol{\mu})'A(C'\mathbf{z} + \boldsymbol{\mu}) = (C'\mathbf{z} + \boldsymbol{\mu})'C^{-1}CAC'(C')^{-1}(C'\mathbf{z} + \boldsymbol{\mu})$$

Apply the rules for transposes to the first term on the right-hand side, multiply by C^{-1}, simplify, and rewrite in transpose form. Multiply the last term on the right-hand side by $(C')^{-1}$ and simplify.)

42. Consider the random vector $\mathbf{u} = \mathbf{z} + (C')^{-1}\boldsymbol{\mu}$ defined in the proof of Theorem 2.3.3. Use the rules for expectation and variance to show that $E[\mathbf{u}] = (C')^{-1}\boldsymbol{\mu}$ and $\text{var}\ \mathbf{u} = I$ as claimed.

43. Show that the noncentrality parameter $\lambda = (\frac{1}{2})[(C')^{-1}\boldsymbol{\mu}]'B[(C')^{-1}\boldsymbol{\mu}]$ given in the proof of

Theorem 2.3.3 can be expressed as $\lambda = (\frac{1}{2})\boldsymbol{\mu}' A \boldsymbol{\mu}$ as claimed. (*Hint*: Use the rules for transposes to simplify $[(C')^{-1}\boldsymbol{\mu}]'$.)

44. Prove Corollary 2.3.3.

45. Prove Corollary 2.3.4. (*Hint*: Argue that V^{-1} exists and apply Theorem 2.3.3.)

46. Let

$$\mathbf{y} = \begin{bmatrix} y_1 \\ y_2 \end{bmatrix}$$

be a multivariate normal random variable with mean

$$\begin{bmatrix} 2 \\ 3 \end{bmatrix}$$

and variance-covariance matrix I. Let

$$A = \begin{bmatrix} \frac{1}{2} & \frac{1}{2} \\ \frac{1}{2} & \frac{1}{2} \end{bmatrix}$$

Write the quadratic form $\mathbf{y}'A\mathbf{y}$ in summation form. What is the distribution of $\mathbf{y}'A\mathbf{y}$?

47. Let

$$\begin{bmatrix} y_1 \\ y_2 \\ y_3 \end{bmatrix}$$

be a multivariate normal random variable with mean

$$\begin{bmatrix} 1 \\ -1 \\ 0 \end{bmatrix}$$

and variance-covariance matrix

$$\begin{bmatrix} 1 & 0 & 2 \\ 0 & 1 & 0 \\ 2 & 0 & 1 \end{bmatrix}$$

Let

$$A = -\frac{1}{3}\begin{bmatrix} 1 & 0 & -2 \\ 0 & -3 & 0 \\ -2 & 0 & 1 \end{bmatrix}$$

Write the quadratic form $\mathbf{y}'A\mathbf{y}$ in summation form. What is the distribution of $\mathbf{y}'A\mathbf{y}$?

48. Let

$$\mathbf{y} = \begin{bmatrix} y_1 \\ y_2 \end{bmatrix}$$

be a multivariate normal random variable with mean $\mathbf{0}$ and variance-covariance matrix I. Let

$$A = \begin{bmatrix} \frac{1}{2} & -\frac{1}{2} \\ -\frac{1}{2} & \frac{1}{2} \end{bmatrix}$$

Write $\mathbf{y}'A\mathbf{y}$ in summation form. What is the distribution of $\mathbf{y}'A\mathbf{y}$? Let

$$B = \begin{bmatrix} 1 & 1 \\ 1 & 1 \end{bmatrix}$$

What can be said about the distribution of $\mathbf{y}'B\mathbf{y}$?

49. Let

$$\begin{bmatrix} y_1 \\ y_2 \end{bmatrix}$$

be a multivariate normal random variable with mean

$$\begin{bmatrix} 1 \\ 1 \end{bmatrix}$$

and variance-covariance matrix

$$V = \begin{bmatrix} 2 & 0 \\ 0 & 2 \end{bmatrix}$$

Let

$$A = \begin{bmatrix} \frac{1}{2} & 0 \\ 0 & \frac{1}{2} \end{bmatrix}$$

Use Corollary 2.3.2 to show that $\mathbf{y}'A\mathbf{y}$ follows a noncentral chi-squared distribution. What is the numerical value of λ?

50. Let $\mathbf{z}$ be an $n \times 1$ multivariate normal random variable. Let $\mathbf{y} = A\mathbf{z}$ where A is an $m \times n$ matrix. Argue that each entry in $\mathbf{y}$ is a linear combination of independent normally distributed random variables and hence that each entry is itself a normally distributed random variable. What is the mean of $\mathbf{y}$? What is its variance?

Section 2.4

51. In the proof of Theorem 2.4.1 it is claimed that $\mathbf{z} = P'(C')^{-1}\mathbf{y}$ is a multivariate normal random variable with mean $P'(C')^{-1}\boldsymbol{\mu}$ and variance I.

a. Prove that each entry in $\mathbf{z}$ is a normally distributed random variable. (*Hint*: See Exercise 50.)
b. Use the rules for expectation and variance to show that the mean and variance of $\mathbf{z}$ are as claimed.

52. In the proof of Theorem 2.4.1, we defined $\mathbf{z} = P'(C')^{-1}\mathbf{y}$, $R = CAC'$, and $S = CBC'$. Show that $\mathbf{y} = C'P\mathbf{z}$, $A = C^{-1}R(C')^{-1}$, and $B = C^{-1}S(C')^{-1}$.

53. Let $\mathbf{y}$ be an $n \times 1$ multivariate normal random variable with mean $\boldsymbol{\mu}$ and variance $\sigma^2 I$. Let $\mathbf{z} = \mathbf{y}/\sigma$.

a. What is the variance of $\mathbf{z}$?
b. Let X be an $n \times p$ matrix of rank p. Verify that

$$\begin{aligned} \mathbf{z}'\mathbf{z} &= \mathbf{z}'[I - X(X'X)^{-1}X']\mathbf{z} + \mathbf{z}'[X(X'X)^{-1}X']\mathbf{z} \\ &= \mathbf{z}'A_1\mathbf{z} + \mathbf{z}'A_2\mathbf{z} \end{aligned}$$

c. Show that $X(X'X)^{-1}X'$ is idempotent of rank p.
d. Show that $I - X(X'X)^{-1}X'$ is idempotent of rank $n - p$.

e. Use Theorem 2.4.3 to determine the distribution of $\mathbf{z}'A_1\mathbf{z}$ and $\mathbf{z}'A_2\mathbf{z}$.
f. What is the distribution of $\mathbf{z}'\mathbf{z}$?

54. Let

$$\begin{bmatrix} y_1 \\ y_2 \end{bmatrix}$$

be a multivariate normal random variable with mean

$$\begin{bmatrix} 2 \\ 4 \end{bmatrix}$$

and variance-covariance matrix

$$V = \begin{bmatrix} 1 & 0 \\ 0 & 1 \end{bmatrix}$$

Let

$$A = \begin{bmatrix} \frac{1}{2} & \frac{1}{2} \\ \frac{1}{2} & \frac{1}{2} \end{bmatrix} \quad \text{and} \quad B = \begin{bmatrix} \frac{1}{2} & -\frac{1}{2} \\ -\frac{1}{2} & \frac{1}{2} \end{bmatrix}$$

a. Write $\mathbf{y}'A\mathbf{y}$ and $\mathbf{y}'B\mathbf{y}$ in summation form.
b. What is the distribution of $\mathbf{y}'A\mathbf{y}$? of $\mathbf{y}'B\mathbf{y}$?
c. Are $\mathbf{y}'A\mathbf{y}$ and $\mathbf{y}'B\mathbf{y}$ independent?
d. What is the distribution of $\mathbf{y}'A\mathbf{y} + \mathbf{y}'B\mathbf{y}$?

55. Let

$$\begin{bmatrix} y_1 \\ y_2 \end{bmatrix}$$

be a multivariate normal random variable with mean $\mathbf{0}$ and variance-covariance matrix

$$V = \begin{bmatrix} 1 & 0 \\ 0 & 1 \end{bmatrix}$$

Let

$$A = \begin{bmatrix} 1 & 0 \\ 0 & 1 \end{bmatrix} \quad \text{and} \quad B = \begin{bmatrix} \frac{1}{2} & -\frac{1}{2} \\ -\frac{1}{2} & \frac{1}{2} \end{bmatrix}$$

a. Write $\mathbf{y}'A\mathbf{y}$ and $\mathbf{y}'B\mathbf{y}$ in summation form.
b. What is the distribution of $\mathbf{y}'A\mathbf{y}$? of $\mathbf{y}'B\mathbf{y}$?
c. Are $\mathbf{y}'A\mathbf{y}$ and $\mathbf{y}'B\mathbf{y}$ independent?
d. Can we use Theorem 2.3.1 to conclude that $\mathbf{y}'A\mathbf{y} + \mathbf{y}'B\mathbf{y}$ follows a noncentral chi-squared distribution? Explain.

In Exercises 56–59, many of the concepts presented in this chapter are reviewed by outlining the proofs of some important results from elementary statistics. These results should be familiar to you but you might not have seen them proved rigorously before. In particular, we review the definition of the *unbiased estimator* and show that the sample variance s^2 is an unbiased estimator for the population variance σ^2; we derive the distribution of the random variable $(n-1)s^2/\sigma^2$, which plays a major role in making inferences on a population variance; we review the definition of the t random variable and show that the statistic $(\bar{y} - \mu_0)/(s/\sqrt{n})$ used to test hypotheses on the mean of a normal distribution does in fact follow a t distribution with $n-1$ degrees of freedom as claimed in beginning courses in statistics. You will see that all of these basic results are easy consequences of the theory developed in the first two chapters of this text.

***56.** Recall that an estimator $\hat{\theta}$ for a parameter is said to be *unbiased* for θ if and only if $E[\hat{\theta}] = \theta$. Usually one of the first unbiased estimators that is presented is s^2, the sample variance. This estimator is defined by

$$s^2 = \frac{1}{n-1} \sum_{i=1}^{n} (y_i - \bar{y})^2 = \frac{1}{n-1} \left[\sum_{i=1}^{n} y_i^2 - \frac{\left(\sum_{i=1}^{n} y_i\right)^2}{n} \right]$$

The fact that s^2 is an unbiased estimator for the population variance σ^2 can be proved easily using the material in this chapter. This exercise outlines the proof for you.

a. Let $\mathbf{y}$ be an $n \times 1$ random vector with mean $\boldsymbol{\mu}' = [\mu \quad \mu \quad \dots \quad \mu]$ and variance-covariance matrix $V = \sigma^2 I$. That is, $y_1, y_2, \dots, y_n$ have common mean μ and common variance σ^2. Let $\mathbf{z}$ be an $n \times 1$ vector of ones. Show that $(\Sigma_{i=1}^{n} y_i)^2/n = (\mathbf{y}'\mathbf{z}\mathbf{z}'\mathbf{y})/n$.

b. Show that $\Sigma_{i=1}^{n} y_i^2 - (\Sigma_{i=1}^{n} y_i)^2/n = \mathbf{y}'[I - (\mathbf{z}\mathbf{z}')/n]\mathbf{y}$.

c. Note that $s^2 = [1/(n-1)]\mathbf{y}'[I - (\mathbf{z}\mathbf{z}')/n]\mathbf{y}$. Use Theorem 2.2.1 to show that $E[s^2] = \sigma^2$, thus showing that the sample variance is an unbiased estimator for the population variance.

***57.** In point estimation problems it is often unnecessary to know the distribution of a particular statistic. For example, in Exercise 56 we were able to show that s^2 is an unbiased estimator for σ^2 without mentioning the family of random variables to which s^2 belongs. However, when constructing confidence intervals or testing hypotheses it is necessary to know the distribution of the statistics involved. In this exercise we prove another result that is probably familiar to you from elementary courses. In particular, we show that if $\mathbf{y}$ is assumed to have a multivariate normal distribution then the random variable

$$\frac{(n-1)s^2}{\sigma^2} = \sum_{i=1}^{n} \frac{(y_i - \bar{y})^2}{\sigma^2}$$

follows a chi-squared distribution with $n - 1$ degrees of freedom.

a. Let $\mathbf{y}$ be an $n \times 1$ multivariate normal random variable with mean $\boldsymbol{\mu}' = [\mu \quad \mu \quad \dots \quad \mu]$ and variance-covariance matrix $V = \sigma^2 I$. Note that from Exercise 56 we know that

$$\sum_{i=1}^{n} (y_i - \bar{y})^2 = \mathbf{y}'[I - (\mathbf{z}\mathbf{z}')/n]\mathbf{y}$$

where $\mathbf{z}$ is an $n \times 1$ matrix of ones. Show that $I - (\mathbf{z}\mathbf{z}')/n$ is symmetric.

b. Show that $I - (\mathbf{z}\mathbf{z}')/n$ is idempotent.

c. Find $\text{tr}[I - (\mathbf{z}\mathbf{z}')/n]$.

d. Find $r[I - (\mathbf{z}\mathbf{z}')/n]$. (*Hint*: See Theorem 1.5.2.)

e. Use Corollary 2.3.2 to show that

$$\frac{(n-1)s^2}{\sigma^2} = \frac{1}{\sigma^2}\mathbf{y}'\left[I - \frac{\mathbf{z}\mathbf{z}'}{n}\right]\mathbf{y}$$

follows a noncentral chi-squared distribution.

f. Show that the noncentrality parameter is zero, thus showing that the distribution is chi-squared with $n - 1$ degrees of freedom as claimed.

***58.** Let $\mathbf{y}$ be an $n \times 1$ multivariate normal random variable with mean $\boldsymbol{\mu}' = [\mu \quad \mu \quad \ldots \quad \mu]$ and variance-covariance matrix $V = \sigma^2 I$. Show that $\bar{y} = (\Sigma_{i=1}^{n} y_i)/n$ and s^2 are independent. (*Hint*: Write $\bar{y}$ in the form

$$\bar{y} = [1/n \quad 1/n \quad \ldots \quad 1/n]\begin{bmatrix} y_1 \\ y_2 \\ \vdots \\ y_n \end{bmatrix}$$

and apply Theorem 2.4.2.)

***59.** One of the most often used probability distributions is the Student's t or just the t distribution. It can be shown that any random variable that can be expressed as the ratio of a standard normal random variable Z to the nonnegative square root of an independent chi-squared random variable, X_k^2, divided by its degrees of freedom follows the t distribution with k degrees of freedom. That is, every t random variable can be written in the form

$$\frac{Z}{\sqrt{X_k^2/k}}$$

This distribution is usually encountered first when conducting t tests on the mean of a normal distribution. The t statistic used in this setting is

$$\frac{\bar{y} - \mu_0}{s/\sqrt{n}}$$

where μ_0 is the null value of the mean. In this exercise we outline the proof that this statistic does in fact follow the t distribution with $n - 1$ degrees of freedom. Let $\mathbf{y}$ be a multivariate normal random variable with mean $\boldsymbol{\mu}' = [\mu_0 \quad \mu_0 \quad \ldots \quad \mu_0]$ and variance $\sigma^2 I$.

a. Show that $\bar{y}$ is normally distributed with mean μ_0 and variance σ^2/n. (*Hint*: Write $\bar{y}$ as $[1/n \quad 1/n \quad \ldots \quad 1/n]\mathbf{y}$ and apply the rules for expectation and variance given in Section 2.2.)

b. Show that

$$z = \frac{\bar{y} - \mu_0}{\sigma/\sqrt{n}}$$

is a standard normal random variable.

c. Based on the results of Exercise 58, argue that z and $q = (n-1)s^2/\sigma^2$ are independent.

d. Use the results of parts b and c to give an argument that

$$\frac{\bar{y} - \mu_0}{s/\sqrt{n}}$$

follows a t distribution with $n - 1$ degrees of freedom.

CHAPTER 3

ESTIMATION IN THE FULL RANK MODEL

In the introduction we briefly described the notion of a linear statistical model. Recall that this is a model of the form

$$y = \beta_0 + \beta_1 x_1 + \beta_2 x_2 + \cdots + \beta_k x_k + \varepsilon$$

where y is a random variable called the response; $x_1, x_2, \ldots, x_k$ are mathematical variables, called regressors, whose values are controlled or at least accurately observed by the experimenter; ε is a random variable that accounts for unexplained random variation in response; and $\beta_0, \beta_1, \ldots, \beta_k$ are constants or parameters whose exact values are not known and hence must be estimated from experimental data. This model is linear in the sense that it is a polynomial of degree one in the *coefficients* $\beta_0, \beta_1, \ldots, \beta_k$. In this chapter we consider problems of estimation relative to this model.

3.1

MATRIX FORMULATION OF THE FULL RANK MODEL

We begin by considering the *full rank* linear model. Assume that it has been postulated that a particular response y depends on the values assumed by k variables $x_1, x_2, \ldots, x_k$ and that a linear model

$$y = \beta_0 + \beta_1 x_1 + \beta_2 x_2 + \cdots + \beta_k x_k + \varepsilon$$

is appropriate. The experiment is performed n times for various values of $x_1, x_2, \ldots, x_k$ where $n \geqslant k+1$ and the responses $y_1, y_2, \ldots, y_n$ are observed. If we let x_{ij} denote the ith value of the variable x_j then the model generates a system of equations linear in $\beta_0, \beta_1, \beta_2, \ldots, \beta_k$ of the form

$$\begin{aligned}
y_1 &= \beta_0 + \beta_1 x_{11} + \beta_2 x_{12} + \cdots + \beta_k x_{1k} + \varepsilon_1 \\
y_2 &= \beta_0 + \beta_1 x_{21} + \beta_2 x_{22} + \cdots + \beta_k x_{2k} + \varepsilon_2 \\
&\vdots \\
y_n &= \beta_0 + \beta_1 x_{n1} + \beta_2 x_{n2} + \cdots + \beta_k x_{nk} + \varepsilon_n
\end{aligned}$$

Now let

$$X = \begin{bmatrix} 1 & x_{11} & x_{12} & \cdots & x_{1k} \\ 1 & x_{21} & x_{22} & \cdots & x_{2k} \\ \vdots & & & & \\ 1 & x_{n1} & x_{n2} & & x_{nk} \end{bmatrix}, \quad \boldsymbol{\beta} = \begin{bmatrix} \beta_0 \\ \beta_1 \\ \vdots \\ \beta_k \end{bmatrix}, \quad \mathbf{y} = \begin{bmatrix} y_1 \\ y_2 \\ \vdots \\ y_n \end{bmatrix}, \quad \text{and} \quad \boldsymbol{\varepsilon} = \begin{bmatrix} \varepsilon_1 \\ \varepsilon_2 \\ \vdots \\ \varepsilon_n \end{bmatrix}$$

and note that the system of linear equations can be expressed using matrix notation as

$$\mathbf{y} = X\boldsymbol{\beta} + \boldsymbol{\varepsilon}$$

In this form, $\mathbf{y}$ and $\boldsymbol{\varepsilon}$ are each $n \times 1$ random vectors. The vector $\mathbf{y}$ is the *response vector* and the vector $\boldsymbol{\varepsilon}$ is a vector of *random errors*. The vector $\boldsymbol{\beta}$ is a $(k+1) \times 1$ vector of unknown parameters and X is an $n \times (k+1)$ matrix of scalars. When we say that the model is *full rank*, we mean that the matrix X is full rank. Recall from Chapter 1 that this means that $r(X) = k+1$ and implies that $X'X$ is nonsingular. Throughout this chapter we assume that the X matrix is of full rank even if this fact is not stated explicitly.

Although Example 3.1.1 is a very oversimplified example of a linear model, it will help illustrate notation and basic concepts.

Example 3.1.1 Suppose we assume that y, the price obtained for a house, depends primarily on two variables: x_1, its age; and x_2, its livable square footage. Suppose we assume a linear model of the form

$$y = \beta_0 + \beta_1 x_1 + \beta_2 x_2 + \varepsilon$$

We observe five randomly selected houses on the market and obtain these data [19]:

Price in Thousands (y)	Age in Years (x_1)	Square Footage in Thousands (x_2)
50	1	1
40	5	1
52	5	2
47	10	2
65	20	3

The system of five linear equations generated by the model is

$$\begin{aligned} 50 &= \beta_0 + 1\beta_1 + 1\beta_2 + \varepsilon_1 \\ 40 &= \beta_0 + 5\beta_1 + 1\beta_2 + \varepsilon_2 \\ 52 &= \beta_0 + 5\beta_1 + 2\beta_2 + \varepsilon_3 \\ 47 &= \beta_0 + 10\beta_1 + 2\beta_2 + \varepsilon_4 \\ 65 &= \beta_0 + 20\beta_1 + 3\beta_2 + \varepsilon_5 \end{aligned}$$

In matrix form, this system is expressed as

$$\mathbf{y} = X\boldsymbol{\beta} + \boldsymbol{\varepsilon}$$

where

$$\mathbf{y} = \begin{bmatrix} 50 \\ 40 \\ 52 \\ 47 \\ 65 \end{bmatrix}, \quad X = \begin{bmatrix} 1 & 1 & 1 \\ 1 & 5 & 1 \\ 1 & 5 & 2 \\ 1 & 10 & 2 \\ 1 & 20 & 3 \end{bmatrix}, \quad \boldsymbol{\beta} = \begin{bmatrix} \beta_0 \\ \beta_1 \\ \beta_2 \end{bmatrix}, \quad \text{and} \quad \boldsymbol{\varepsilon} = \begin{bmatrix} \varepsilon_1 \\ \varepsilon_2 \\ \varepsilon_3 \\ \varepsilon_4 \\ \varepsilon_5 \end{bmatrix}$$

Our job eventually is to find methods by which β_0, β_1, and β_2 can be estimated from this data set.

Most elementary statistics texts include a discussion of at least one linear model, the model for *simple linear regression.* In such texts the model is often expressed and treated without referring to matrix algebra. Example 3.1.2 reviews the notion of simple linear regression and formulates the model in the matrix setting.

Example 3.1.2 In simple linear regression it is assumed that the response y depends only on one variable x. The model is linear in the parameters and is given by

$$y = \beta_0 + \beta_1 x + \varepsilon$$

A set of n responses gives rise to this system of linear equations:

$$\begin{aligned} y_1 &= \beta_0 + \beta_1 x_1 + \varepsilon_1 \\ y_2 &= \beta_0 + \beta_1 x_2 + \varepsilon_2 \\ &\vdots \\ y &= \beta_0 + \beta_1 x_n + \varepsilon_n \end{aligned}$$

The response, parameter, and random error vectors and the X matrix are given by

$$\mathbf{y} = \begin{bmatrix} y_1 \\ y_2 \\ \vdots \\ y_n \end{bmatrix}, \quad \boldsymbol{\beta} = \begin{bmatrix} \beta_0 \\ \beta_1 \end{bmatrix}, \quad \boldsymbol{\varepsilon} = \begin{bmatrix} \varepsilon_1 \\ \varepsilon_2 \\ \vdots \\ \varepsilon_n \end{bmatrix}, \quad \text{and} \quad X = \begin{bmatrix} 1 & x_1 \\ 1 & x_2 \\ \vdots & \vdots \\ 1 & x_n \end{bmatrix}$$

respectively. In elementary courses the estimators for β_0 and β_1 are found by solving a system of two equations in the two unknowns β_0 and β_1 algebraically. These solutions will be found using matrix techniques later.

Remember that a linear model is a linear function of the *parameters* $\beta_0, \beta_1, \ldots, \beta_k$. The variables $x_1, x_2, \ldots, x_k$ can appear in the model in a nonlinear fashion. Thus expressions such as

$$\begin{aligned} y &= \beta_0 + \beta_1 e^x + \varepsilon \\ y &= \beta_0 + \beta_1 x_1^2 + \beta_2 x_2^2 + \varepsilon \\ y &= \beta_0 + \beta_1 x_1 + \beta_2 x_2 + \beta_{12} x_1 x_2 + \varepsilon \\ y &= \beta_0 + \beta_1 \ln(x) + \varepsilon \end{aligned}$$

represent linear models; expressions such as $y = e^{\beta_0 x} + \varepsilon$ and $y = 1/(1 + e^{-\beta_0 - x\beta_1 + \varepsilon})$ do not represent linear models.

3.2

LEAST SQUARES ESTIMATORS OF THE MODEL PARAMETERS

As mentioned earlier, our first task is to estimate the coefficients $\beta_0, \beta_1, \ldots, \beta_k$ in a logical way. We want to do this in a relatively unrestricted setting. The method presented in this section is the *method of least squares*. In using this method here, we require only that the vector $\boldsymbol{\varepsilon}$ of random errors has mean $\mathbf{0}$ and variance $\sigma^2 I$. This will be assumed throughout this section even if not explicitly stated. Using the rules for expectation and variance, this implies that the random vector $\mathbf{y}$ has mean $X\boldsymbol{\beta}$ and variance-covariance matrix $\sigma^2 I$. Since normality has not been mentioned, we are not assuming that $y_1, y_2, \ldots, y_n$ are independent; however, we are requiring that they be uncorrelated with common variance σ^2. To say that $E[\mathbf{y}] = X\boldsymbol{\beta}$ implies that the *mean* value of a response for a given set of x values depends on these values as well as on the values of the parameters $\beta_0, \beta_1, \ldots, \beta_k$ and that this dependence is a linear function of the β's. To understand the method of least squares, we must take a close look at the last statement. Consider the response y_i corresponding to the x values $x_{i1}, x_{i2}, \ldots, x_{ik}$. The linear model indicates that

$$y_i = \beta_0 + \beta_1 x_{i1} + \beta_2 x_{i2} + \cdots + \beta_k x_{ik} + \varepsilon_i$$

Taking the expectation,

$$E[y_i] = \beta_0 + \beta_1 x_{i1} + \beta_2 x_{i2} + \cdots + \beta_k x_{ik} + E[\varepsilon_i]$$

Since $E[\varepsilon_i]$ is assumed to be zero, it is easy to see that the mean value of the response y_i is given by

$$E[y_i] = \beta_0 + \beta_1 x_{i1} + \beta_2 x_{i2} + \cdots + \beta_k x_{ik}$$

As you can see, this mean response depends on the values of the x variables and is a linear function of the parameters as claimed.

To develop the method of least squares, the model is rewritten in the form

$$\begin{aligned} \varepsilon_i &= y_i - (\beta_0 + \beta_1 x_{i1} + \beta_2 x_{i2} + \cdots + \beta_k x_{ik}) \\ &= y_i - E[y_i] \end{aligned}$$

From this it can be seen that the ith random error is the difference between the ith response and its mean, or *theoretical*, average value. Since this average response depends on $\beta_0, \beta_1, \ldots, \beta_k$ whose values are not known, random errors cannot be observed from a sample. However, we can observe *residuals*. Imagine that $b_0, b_1, \ldots, b_k$ are estimators for $\beta_0, \beta_1, \ldots, \beta_k$, respectively. These estimators can be used to approximate $E[y_i]$ in a logical way. Simply replace the model parameters with these estimators to conclude that

$$\widehat{E[y_i]} = b_0 + b_1 x_{i1} + b_2 x_{i2} + \cdots + b_k x_{ik}$$

where the "hat" ($\hat{\ }$) is used to indicate an estimator. The ith residual is defined to be the difference between the ith response and its *estimated* expected value. The ith residual is denoted by e_i and is given by

$$\begin{aligned} e_i &= y_i - (b_0 + b_1 x_{i1} + b_2 x_{i2} + \cdots + b_k x_{ik}) \\ &= y_i - \widehat{E[y_i]} \end{aligned}$$

It is hoped that the behavior of the unobservable random errors is mirrored by the observable residuals. In particular, if the linear model is an accurate description of the relationship between y and the x variables, then most of the variation in response can be attributed to changes in the x's; little of the variation in y is left unexplained. This means that the random errors, which represent unexplained variation, should be small. To mirror this characteristic, estimators $b_0, b_1, \ldots, b_k$ must be found that force the residuals to be as small as possible. It is natural to suggest that the residuals be summed and the values of $b_0, b_1, \ldots, b_k$ be chosen so as to minimize this sum. However, care must be taken. Residuals can be either positive, negative, or zero in value. In summing, positive and negative residuals of the same magnitude will add to 0. In fact, it can be shown that the sum of the residuals is always 0 (see Exercise 5). This leads to a misrepresentation of the variability of the responses about the estimated line of regression. A negative residual could be handled by considering its absolute value. However, this is rather inconvenient mathematically because the absolute value function is not differentiable everywhere. To avoid the problems posed by negative residuals, we work not with the residuals or their absolute values but with their squares. The idea behind the method of least squares is to choose estimators in such a way that the *sum of the squares of the residuals is minimized.*

To obtain the least squares estimators for $\beta_0, \beta_1, \ldots, \beta_k$ in the linear model, we first express the responses in terms of residuals as shown:

$$\begin{aligned}
y_1 &= b_0 + b_1x_{11} + b_2x_{12} + \cdots + b_kx_{1k} + e_1 \\
y_2 &= b_0 + b_1x_{21} + b_2x_{22} + \cdots + b_kx_{2k} + e_2 \\
&\vdots \\
y_n &= b_0 + b_1x_{n1} + b_2x_{n2} + \cdots + b_kx_{nk} + e_n
\end{aligned}$$

Now let

$$X = \begin{bmatrix} 1 & x_{11} & x_{12} & \cdots & x_{1k} \\ 1 & x_{21} & x_{22} & \cdots & x_{2k} \\ \vdots & \vdots & \vdots & & \vdots \\ 1 & x_{n1} & x_{n2} & & x_{nk} \end{bmatrix}, \quad \mathbf{b} = \begin{bmatrix} b_0 \\ b_1 \\ \vdots \\ b_k \end{bmatrix}, \quad \mathbf{y} = \begin{bmatrix} y_1 \\ y_2 \\ \vdots \\ y_n \end{bmatrix}, \quad \text{and} \quad \mathbf{e} = \begin{bmatrix} e_1 \\ e_2 \\ \vdots \\ e_n \end{bmatrix}$$

and write the system in matrix form as

$$\mathbf{y} = X\mathbf{b} + \mathbf{e}$$

Here $\mathbf{y}$ is the $n \times 1$ vector of responses, $\mathbf{e}$ is the $n \times 1$ random vector of residuals, and $\mathbf{b}$ is the $(k+1) \times 1$ vector of estimators for the coefficients $\beta_0, \beta_1, \ldots, \beta_k$. X is an $n \times (k+1)$ full rank matrix of real numbers. In the method of least squares, we want to minimize the sum of squares of the residuals. That is, we want to find estimators $b_0, b_1, \ldots, b_k$ such that

$$\mathbf{e}'\mathbf{e} = \sum_{i=1}^{n} e_i^2$$

is as small as possible. Theorem 3.2.1 gives these estimators.

Theorem 3.2.1 Let $\mathbf{y} = X\boldsymbol{\beta} + \boldsymbol{\varepsilon}$ where X is an $n \times (k+1)$ matrix of full rank, $\boldsymbol{\beta}$ is a $(k+1) \times 1$ vector of unknown parameters, and $\boldsymbol{\varepsilon}$ is an $n \times 1$ random vector with mean $\mathbf{0}$ and variance $\sigma^2 I$. The least squares estimator for $\boldsymbol{\beta}$ is denoted by $\mathbf{b}$ and is given by

$$\mathbf{b} = (X'X)^{-1}X'\mathbf{y}$$

Proof

The vector of residuals $\mathbf{e}$ can be written as $\mathbf{e} = \mathbf{y} - X\mathbf{b}$ and hence

$$\mathbf{e}'\mathbf{e} = (\mathbf{y} - X\mathbf{b})'(\mathbf{y} - X\mathbf{b})$$

Simplifying,

$$\begin{aligned}\mathbf{e}'\mathbf{e} &= (\mathbf{y}' - \mathbf{b}'X')(\mathbf{y} - X\mathbf{b}) \\ &= \mathbf{y}'\mathbf{y} - \mathbf{y}'X\mathbf{b} - \mathbf{b}'X'\mathbf{y} + \mathbf{b}'X'X\mathbf{b}\end{aligned}$$

Since $\mathbf{b}'X'\mathbf{y}$ is 1×1, $\mathbf{b}'X'\mathbf{y} = (\mathbf{b}'X'\mathbf{y})' = \mathbf{y}'X\mathbf{b}$. By substitution,

$$\begin{aligned}\mathbf{e}'\mathbf{e} &= \mathbf{y}'\mathbf{y} - 2\mathbf{y}'X\mathbf{b} + \mathbf{b}'(X'X)\mathbf{b} \\ &= \mathbf{y}'\mathbf{y} - 2(X'\mathbf{y})'\mathbf{b} + \mathbf{b}'(X'X)\mathbf{b}\end{aligned}$$

To minimize $\mathbf{e}'\mathbf{e}$ as a function of $\mathbf{b}$, this expression is differentiated with respect to $\mathbf{b}$, the derivative is set equal to zero, and the resulting equation, called the *normal equation*, is solved for $\mathbf{b}$. Using the rules for differentiation given in Section 2.2,

$$\begin{aligned}\frac{\partial \mathbf{e}'\mathbf{e}}{\partial \mathbf{b}} &= -2X'\mathbf{y} + (X'X)\mathbf{b} + (X'X)'\mathbf{b} \\ &= -2X'\mathbf{y} + 2(X'X)\mathbf{b}\end{aligned}$$

Setting this derivative equal to zero, we obtain

$$-2X'\mathbf{y} + 2(X'X)\mathbf{b} = \mathbf{0}$$

or

$$(X'X)\mathbf{b} = X'\mathbf{y} \qquad \text{(Normal equation)}$$

To solve for $\mathbf{b}$, multiply each side of the equation by $(X'X)^{-1}$ to obtain the least squares estimators

$$\mathbf{b} = (X'X)^{-1}X'\mathbf{y}$$

as claimed. ■

Before illustrating this theorem in an applied setting, we must ask one question. Is $\mathbf{b}$ an unbiased estimator for $\boldsymbol{\beta}$? That is, is $E[\mathbf{b}] = \boldsymbol{\beta}$? Theorem 3.2.2 answers this question. Its proof is an easy application of the rules of expectation and variance and is left as an exercise (see Exercise 4).

Theorem 3.2.2 Let $\mathbf{y} = X\boldsymbol{\beta} + \boldsymbol{\varepsilon}$ where X is an $n \times (k+1)$ matrix of full rank, $\boldsymbol{\beta}$ is a $(k+1) \times 1$ vector of unknown parameters, and $\boldsymbol{\varepsilon}$ is an $n \times 1$ random vector with mean $\mathbf{0}$ and variance $\sigma^2 I$. The least squares estimator $\mathbf{b} = (X'X)^{-1}X'\mathbf{y}$ is an unbiased estimator for $\boldsymbol{\beta}$. Furthermore, $\operatorname{var} \mathbf{b} = (X'X)^{-1}\sigma^2$. ■

As indicated earlier, we view the residuals as entities that mirror the behavior of random errors. Although this reflection is partially true, it is not perfect. Ideally, $\mathbf{e}$ should have mean $\mathbf{0}$ and variance $\sigma^2 I$, as does the vector of random errors. However, this is not the case. In Exercise 5 you are asked to show that $E[\mathbf{e}] = \mathbf{0}$ as desired but that the residuals are correlated so that var $\mathbf{e} \neq \sigma^2 I$. Since $E[\mathbf{e}] = \mathbf{0}$ is the property relied upon, this will not present a problem.

To illustrate the use of Theorem 3.2.1, we reconsider the data of Example 3.1.1.

Example 3.2.1 The assumed model is

$$y = \beta_0 + \beta_1 x_1 + \beta_2 x_2 + \varepsilon$$

where y is the price obtained for a house in thousands of dollars, x_1 is its age in years, and x_2 is its livable square footage in thousands. Data available are

$$\mathbf{y} = \begin{bmatrix} 50 \\ 40 \\ 52 \\ 47 \\ 65 \end{bmatrix} \quad \text{and} \quad X = \begin{bmatrix} 1 & 1 & 1 \\ 1 & 5 & 1 \\ 1 & 5 & 2 \\ 1 & 10 & 2 \\ 1 & 20 & 3 \end{bmatrix}$$

For these data,

$$X'X = \begin{bmatrix} 5 & 41 & 9 \\ 41 & 551 & 96 \\ 9 & 96 & 19 \end{bmatrix} \quad \text{and} \quad X'\mathbf{y} = \begin{bmatrix} 254 \\ 2280 \\ 483 \end{bmatrix}$$

Since X is of full rank, $(X'X)^{-1}$ exists. The inverse is found via the computer to be approximately

$$(X'X)^{-1} = \begin{bmatrix} 2.307551 & 0.1565378 & -1.88398 \\ 0.1565378 & 0.02578269 & -0.20442 \\ -1.88398 & -0.20442 & 1.977901 \end{bmatrix}$$

(See Computing Supplement A at the end of this chapter.) The least squares estimators are given by

$$(X'X)^{-1}X'\mathbf{y} = \begin{bmatrix} 2.307551 & 0.1565378 & -1.88398 \\ 0.1565378 & 0.02578269 & -0.20442 \\ -1.88398 & -0.20442 & 1.977901 \end{bmatrix} \begin{bmatrix} 254 \\ 2280 \\ 483 \end{bmatrix}$$

$$\doteq \begin{bmatrix} 33.06 \\ -0.189 \\ 10.718 \end{bmatrix}$$

The estimated model is

$$y = b_0 + b_1x_1 + b_2x_2 + e$$

or

$$y = 33.06 - 0.189x_1 + 10.718x_2 + e$$

The simple linear regression model is $y = \beta_0 + \beta_1 x + \varepsilon$, which implies that $E[y] = \beta_0 + \beta_1 x$. That is, the simple linear regression model postulates that the average response is a linear function of x alone. The graph of this function is a straight line whose theoretical intercept is β_0 and whose theoretical slope is β_1. It is easy to obtain formulas for b_0, the estimated intercept, and b_1, the estimated slope. Although these formulas can be found without using matrix algebra, their derivation via matrix techniques is very straightforward, as is demonstrated in Example 3.2.2.

Example 3.2.2 The estimated simple linear regression model is given by

$$y = b_0 + b_1x + e$$

In matrix notation,

$$\mathbf{y} = \begin{bmatrix} y_1 \\ y_2 \\ \vdots \\ y_n \end{bmatrix}, \quad \mathbf{b} = \begin{bmatrix} b_0 \\ b_1 \end{bmatrix}, \quad \text{and} \quad X = \begin{bmatrix} 1 & x_1 \\ 1 & x_2 \\ \vdots & \vdots \\ 1 & x_n \end{bmatrix}$$

Here

$$X'X = \begin{bmatrix} n & \sum_{i=1}^{n} x_i \\ \sum_{i=1}^{n} x_i & \sum_{i=1}^{n} x_i^2 \end{bmatrix} \quad \text{and} \quad X'\mathbf{y} = \begin{bmatrix} \sum_{i=1}^{n} y_i \\ \sum_{i=1}^{n} x_i y_i \end{bmatrix}$$

The inverse of the 2×2 matrix $X'X$ is

$$(X'X)^{-1} = \frac{1}{n\sum_{i=1}^{n} x_i^2 - \left(\sum_{i=1}^{n} x_i\right)^2} \begin{bmatrix} \sum_{i=1}^{n} x_i^2 & -\sum_{i=1}^{n} x_i \\ -\sum_{i=1}^{n} x_i & n \end{bmatrix}$$

This can be checked by direct multiplication. By Theorem 3.2.1, the least squares estimator for $\boldsymbol{\beta}$ is

$$\mathbf{b} = (X'X)^{-1}X'\mathbf{y}$$

$$= \frac{1}{n\sum_{i=1}^{n} x_i^2 - \left(\sum_{i=1}^{n} x_i\right)^2} \begin{bmatrix} \sum_{i=1}^{n} x_i^2 & -\sum_{i=1}^{n} x_i \\ -\sum_{i=1}^{n} x_i & n \end{bmatrix} \begin{bmatrix} \sum_{i=1}^{n} y_i \\ \sum_{i=1}^{n} x_i y_i \end{bmatrix}$$

$$= \frac{1}{n\sum_{i=1}^{n} x_i^2 - \left(\sum_{i=1}^{n} x_i\right)^2} \begin{bmatrix} \sum_{i=1}^{n} x_i^2 \sum_{i=1}^{n} y_i - \sum_{i=1}^{n} x_i \sum_{i=1}^{n} x_i y_i \\ n\sum_{i=1}^{n} x_i y_i - \sum_{i=1}^{n} x_i \sum_{i=1}^{n} y_i \end{bmatrix}$$

From this, it is easy to see that the slope of the regression line is given by

$$b_1 = \frac{n\sum_{i=1}^{n} x_i y_i - \sum_{i=1}^{n} x_i \sum_{i=1}^{n} y_i}{n\sum_{i=1}^{n} x_i^2 - \left(\sum_{i=1}^{n} x_i\right)^2}$$

This formula should be familiar to you from your earlier courses. The estimator for the intercept of the regression line is

$$b_0 = \frac{\sum_{i=1}^{n} x_i^2 \sum_{i=1}^{n} y_i - \sum_{i=1}^{n} x_i \sum_{i=1}^{n} x_i y_i}{n\sum_{i=1}^{n} x_i^2 - \left(\sum_{i=1}^{n} x_i\right)^2}$$

This probably does not look familiar! In most elementary courses, the estimator for the intercept is given in a more compact form as $\bar{y} - b_1\bar{x}$. It is left as an exercise for you to verify that these two estimators are identical (see Exercise 6).

GAUSS–MARKOFF THEOREM

The unbiased least squares estimators developed here are examples of *linear estimators.* These are estimators of the form $L\mathbf{y}$ where L is a matrix of real numbers. Here $\mathbf{b}$ assumes this form with $L = (X'X)^{-1}X'$. Unfortunately, unbiasedness does not guarantee uniqueness. There might be more than one set of unbiased linear estimators for $\beta_0, \beta_1, \ldots, \beta_k$. Theorem 3.2.3, the *Gauss–Markoff theorem,* guarantees that among the class of linear unbiased estimators for $\boldsymbol{\beta}$ the estimator $\mathbf{b}$ is best in the sense that the variances of $b_0, b_1, b_2, \ldots, b_k$ are minimized. For this reason, the least squares estimator $\mathbf{b}$ is called a BLUE (best linear unbiased estimator) estimator.

Theorem 3.2.3 Let $\mathbf{y} = X\boldsymbol{\beta} + \boldsymbol{\varepsilon}$ where X is an $n \times (k+1)$ matrix of full rank, $\boldsymbol{\beta}$ is a $(k+1) \times 1$ vector of unknown parameters, and $\boldsymbol{\varepsilon}$ is an $n \times 1$ random vector with mean $\mathbf{0}$ and variance $\sigma^2 I$. The least squares estimator $\mathbf{b}$ is the best linear unbiased estimator for $\boldsymbol{\beta}$.

Proof

Let $\mathbf{b}^*$ denote any other linear unbiased estimator for $\boldsymbol{\beta}$. This estimator can be expressed in the form

$$\mathbf{b}^* = [(X'X)^{-1}X' + B]\mathbf{y}$$

where B is a $(k+1) \times n$ matrix of real numbers. Taking expectations,

$$\begin{aligned} E[\mathbf{b}^*] &= [(X'X)^{-1}X' + B]E[\mathbf{y}] \\ &= [(X'X)^{-1}X' + B]X\boldsymbol{\beta} \\ &= [I + BX]\boldsymbol{\beta} \end{aligned}$$

Since $\mathbf{b}^*$ is an unbiased estimator for $\boldsymbol{\beta}$, $E[\mathbf{b}^*] = \boldsymbol{\beta}$ and hence $[I + BX]\boldsymbol{\beta} = \boldsymbol{\beta}$. Thus $I + BX$ is the identity matrix, which implies that $BX = 0$. By the rules for variance,

$$\begin{aligned} \text{var } \mathbf{b}^* &= \text{var } [(X'X)^{-1}X' + B]\mathbf{y} \\ &= [(X'X)^{-1}X' + B]\sigma^2 I[(X'X)^{-1}X' + B]' \\ &= \sigma^2[(X'X)^{-1}X' + B][X(X'X)^{-1} + B'] \\ &= \sigma^2[(X'X)^{-1}X'X(X'X)^{-1} + (X'X)^{-1}X'B' + BX(X'X)^{-1} + BB'] \end{aligned}$$

Since $BX = 0$, $X'B' = 0$, and var $\mathbf{b}^*$ can be written as

$$\begin{aligned} \text{var } \mathbf{b}^* &= \sigma^2[(X'X)^{-1} + BB'] \\ &= (X'X)^{-1}\sigma^2 + BB'\sigma^2 \\ &= \text{var } \mathbf{b} + BB'\sigma^2 \end{aligned}$$

The ith entry of the main diagonal of BB' is

$$\sum_{j=1}^{n} b_{ij}^2 \geqslant 0 \quad i = 1, 2, 3, \ldots, n$$

Therefore, the entries on the main diagonal are minimized when $B = 0$. From this it is easy to see that var b_0, var b_1, ..., var b_k are minimized by letting $B = 0$. In this case, $\mathbf{b}^* = \mathbf{b}$ as claimed. ■

ESTIMATING LINEAR FUNCTIONS OF $\beta_0, \beta_1, \ldots, \beta_k$

Thus far, attention has centered on the least squares estimators of the coefficients of the linear model. Suppose we need to estimate some linear function of these coefficients. This function can be written as $\mathbf{t}'\boldsymbol{\beta}$ where $\mathbf{t}'$ is a $1 \times (k+1)$ vector of real numbers. The logical estimator is $\mathbf{t}'\mathbf{b}$ where $\mathbf{b}$ is the least squares estimator for $\boldsymbol{\beta}$. Theorem 3.2.4 extends Theorem 3.2.3 to this more general estimation problem.

Theorem 3.2.4 Let $\mathbf{y} = X\boldsymbol{\beta} + \boldsymbol{\varepsilon}$ where X is an $n \times (k+1)$ matrix of full rank, $\boldsymbol{\beta}$ is a $(k+1) \times 1$ vector of unknown parameters, and $\boldsymbol{\varepsilon}$ is an $n \times 1$ random vector with mean $\mathbf{0}$ and variance $\sigma^2 I$. Let $\mathbf{t}'$ be a $1 \times (k+1)$ nonzero vector of real numbers. The best linear unbiased estimator for $\mathbf{t}'\boldsymbol{\beta}$ is $\mathbf{t}'\mathbf{b}$ where $\mathbf{b}$ is the least squares estimator for $\boldsymbol{\beta}$. ■

The proof of this theorem parallels that of Theorem 3.2.3 and is outlined in Exercise 9. One important use of this theorem is to find the best linear unbiased estimator for the mean response for a given set of values of the x's. This idea is illustrated in Example 3.2.3.

Example 3.2.3 Consider the model

$$y = \beta_0 + \beta_1 x_1 + \beta_2 x_2 + \varepsilon$$

of Example 3.2.1. Here y is the price obtained for a house in thousands of dollars, x_1 is its age in years, and x_2 is its livable square footage. Suppose we want to find the best linear unbiased estimator for the average price of a house when x_1 and x_2 assume some specific values x_{*1} and x_{*2}, respectively. We know that

$$E[y] = \beta_0 + \beta_1 x_{*1} + \beta_2 x_{*2}$$

Thus we want to estimate the linear function

$$E[y] = \mathbf{t}'\boldsymbol{\beta}$$

where $\mathbf{t}' = [1 \quad x_{*1} \quad x_{*2}]$. By Theorem 3.2.4, an unbiased estimator for $E[y]$ is

$$\mathbf{t}'\mathbf{b} = [1 \quad x_{*1} \quad x_{*2}]\mathbf{b} = b_0 + b_1 x_{*1} + b_2 x_{*2}$$

where b_0, b_1, and b_2 are the least squares estimators for β_0, β_1, and β_2, respectively. For example, to estimate the average selling price of 15-year-old houses with 2500 square feet of living space we let $x_{*1} = 15$, $x_{*2} = 2.5$, and $\mathbf{t}' = [1 \quad 15 \quad 2.5]$. The estimated average price of these houses is

$$\begin{aligned}\widehat{E[y]} &= \mathbf{t}'\mathbf{b} = [1 \quad 15 \quad 2.5]\mathbf{b} \\ &= b_0 + 15b_1 + 2.5b_2\end{aligned}$$

Using the least squares estimates from Example 3.2.1,

$$\widehat{E[y]} = 33.06 - 0.189(15) + 10.718(2.5) = 57.02$$

The estimated average selling price of such houses is \$57,020.

3.3

ESTIMATING THE VARIANCE

Eventually we will want to find confidence intervals on $\beta_0, \beta_1, \ldots, \beta_k$ and to test hypotheses concerning the values of these parameters. To do so it is necessary to be able to estimate σ^2, the common variance of the random variables $y_1, y_2, \ldots, y_n$. Recall that the variance of a random variable is a measure of its variability. It is the theoretical average or expected value of the square of the difference between the random variable and its *population* mean. To define an estimator for σ^2 in a logical way, we parallel this definition as closely as possible. In particular, we subtract the vector of *estimated* means, $X\mathbf{b}$, from the vector of responses, $\mathbf{y}$, to form the vector of differences, $\mathbf{y} - X\mathbf{b}$. The sum of squares of these differences is given by $(\mathbf{y} - X\mathbf{b})'(\mathbf{y} - X\mathbf{b})$, and the average value of the squared differences is found by dividing by n, the number of responses available. Thus a logical estimator for σ^2 is

$$\widehat{\sigma^2} = \frac{(\mathbf{y} - X\mathbf{b})'(\mathbf{y} - X\mathbf{b})}{n}$$

Notice that the numerator of this estimator is the *sum of the squares of the residuals* encountered earlier in the chapter. Since this residual sum of squares is known to reflect random or unexplained variation in response, this should not be surprising. Is this estimator unbiased? To answer this question, $E[\widehat{\sigma^2}]$ is found as follows:

$$\begin{aligned} E[\widehat{\sigma^2}] &= \frac{1}{n} E[\mathbf{y} - X\mathbf{b})'(\mathbf{y} - X\mathbf{b})] \\ &= \frac{1}{n} E[(\mathbf{y} - X(X'X)^{-1}X'\mathbf{y})'(\mathbf{y} - X(X'X)^{-1}X'\mathbf{y})] \\ &= \frac{1}{n} E[\mathbf{y}'(I - X(X'X)^{-1}X')(I - X(X'X)^{-1}X')\mathbf{y}] \end{aligned}$$

It is easy to show that $I - X(X'X)^{-1}X'$ is idempotent (see Exercise 13). Hence

$$E[\widehat{\sigma^2}] = \frac{1}{n} E[\mathbf{y}'(I - X(X'X)^{-1}X')\mathbf{y}]$$

Applying Theorem 2.2.1 to the right-hand side with $\boldsymbol{\mu} = E[\mathbf{y}] = X\boldsymbol{\beta}$,

$$E[\widehat{\sigma^2}] = \frac{1}{n}[\operatorname{tr}(I - X(X'X)^{-1}X')\sigma^2 I + (X\boldsymbol{\beta})'(I - X(X'X)^{-1}X')X\boldsymbol{\beta}]$$

Since the trace of a difference is the difference in traces, we obtain

$$\begin{aligned} E[\widehat{\sigma^2}] = \frac{1}{n}\{\sigma^2[\operatorname{tr}(I) - \operatorname{tr}(X(X'X)^{-1}X')] \\ + \boldsymbol{\beta}'X'X\boldsymbol{\beta} - \boldsymbol{\beta}'X'X(X'X)^{-1}X'X\boldsymbol{\beta}\} \end{aligned}$$

Since I is an $n \times n$ matrix, its trace is n. Furthermore, by property 3 of the trace given in Section 1.5,

$$\begin{aligned} \operatorname{tr}[X(X'X)^{-1}X'X] &= \operatorname{tr}(X'X)(X'X)^{-1} \\ &= \operatorname{tr}(I) \end{aligned}$$

where, in this case, I is a $(k+1) \times (k+1)$ matrix. Thus,

$$\begin{aligned} E[\widehat{\sigma^2}] &= \frac{1}{n}\sigma^2[n-(k+1)] \\ &= \frac{n-(k+1)}{n}\sigma^2 \end{aligned}$$

As you can see, $E[\widehat{\sigma^2}] \neq \sigma^2$, and hence the proposed estimator for σ^2 is not unbiased. Since

$$\frac{n-(k+1)}{n} < 1$$

the estimates generated by $\widehat{\sigma^2}$ tend on the average to underestimate σ^2. This is not really a problem. Unbiasedness is not an essential property of a good estimator. However, it is usually considered to be desirable. Here it is easy to adjust the proposed estimator to obtain an unbiased estimator for σ^2. Simply multiply $\widehat{\sigma^2}$ by

$$\frac{n}{n-(k+1)}$$

to obtain an estimator, s^2, that we shall call the *sample variance*. This estimator is given by

$$s^2 = \frac{n}{n-(k+1)}\left[\frac{(\mathbf{y}-X\mathbf{b})'(\mathbf{y}-X\mathbf{b})}{n}\right] = \frac{(\mathbf{y}-X\mathbf{b})'(\mathbf{y}-X\mathbf{b})}{n-(k+1)}$$

A quick check will show that this estimator is unbiased for σ^2 (see Exercise 14). For simplicity of notation later, the number of model parameters to be estimated will be denoted in general by p. Here $p = k+1$. The residual sum of squares will be denoted by SS_{Res}. The sample variance can now be written in a form quite easily recalled, namely

$$\boxed{\text{Unbiased estimator for } \sigma^2 = s^2 = \frac{SS_{\text{Res}}}{n-p}}$$

In the work to come, this estimator will appear in various alternative forms. You are asked to derive some of these in Exercise 16. Example 3.3.1 demonstrates the computation of s^2 from its definition.

Example 3.3.1 In industry, useful information is often obtained from "accelerated" tests. In such tests, the product under study is subjected in a short time to conditions similar to those that will be encountered in the working environment. A study is designed to investigate the ability to predict the extent of cracking of latex paint in the natural environment based on accelerated tests of the paint in the laboratory. Two variables are used: y, the cracking rate in the field; and x, the accelerated cracking rate obtained in the laboratory. Each of these rates is nonnegative. A simple linear regression model is assumed and these data are obtained:

x	y
2.0	1.9
3.0	2.7
4.0	4.2
5.0	4.8
6.0	4.8
7.0	5.1

You can use the formulas developed in Example 3.2.2 to verify that

$$\mathbf{b} = \begin{bmatrix} b_0 \\ b_1 \end{bmatrix} = \begin{bmatrix} 0.97 \\ 0.65 \end{bmatrix}$$

Here

$$\mathbf{y} = \begin{bmatrix} 1.9 \\ 2.7 \\ 4.2 \\ 4.8 \\ 4.8 \\ 5.1 \end{bmatrix}, \quad X = \begin{bmatrix} 1 & 2 \\ 1 & 3 \\ 1 & 4 \\ 1 & 5 \\ 1 & 6 \\ 1 & 7 \end{bmatrix}, \quad \text{and} \quad X\mathbf{b} = \begin{bmatrix} 2.27 \\ 2.92 \\ 3.57 \\ 4.22 \\ 4.87 \\ 5.52 \end{bmatrix}$$

By definition,

$$s^2 = \frac{(\mathbf{y} - X\mathbf{b})'(\mathbf{y} - X\mathbf{b})}{n-p} = \frac{SS_{\mathrm{Res}}}{n-p}$$

Substituting,

$$\mathbf{y} - X\mathbf{b} = \begin{bmatrix} 1.9 \\ 2.7 \\ 4.2 \\ 4.8 \\ 4.8 \\ 5.1 \end{bmatrix} - \begin{bmatrix} 2.27 \\ 2.92 \\ 3.57 \\ 4.22 \\ 4.87 \\ 5.52 \end{bmatrix} = \begin{bmatrix} -0.37 \\ -0.22 \\ 0.63 \\ 0.58 \\ -0.07 \\ -0.42 \end{bmatrix} = \mathbf{e}$$

and

$$s^2 = \frac{\mathbf{e'e}}{n-p} = \frac{\mathbf{e'e}}{6-2} = \sum_{i=1}^{6} e_i^2/4$$
$$= [(-0.37)^2 + (-0.22)^2 + \cdots + (-0.42)^2]/4$$
$$= 0.2749$$

The estimated variance in the field cracking rate is approximately 0.2749. (Based on a study reported in *Journal of Coating Technology* 55, 1983.)

The above example demonstrates one way to compute s^2. Since SS_{Res} can be written in several alternative forms, this is not the only method used to find the estimated variance in response. You are asked to find s^2 using some of these alternatives in Exercises 18 and 20.

REGRESSION THROUGH THE ORIGIN

Thus far we have considered the multiple linear regression model with an intercept term. This means that the model contains k parameters, $\beta_1, \beta_2, \ldots, \beta_k$, that are associated with the k variables, $x_1, x_2, \ldots, x_k$, respectively. It also contains a parameter, β_0, that stands alone. This parameter is called the *intercept*. Hence the number of model parameters estimated is $p = k + 1$, as mentioned earlier. Occasionally it is reasonable to assume from physical considerations that β_0 is equal to zero; that is, that no intercept term is needed in the model. Such a model assumes the form

$$y = \beta_1 x_1 + \beta_2 x_2 + \cdots + \beta_k x_k + \varepsilon$$

and is called *regression through the origin*. The formulas already derived for estimating the model parameters as well as that for estimating σ^2 hold as stated. Thus

$$\boldsymbol{\beta} = \begin{bmatrix} \beta_1 \\ \beta_2 \\ \vdots \\ \beta_k \end{bmatrix}$$

is estimated by

$$\mathbf{b} = (X'X)^{-1}X'\mathbf{y}$$

and the estimator for σ^2 is

$$\widehat{\sigma^2} = s^2 = \frac{SS_{\text{Res}}}{n-p}$$

where p, the number of model parameters, is k rather than $k + 1$. The only other change needed is to drop the first column of ones in the X matrix. This model is useful when it is reasonable to assume that the response is zero whenever each of the variables $x_1, x_2, \ldots, x_k$ assumes the value zero. Exercise 23 clarifies this idea.

3.4

MAXIMUM LIKELIHOOD ESTIMATORS (ADVANCED)

In Section 3.2 we developed the least squares estimators for $\beta_1, \beta_2, \ldots, \beta_k$. To do so the vector $\boldsymbol{\varepsilon}$ of random errors was required to have mean $\mathbf{0}$ and variance $\sigma^2 I$. If, in addition, it is assumed that $\varepsilon_1, \varepsilon_2, \ldots, \varepsilon_n$ are normally distributed, then maximum likelihood estimators can be found for these parameters. By assuming normality and zero covariance, we are in fact assuming that the random errors are independent. That is, we are assuming that $\varepsilon_1, \varepsilon_2, \ldots, \varepsilon_n$ are independent normally distributed random variables each with mean 0 and variance σ^2.

You may be familiar with the general technique for deriving maximum likelihood estimators:

1. Write the expression for the density, $f(\varepsilon_i)$ of the ith random error.
2. The likelihood function L is defined to be the joint density of the random errors. Since these are assumed to be independent, the joint density is the product of the marginals and hence L is given by

$$L(\varepsilon_1, \varepsilon_2, \ldots, \varepsilon_n; \sigma^2) = \prod_{i=1}^{n} f(\varepsilon_i)$$

3. Express L as a function of $\boldsymbol{\beta}$ and σ^2.
4. Find $\ln(L)$.
5. Maximize $\ln(L)$ with respect to $\boldsymbol{\beta}$ to obtain the maximum likelihood estimators for $\beta_0, \beta_1, \ldots, \beta_k$.
6. Maximize $\ln(L)$ with respect to σ^2 to obtain the maximum likelihood estimator for σ^2.

The technique can be demonstrated by finding the maximum likelihood estimator for $\boldsymbol{\beta}$:

1. Since ε_i is assumed to be a normally distributed random variable with mean 0 and variance σ^2, its density is given by

$$f(\varepsilon_i) = \frac{1}{\sqrt{2\pi}\sigma} e^{(-1/2)[\varepsilon_i/\sigma]^2}$$

2. The likelihood function is given by

$$L(\varepsilon_1, \varepsilon_2, \ldots, \varepsilon_n; \sigma^2) = \prod_{i=1}^{n} f(\varepsilon_i)$$
$$= \prod_{i=1}^{n} \frac{1}{\sqrt{2\pi}\sigma} e^{(-1/2)[\varepsilon_i/\sigma]^2}$$
$$= \frac{1}{(2\pi)^{n/2}(\sigma^2)^{n/2}} e^{-1/(2\sigma^2)\sum_{i=1}^{n}\varepsilon_i^2}$$

3. Since $\boldsymbol{\varepsilon} = \mathbf{y} - X\boldsymbol{\beta}$, $\Sigma_{i=1}^{n}\varepsilon_i^2 = \boldsymbol{\varepsilon}'\boldsymbol{\varepsilon} = (\mathbf{y} - X\boldsymbol{\beta})'(\mathbf{y} - X\boldsymbol{\beta})$. Substituting,

$$L(\varepsilon_1, \varepsilon_2, \ldots, \varepsilon_n; \sigma^2) = \frac{1}{(2\pi)^{n/2}(\sigma^2)^{n/2}} e^{-1/(2\sigma^2)[(\mathbf{y}-X\boldsymbol{\beta})'(\mathbf{y}-X\boldsymbol{\beta})]}$$

4. Taking the natural logarithm of each side and simplifying,

$$\ln L(\varepsilon_1, \varepsilon_2, \ldots, \varepsilon_n; \sigma^2) = -\frac{n}{2}\ln 2\pi - \frac{n}{2}\ln \sigma^2$$
$$-1/(2\sigma^2)[(\mathbf{y} - X\boldsymbol{\beta})'(\mathbf{y} - X\boldsymbol{\beta})]$$

5. To maximize $\ln(L)$, we must differentiate with respect to $\boldsymbol{\beta}$, set the derivative equal to zero, and solve for $\boldsymbol{\beta}$. Note that the derivative of the first two terms of $\ln(L)$ is zero. Thus the real problem is to use the rules for differentiation given in Section 2.2 to differentiate $(\mathbf{y} - X\boldsymbol{\beta})'(\mathbf{y} - X\boldsymbol{\beta})$. Note that

$$(\mathbf{y} - X\boldsymbol{\beta})'(\mathbf{y} - X\boldsymbol{\beta}) = \mathbf{y}'\mathbf{y} - 2(X'\mathbf{y})\boldsymbol{\beta} + \boldsymbol{\beta}'(X'X)\boldsymbol{\beta} = Q$$

and

$$\frac{\partial Q}{\partial \boldsymbol{\beta}} = -2(X'\mathbf{y}) + 2(X'X)\boldsymbol{\beta}$$

To maximize, the equation

$$-2(X'\mathbf{y}) + 2(X'X)\boldsymbol{\beta} = 0$$

is solved for $\boldsymbol{\beta}$ to obtain

$$\hat{\boldsymbol{\beta}} = (X'X)^{-1}X'\mathbf{y}$$

That is, the maximum likelihood estimator for $\boldsymbol{\beta}$ is $(X'X)^{-1}X'\mathbf{y}$. This is also the least squares estimator for $\boldsymbol{\beta}$. Theorem 3.4.1 formalizes this result for easy reference.

Theorem 3.4.1 Let $\mathbf{y} = X\boldsymbol{\beta} + \boldsymbol{\varepsilon}$ where X is an $n \times (k+1)$ matrix of full rank, $\boldsymbol{\beta}$ is a $(k+1) \times 1$ vector of unknown parameters, and $\boldsymbol{\varepsilon}$ is an $n \times 1$ normally distributed random vector with mean $\mathbf{0}$ and variance $\sigma^2 I$. The *maximum likelihood* estimator for $\boldsymbol{\beta}$ is denoted by $\tilde{\boldsymbol{\beta}}$

and is given by

$$\tilde{\boldsymbol{\beta}} = (X'X)^{-1}X'\mathbf{y}$$ ■

Do least squares estimators and maximum likelihood estimators always coincide? As Theorem 3.4.2 will show, the answer to this question is no. Theorem 3.4.2 shows that s^2, the least squares estimator for σ^2, differs from that obtained via the method of maximum likelihood. The proof of the theorem is outlined in Exercise 25.

Theorem 3.4.2 Let $\mathbf{y} = X\boldsymbol{\beta} + \boldsymbol{\varepsilon}$ where X is an $n \times (k+1)$ matrix of full rank, $\boldsymbol{\beta}$ is a $(k+1) \times 1$ vector of unknown parameters, and $\boldsymbol{\varepsilon}$ is an $n \times 1$ normally distributed random vector with mean $\mathbf{0}$ and variance $\sigma^2 I$. The maximum likelihood estimator for σ^2 is denoted by $\widetilde{\sigma^2}$ and is given by

$$\widetilde{\sigma^2} = \frac{SS_{\text{Res}}}{n}$$ ■

Since $s^2 = SS_{\text{Res}}/(n-p)$ and $\widetilde{\sigma^2} = SS_{\text{Res}}/n$ differ, there is a choice to be made when estimating σ^2. Recall that in the full rank model, $n \geqslant p$. If n (the sample size) is greater than p (the number of model parameters) and is large relative to p, then s^2 and $\widetilde{\sigma^2}$ will yield estimates that are very close in value. Either provides a suitable estimate for σ^2. However, as you shall see later, there are strong theoretical reasons for preferring s^2 to $\widetilde{\sigma^2}$ in problems that entail more than simple point estimation. Unfortunately, if $n = p$ then s^2 does not exist and $\widetilde{\sigma^2}$ always assumes the value 0 (see Exercises 31 and 32). In this case we can estimate $\boldsymbol{\beta}$ but we cannot estimate σ^2. Such a model is said to be *saturated.*

3.5

FURTHER PROPERTIES OF LEAST SQUARES ESTIMATORS (ADVANCED)

In Section 3.2 we proved the Gauss–Markoff theorem. This theorem shows that among the class of linear unbiased estimators for $\boldsymbol{\beta}$, the least squares estimator $\mathbf{b}$ is best in the sense that the variances of $b_0, b_1, b_2, \ldots, b_k$ are minimized. That is, $\mathbf{b}$ is a BLUE estimator. To prove this, we assumed only that the vector of random errors has mean $\mathbf{0}$ and variance $\sigma^2 I$. If we further assume that $\varepsilon_1, \varepsilon_2, \ldots, \varepsilon_n$ are normally distributed, then a stronger result can be obtained. To derive this result, we must review the notion of a sufficient statistic.

Consider a density characterized by a single parameter θ whose value is unknown. Typically we draw a random sample from this distribution and attempt to develop a statistic by which θ can be estimated from this sample. Here we want to find a statistic that is "sufficient" for θ. Roughly speaking, to say that a statistic is *sufficient* for a parameter θ means that the statistic exhausts all of the useful information about θ that is contained in the sample [**9**]. That is, the statistic condenses the sample in such a way that no information about θ is lost. The theoretical definition of sufficiency is given in terms of conditional distributions and can be found in [**9**] as well as in most advanced texts on mathematical statistics. For our purposes, we need only know how to recognize a sufficient statistic from knowledge of the form of the joint distribution of the sample. The *Fisher–Neyman factorization theorem* is useful in this regard.

Theorem 3.5.1 (Fisher–Neyman Factorization) Let X denote a random variable whose density depends on a single parameter θ. Let $X_1, X_2, \ldots, X_n$ be a random sample drawn from this distribution with joint density $f(x_1, x_2, \ldots, x_n; \theta)$. The statistic $Y = u(X_1, X_2, \ldots, X_n)$ is sufficient for θ if and only if

$$f(x_1, x_2, \ldots, x_n; \theta) = g[Y; \theta]h(x_1, x_2, \ldots, x_n)$$

where g depends on the x's only through Y, and h does not depend on θ. ■

Basically, this theorem states that a statistic Y is sufficient for θ if and only if the joint density of the sample can be factored into two components: the first dependent only on the statistic and the parameter and the second independent of the parameter. An example will illustrate the idea.

Example 3.5.1 Let $X_1, X_2, \ldots, X_n$ denote a random sample from a Poisson distribution with parameter θ. Find a sufficient statistic for estimating θ. The density for X_i, $i = 1, 2, \ldots, n$, is given by

$$f(x_i; \theta) = \frac{e^{-\theta}\theta^{x_i}}{x_i!}$$

Since $X_1, X_2, \ldots, X_n$ constitute a random sample, these random variables are independent. Hence the joint density for the sample is given by

$$\begin{aligned} f(x_1, x_2, \ldots, x_n; \theta) &= f(x_1; \theta)f(x_2; \theta) \ldots f(x_n; \theta) \\ &= \prod_{i=1}^{n} \frac{e^{-\theta}\theta^{x_i}}{x_i!} \\ &= \frac{e^{-n\theta}\theta^{\Sigma x_i}}{\prod_{i=1}^{n} x_i!} \end{aligned}$$

Rewriting this density,

$$f(x_1, x_2, \dots, x_n; \theta) = [e^{-n\theta}\theta^{\Sigma x_i}]\frac{1}{[x_1!x_2! \dots x_n!]}$$

If we let $Y = \Sigma X_i$, then f can be factored as

$$f(x_1, x_2, \dots, x_n; \theta) = g[Y; \theta]h(x_1, x_2, \dots, x_n)$$

where

$$g[Y; \theta] = e^{-n\theta}\theta^Y$$

and

$$h(x_1, x_2, \dots, x_n) = \frac{1}{x_1!x_2! \dots x_n!}$$

By the Fisher–Neyman factorization theorem, the statistic $Y = \Sigma_{i=1}^n X_i$ is sufficient for θ.

Sufficient statistics are not unique. For instance, in Example 3.5.1 f could have been expressed in the form

$$f(x_1, x_2, \dots, x_n; \theta) = e^{-n\theta}\theta^{n\bar{X}}\frac{1}{x_1!x_2! \dots x_n!}$$

Here the statistic $\bar{X} = \Sigma_{i=1}^n X_i/n$ is shown to be sufficient for θ also. In fact, any one-to-one function of $Y = \Sigma_{i=1}^n X_i$ not involving θ will be sufficient for θ.

In many problems, the joint density for the sample depends on more than one parameter. In this case, we want to find a set of statistics that are "jointly sufficient" for these parameters. That is, given a vector of parameters $\boldsymbol{\theta}' = [\theta_1, \theta_2, \dots, \theta_m]$ we want to find a set of statistics that condense the sample in such a way that no information concerning any of the parameters is lost. The Fisher–Neyman factorization theorem can be extended easily to include this case.

Theorem 3.5.2 Let $X_1, X_2, \dots, X_n$ be a random sample with joint density $f(x_1, x_2, \dots, x_n; \boldsymbol{\theta})$ where $\boldsymbol{\theta}$ is a vector of parameters. The vector of statistics $\mathbf{T}$ is sufficient for $\boldsymbol{\theta}$ if and only if

$$f(x_1, x_2, \dots, x_n; \boldsymbol{\theta}) = g(\mathbf{T}; \boldsymbol{\theta})\,h(x_1, x_2, \dots, x_n)$$

where g depends on the x's only through the random vector $\mathbf{T}$ and h does not depend on $\boldsymbol{\theta}$. ■

The Fisher–Neyman theorem is illustrated in the linear models context in Theorem 3.5.3.

Theorem 3.5.3 Let $\mathbf{y} = X\boldsymbol{\beta} + \boldsymbol{\varepsilon}$ where X is an $n \times p$ matrix of full rank, $\boldsymbol{\beta}$ is a $p \times 1$ vector of unknown parameters, and $\boldsymbol{\varepsilon}$ is an $n \times 1$ normally distributed random vector with mean $\mathbf{0}$ and variance $\sigma^2 I$. Then $\mathbf{b} = (X'X)^{-1}X'\mathbf{y}$ and $s^2 = \mathbf{e}'\mathbf{e}/(n-p)$ are jointly sufficient for $\boldsymbol{\beta}$ and σ^2.

Proof

Since each of the random errors $\varepsilon_1, \varepsilon_2, \ldots, \varepsilon_n$ is normally distributed with mean 0 and variance σ^2, the density for ε_i, $i = 1, 2, \ldots, n$, is given by

$$f(\varepsilon_i) = \frac{1}{\sqrt{2\pi}\sigma} e^{-1/2[\varepsilon_i/\sigma]^2}$$

Since covariance 0 implies independence when sampling from a normal distribution, the joint density for $\varepsilon_1, \varepsilon_2, \ldots, \varepsilon_n$ is given by

$$\begin{aligned} f(\varepsilon_1, \varepsilon_2, \ldots, \varepsilon_n; \boldsymbol{\beta}, \sigma^2) &= \prod_{i=1}^{n} f(\varepsilon_i) \\ &= \frac{1}{(2\pi)^{n/2}(\sigma^2)^{n/2}} \exp\left[-1/(2\sigma^2) \sum_{i=1}^{n} \varepsilon_i^2\right] \end{aligned}$$

Now $\boldsymbol{\varepsilon} = \mathbf{y} - X\boldsymbol{\beta}$ and hence

$$\varepsilon_i^2 = \boldsymbol{\varepsilon}'\boldsymbol{\varepsilon} = (\mathbf{y} - X\boldsymbol{\beta})'(\mathbf{y} - X\boldsymbol{\beta})$$

Substituting,

$$f(\varepsilon_1, \varepsilon_2, \ldots, \varepsilon_n; \boldsymbol{\beta}, \sigma^2) = \frac{1}{(2\pi\sigma^2)^{n/2}} \exp\left[(-1/2\sigma^2)(\mathbf{y} - X\boldsymbol{\beta})'(\mathbf{y} - X\boldsymbol{\beta})\right]$$

This density must now be factored as indicated in the Fisher–Neyman theorem. Note that the term $(\mathbf{y} - X\boldsymbol{\beta})'(\mathbf{y} - X\boldsymbol{\beta})$ can be partitioned as follows (see Exercise 36):

$$(\mathbf{y} - X\boldsymbol{\beta})'(\mathbf{y} - X\boldsymbol{\beta}) = (\mathbf{y} - X\mathbf{b})'(\mathbf{y} - X\mathbf{b}) + (\mathbf{b} - \boldsymbol{\beta})'X'X(\mathbf{b} - \boldsymbol{\beta})$$

Now $(\mathbf{y} - X\mathbf{b})'(\mathbf{y} - X\mathbf{b}) = SS_{\text{Res}} = (n-p)s^2$. Substituting, the density f can be expressed as

$$f(\varepsilon_1, \varepsilon_2, \ldots, \varepsilon_n; \boldsymbol{\beta}, \sigma^2) = \frac{1}{(2\pi\sigma^2)^{n/2}} \exp\{(-1/2\sigma^2)[(n-p)s^2 + (\mathbf{b} - \boldsymbol{\beta})'X'X(\mathbf{b} - \boldsymbol{\beta})]\}$$

Denote the right-hand side of this equation by $g(\mathbf{b}, s^2; \boldsymbol{\beta}, \sigma^2)$ and let $h(\varepsilon_1, \varepsilon_2, \ldots, \varepsilon_n) = 1$. Then

$$f(\varepsilon_1, \varepsilon_2, \ldots, \varepsilon_n; \boldsymbol{\beta}, \sigma^2) = g(\mathbf{b}, s^2; \boldsymbol{\beta}, \sigma^2)\, h(\varepsilon_1, \varepsilon_2, \ldots, \varepsilon_n)$$

By the Fisher–Neyman factorization theorem, $\mathbf{b}$ and s^2 are jointly sufficient for $\boldsymbol{\beta}$ and σ^2 as claimed. ■

At this point we know that the least squares estimators are the best *linear* unbiased estimators for $\boldsymbol{\beta}$ and σ^2 (BLUE). Furthermore, if it is assumed that the vector $\boldsymbol{\varepsilon}$ of random errors is normally distributed, then $\mathbf{b}$ and s^2 are sufficient for $\boldsymbol{\beta}$ and σ^2.

An important result from the field of statistical inference allows us to exploit the joint sufficiency and unbiasedness of $\mathbf{b}$ and s^2 to derive a stronger result than BLUE in the normal theory case. In particular, the Lehmann–Scheffé theorem can be applied to show that in this case $\mathbf{b}$ and s^2 are the *uniformly minimum variance unbiased estimators* (UMVUE) for $\boldsymbol{\beta}$ and σ^2, respectively. That is, it can be shown that if $\boldsymbol{\varepsilon}$ is a normally distributed random vector with mean $\mathbf{0}$ and variance $\sigma^2 I$, then not only are the least squares estimators best in the sense of minimum variance among *linear* estimators but they are best among *all* unbiased estimators for $\boldsymbol{\beta}$ and σ^2. The full development of this result is beyond the scope of this text. Refer to [**14**] for details.

3.6

INTERVAL ESTIMATION OF THE COEFFICIENTS

Once a point estimate is obtained for a parameter the usual question asked by a researcher is, How good is this estimate? By "good" he or she usually means that the estimate is close in value to the true value of the parameter under study. Unfortunately, neither the unbiasedness property nor the notion of maximum likelihood guarantees that a *single* estimate will closely approximate the true value of the parameter. Thus to answer the researcher's question, we must turn from point to interval estimation. We create an interval of values in such a way that we can attach a measure of confidence to the interval. This measure gives us an idea of the faith that we have that the interval actually contains the true value of the parameter of interest.

The probability distribution that will be used to create confidence intervals on the coefficients $\beta_0, \beta_1, \ldots, \beta_k$ of the full rank linear model is the Student's t distribution. This continuous distribution was first described by W. S. Gosset in 1908. The graph of the density of a t random variable is a symmetric bell-shaped curve centered at 0. The exact shape of the bell is determined by one parameter, γ, called *degrees of freedom.* This parameter is always a positive integer whose numerical value depends in part on the sample size involved. The bell-shaped curves associated with the t distribution for small samples are rather flat, indicative of a large variance. As the sample size and degrees of freedom increase, the graph of the t density approaches that of the standard normal curve until eventually they are virtually identical. For this reason, the standard normal curve is used to approximate t points for large samples. The t distribution can be defined by stating the form assumed by its density. However, for our purposes it is more useful to define it in terms of a ratio of two other random variables whose distributions are already familiar.

Definition 3.6.1

Let Z be a standard normal random variable and let X_γ^2 be an independent chi-squared random variable with γ degrees of freedom. The random variable

$$\frac{Z}{\sqrt{X_\gamma^2/\gamma}}$$

is said to follow a t distribution with γ degrees of freedom.

Three things about this definition are important. First, the number of degrees of freedom associated with the t random variable is equal to that associated with the chi-squared random variable that appears in the denominator; second, the numerator and denominator must be independent random variables; and third, to show that a particular random variable follows a t distribution, it is sufficient to show that it can be written as the ratio of a standard normal random variable to the square root of an independent chi-squared random variable divided by its degrees of freedom.

Recall that to obtain the least squares estimators for the coefficients, it is not necessary to make any assumption concerning the family of random variables to which the random errors $\varepsilon_1, \varepsilon_2, \ldots, \varepsilon_n$ belong. It is assumed only that $E[\boldsymbol{\varepsilon}] = \mathbf{0}$ and $\text{var}\ \boldsymbol{\varepsilon} = \sigma^2 I$. With these assumptions it is possible to show that $\mathbf{b}$ is an unbiased estimator for $\boldsymbol{\beta}$ and that $\text{var}\ \mathbf{b} = (X'X)^{-1}\sigma^2$. To create confidence intervals on $\beta_0, \beta_1, \ldots, \beta_k$, an additional assumption is necessary. In particular, we assume that $\boldsymbol{\varepsilon}$ is *normally* distributed with mean $\mathbf{0}$ and variance $\sigma^2 I$. This assumption makes it possible to consider fully the distribution of $\mathbf{b}$. Since $\mathbf{y} = X\boldsymbol{\beta} + \boldsymbol{\varepsilon}$, we can write y_i in the form

$$y_i = \beta_0 + x_{i1}\beta_1 + x_{i2}\beta_2 + \ \ldots \ + x_{ik}\beta_k + \varepsilon_i$$

Since we are now assuming that ε_i is normally distributed, it is evident that we are also assuming that y_i is normally distributed for $i = 1, 2, \ldots, n$. Hence $\mathbf{y}$ is a normally distributed random vector with mean $X\boldsymbol{\beta}$ and variance $\sigma^2 I$. This means, of course, that $y_1, y_2, \ldots, y_n$ are independent normally distributed random variables. Since $\mathbf{b} = (X'X)^{-1}X'\mathbf{y}$, each element of $\mathbf{b}$ is a linear combination of $y_1, y_2, \ldots, y_n$. It is known that any linear combination of independent normally distributed random variables is also normally distributed. Thus we can conclude that $\mathbf{b}$ is a normally distributed random vector with mean $\boldsymbol{\beta}$ and variance $(X'X)^{-1}\sigma^2$. This result is summarized in Theorem 3.6.1.

Theorem 3.6.1 Let $\mathbf{y} = X\boldsymbol{\beta} + \boldsymbol{\varepsilon}$ where X is an $n \times (k+1)$ matrix of full rank, $\boldsymbol{\beta}$ is a $(k+1) \times 1$ vector of unknown parameters, and $\boldsymbol{\varepsilon}$ is an $n \times 1$ normally distributed random vector with mean $\mathbf{0}$ and variance $\sigma^2 I$. Then the least squares estimator for $\boldsymbol{\beta}$, $\mathbf{b}$, is normally distributed with mean $\boldsymbol{\beta}$ and variance $(X'X)^{-1}\sigma^2$. ■

Recall that the least squares estimator for σ^2 is s^2 where

$$s^2 = \frac{SS_{\mathrm{Res}}}{n-p}$$

We already know that s^2 is an unbiased estimator for σ^2, but nothing else has been said about its distribution. To create confidence intervals on the coefficients, it is necessary to know the distribution of the random variable

$$\frac{(n-p)s^2}{\sigma^2} = \frac{SS_{\mathrm{Res}}}{\sigma^2}$$

Theorem 3.6.2 gives this distribution.

Theorem 3.6.2 Let $\mathbf{y} = X\boldsymbol{\beta} + \boldsymbol{\varepsilon}$ where X is an $n \times (k+1) = n \times p$ matrix of full rank, $\boldsymbol{\beta}$ is a $(k+1) \times 1$ vector of unknown parameters, and $\boldsymbol{\varepsilon}$ is an $n \times 1$ normally distributed random vector with mean $\mathbf{0}$ and variance $\sigma^2 I$. Then

$$\frac{(n-p)s^2}{\sigma^2} = \frac{SS_{\mathrm{Res}}}{\sigma^2}$$

follows a chi-squared distribution with $n-p$ degrees of freedom.

Proof

The residual sum of squares can be expressed as a quadratic form (see Exercise 16):

$$SS_{\mathrm{Res}} = \mathbf{y}'[I - X(X'X)^{-1}X']\mathbf{y}$$

Note that $I - X(X'X)^{-1}X'$ is a symmetric, idempotent matrix of rank $n-p$ (Exercise 37) and that $\mathbf{y}$ is normally distributed with mean $X\boldsymbol{\beta}$ and variance $\sigma^2 I$. By Corollary 2.3.2,

$$\frac{SS_{\mathrm{Res}}}{\sigma^2} = \frac{\mathbf{y}'[I - X(X'X)^{-1}X']\mathbf{y}}{\sigma^2} = \mathbf{y}'A\mathbf{y}$$

follows a noncentral chi-squared distribution with $n-p$ degrees of freedom and noncentrality parameter

$$\lambda = \frac{1}{2\sigma^2}\boldsymbol{\mu}'A\boldsymbol{\mu}$$

Here $\boldsymbol{\mu} = X\boldsymbol{\beta}$ and hence

$$\begin{aligned}\lambda &= \frac{1}{2\sigma^2}(X\boldsymbol{\beta})'[I - X(X'X)^{-1}X']X\boldsymbol{\beta} \\ &= \frac{1}{2\sigma^2}[\boldsymbol{\beta}'X'X\boldsymbol{\beta} - \boldsymbol{\beta}'X'X(X'X)^{-1}X'X\boldsymbol{\beta}] = 0\end{aligned}$$

By definition, a noncentrality parameter of 0 implies that the distribution is chi-squared as claimed. ■

Remember that our aim is to create confidence intervals on the coefficients via the t distribution. We have the basic ingredients needed to do so, namely, a normally distributed random vector $\mathbf{b}$ that can be standardized easily and a chi-squared random variable SS_{Res}/σ^2. We need only show that $\mathbf{b}$ and SS_{Res}/σ^2 are independent, and the structure required to form a t random variable will be available. Theorem 3.6.3 shows that these random variables are in fact independent.

Theorem 3.6.3 Let $\mathbf{y} = X\boldsymbol{\beta} + \boldsymbol{\varepsilon}$ where X is an $n \times p$ matrix of full rank, $\boldsymbol{\beta}$ is a $p \times 1$ vector of unknown parameters, and $\boldsymbol{\varepsilon}$ is an $n \times 1$ normally distributed random vector with mean $\mathbf{0}$ and variance $\sigma^2 I$. Then $\mathbf{b}$ and SS_{Res}/σ^2 are independent.

Proof

Note that

$$\mathbf{b} = (X'X)^{-1}X'\mathbf{y} = B\mathbf{y}$$

and that

$$\frac{SS_{\text{Res}}}{\sigma^2} = \frac{\mathbf{y}'[I - X(X'X)^{-1}X']\mathbf{y}}{\sigma^2} = \mathbf{y}'A\mathbf{y}$$

Furthermore, $\mathbf{y}$ is a multivariate normal random variable with mean $X\boldsymbol{\beta}$ and variance $V = \sigma^2 I$, and the matrix A is symmetric. Consider the matrix BVA. Here

$$\begin{aligned}BVA &= (X'X)^{-1}X'\sigma^2\frac{I[I - X(X'X)^{-1}X']}{\sigma^2} \\ &= (X'X)^{-1}X' - (X'X)^{-1}X'X(X'X)^{-1}X' \\ &= 0\end{aligned}$$

By Theorem 2.4.2, $\mathbf{b}$ and SS_{Res}/σ^2 are independent as claimed. ■

It is now easy to derive the bounds for a confidence interval on a coefficient, β_j, of the full rank linear model. To do so, note that the variance-covariance matrix for $\mathbf{b}$

is $(X'X)^{-1}\sigma^2$. Think of this matrix in expanded form as

$$(X'X)^{-1}\sigma^2 = \begin{bmatrix} c_{00} & c_{01} & \cdots & c_{0k} \\ c_{10} & c_{11} & \cdots & c_{1k} \\ \vdots & \vdots & & \vdots \\ c_{k0} & c_{k1} & & c_{kk} \end{bmatrix} \sigma^2$$

By definition, this matrix displays the variances of $b_0, b_1, b_2, \ldots, b_k$ along the main diagonal. Thus, the variance of b_j is denoted by $c_{jj}\sigma^2$. Since b_j is normally distributed with mean β_j and variance $c_{jj}\sigma^2$, we can standardize to conclude that the random variable

$$\frac{b_j - \beta_j}{\sigma\sqrt{c_{jj}}}$$

follows a standard normal distribution. Now consider the random variable

$$\frac{(b_j - \beta_j)/\sigma\sqrt{c_{jj}}}{\sqrt{(SS_{\text{Res}}/\sigma^2)/(n-p)}}$$

Since SS_{Res}/σ^2 is independent of **b**, by definition it is independent of each entry of **b**. Since the standard normal random variable in the numerator depends only on b_j, the numerator and denominator of this ratio are independent. It can be concluded that the random variable

$$\frac{(b_j - \beta_j)/\sigma\sqrt{c_{jj}}}{\sqrt{(SS_{\text{Res}}/\sigma^2)/(n-p)}} = \frac{(b_j - \beta_j)/\sigma\sqrt{c_{jj}}}{\sqrt{s^2/\sigma^2}} = \frac{b_j - \beta_j}{s\sqrt{c_{jj}}}$$

follows a t distribution with $n - p$ degrees of freedom. Letting $t_{\alpha/2}$ denote the point associated with a t distribution with $n - p$ degrees of freedom with area $\alpha/2$ to its right, the usual procedure for deriving confidence bounds can be used to derive the bounds for a $100(1-\alpha)\%$ confidence interval on β_j as follows:

$$P[-t_{\alpha/2} \leqslant (b_j - \beta_j)/s\sqrt{c_{jj}} \leqslant t_{\alpha/2}] = 1 - \alpha$$
$$P[-t_{\alpha/2}s\sqrt{c_{jj}} \leqslant b_j - \beta_j \leqslant t_{\alpha/2}s\sqrt{c_{jj}}] = 1 - \alpha$$
$$P[b_j - t_{\alpha/2}s\sqrt{c_{jj}} \leqslant \beta_j \leqslant b_j + t_{\alpha/2}s\sqrt{c_{jj}}] = 1 - \alpha$$

From this computation we can see that the desired bounds are given by

Confidence bounds on β_j ($n-p$ df): $b_j \pm t_{\alpha/2}s\sqrt{c_{jj}}$

Example 3.6.1 It is known that y, the amount of a chemical that will dissolve in a fixed volume of water, depends in part on the water temperature, x. These data are found for a particular chemical.

Temperature °C (x)	g/liter (y)
0	2.1
10	4.5
20	6.1
30	11.2
40	13.8
50	17.0

The model employed is the simple linear regression model

$$y = \beta_0 + \beta_1 x + \varepsilon$$

A sample of size 6 yields the following pertinent statistics and point estimates for β_0, β_1, and σ^2:

$$\hat{\beta}_0 = b_0 \doteq 1.438$$
$$\hat{\beta}_1 = b_1 \doteq 0.307$$
$$\widehat{\sigma^2} = s^2 \doteq 0.745$$
$$\hat{\sigma} = s \doteq 0.863$$
$$\sum_{i=1}^{6} x_i = 150$$
$$\sum_{i=1}^{6} x_i^2 = 5500$$

(s^2 was found using a computer.)

To find a 95% confidence interval on β_0, the intercept of the regression line, we need to find the matrix $(X'X)^{-1}$. In Example 3.2.2, we found that the general form for $(X'X)^{-1}$ in a simple linear regression model is

$$(X'X)^{-1} = \begin{bmatrix} c_{00} & c_{01} \\ c_{10} & c_{11} \end{bmatrix} = \frac{1}{n\sum_{i=1}^{n} x_i^2 - \left(\sum_{i=1}^{n} x_i\right)^2} \begin{bmatrix} \sum_{i=1}^{n} x_i^2 & -\sum_{i=1}^{n} x_i \\ -\sum_{i=1}^{n} x_i & n \end{bmatrix}$$

From this, it is easy to see that c_{00} is given by

$$c_{00} = \frac{\sum_{i=1}^{n} x_i^2}{n\sum_{i=1}^{n} x_i^2 - \left(\sum_{i=1}^{n} x_i\right)^2} = \frac{5500}{6(5500) - (150)^2} \doteq 0.524$$

Here the number of degrees of freedom is $n - p = 6 - 2 = 4$. For a 95% confidence interval, $\alpha = 0.05$ and $\alpha/2 = 0.025$. The required t point is found to be $t_{0.025} = 2.776$. Substituting, the 95% confidence bounds on β_0 are

$$\begin{aligned} b_0 \pm t_{\alpha/2} s\sqrt{c_{00}} &= 1.438 \pm 2.776(0.863)\sqrt{0.524} \\ &= 1.438 \pm 1.734 \end{aligned}$$

What does this mean? We know that

$$\widehat{E[y]} = b_0 + b_1 x = 1.438 + 0.307x$$

The estimated average amount of chemical dissolved is $1.438 + 0.307x$ where x is the temperature of the water in °C. If $x = 0$, then $b_0 = 1.438$ is the estimated average amount of chemical dissolved when the temperature is 0°C. We know that this point estimate is probably not exactly equal to β_0. The confidence interval allows us to conclude that the true value of β_0 is likely to lie between $1.438 - 1.734 = -0.296$ and $1.438 + 1.734 = 3.172$. Since physically the amount dissolved is nonnegative, we should say that we are 95% confident that the average amount of chemical dissolved at 0°C lies between 0 and 3.172 g/liter.

Once again it is obvious that there is a computational problem in any but the simplest models. In particular, we must find the matrix $(X'X)^{-1}$ in order to obtain the values of $c_{00}, c_{11}, \ldots, c_{kk}$. In models other than the location model or simple linear regression, this is not done by hand. Rather, we rely on the computer to find and display this matrix. Computer Supplement A at the end of the chapter demonstrates how this is done using SAS.

3.7

INTERVAL ESTIMATION OF LINEAR FUNCTIONS OF THE COEFFICIENTS

In Section 3.2 we considered the problem of point estimation of a linear function of the coefficients $\beta_0, \beta_1, \ldots, \beta_k$. Recall that such a function can be expressed in the form $\mathbf{t}'\boldsymbol{\beta}$ where $\mathbf{t}'$ is a $1 \times (k+1)$ vector of scalars. The BLUE estimator for $\mathbf{t}'\boldsymbol{\beta}$ is $\mathbf{t}'\mathbf{b}$ where $\mathbf{b}$ is the least squares estimator for $\boldsymbol{\beta}$. Here we want to derive bounds for a $100(1-\alpha)\%$ confidence interval on $\mathbf{t}'\boldsymbol{\beta}$. To do so, we need only consider the distribution of $\mathbf{t}'\mathbf{b}$ under the assumption that $\mathbf{y}$ is a normally distributed random vector with mean $X\boldsymbol{\beta}$ and variance $\sigma^2 I$. Since $\mathbf{t}'\mathbf{b} = \mathbf{t}'(X'X)^{-1}X'\mathbf{y}$, $\mathbf{t}'\mathbf{b}$ is a linear function of the independent normally distributed random variables $y_1, y_2, \ldots, y_n$ and is therefore itself normally distributed. We already know that $E[\mathbf{t}'\mathbf{b}] = \mathbf{t}'\boldsymbol{\beta}$. Applying the rules for variance,

$$\begin{aligned}
\text{var } \mathbf{t}'\mathbf{b} &= \text{var } \mathbf{t}'(X'X)^{-1}X'\mathbf{y} \\
&= \text{var } [X(X'X)^{-1}\mathbf{t}]'\mathbf{y} \\
&= \mathbf{t}'(X'X)^{-1}X'\sigma^2 I X(X'X)^{-1}\mathbf{t} \\
&= \mathbf{t}'(X'X)^{-1}\mathbf{t}\sigma^2
\end{aligned}$$

Standardizing, the random variable

$$\frac{\mathbf{t}'\mathbf{b} - \mathbf{t}'\boldsymbol{\beta}}{\sqrt{\mathbf{t}'(X'X)^{-1}\mathbf{t}\sigma^2}}$$

follows a standard normal distribution. It can be shown that the chi-squared random variable SS_{Res}/σ^2 is independent of $\mathbf{t}'\mathbf{b}$ (see Exercise 45). Thus, by definition, the ratio

$$\frac{(\mathbf{t}'\mathbf{b} - \mathbf{t}'\boldsymbol{\beta})/\sqrt{\mathbf{t}'(X'X)^{-1}\mathbf{t}\sigma^2}}{\sqrt{\dfrac{SS_{\text{Res}}}{\sigma^2(n-p)}}} = \frac{\mathbf{t}'\mathbf{b} - \mathbf{t}'\boldsymbol{\beta}}{s\sqrt{\mathbf{t}'(X'X)^{-1}\mathbf{t}}}$$

follows a t distribution with $n-p$ degrees of freedom. A derivation similar to that given in the last section yields these confidence bounds:

Confidence bounds on $\mathbf{t}'\boldsymbol{\beta}$ ($n-p$ df): $\mathbf{t}'\mathbf{b} \pm t_{\alpha/2}s\sqrt{\mathbf{t}'(X'X)^{-1}\mathbf{t}}$

Although these bounds can be used to find confidence intervals on any linear function of the coefficients, they are employed most often to find bounds on the mean response for a particular set of observations on the x variables. To see how this is done, let $x_{*1}, x_{*2}, x_{*3}, \ldots, x_{*k}$ denote some specific values of the variables $x_1, x_2, x_3, \ldots, x_k$, respectively. We already know that the true average response for this set of observations is

$$E[y] = \beta_0 + \beta_1 x_{*1} + \beta_2 x_{*2} + \cdots + \beta_k x_{*k}$$

To form a confidence interval on this mean, we write it in the form $\mathbf{t}'\boldsymbol{\beta}$ where $\mathbf{t}' = [1 \quad x_{*1} \quad x_{*2} \quad \ldots \quad x_{*k}] = \mathbf{x}'_*$. Substituting, the $100(1-\alpha)\%$ confidence bounds are given by

Confidence bounds on the mean response for a particular set of x values ($n-p$ df): $\mathbf{x}'_*\mathbf{b} \pm t_{\alpha/2}s\sqrt{\mathbf{x}'_*(X'X)^{-1}\mathbf{x}_*}$

An example should clarify the point.

Example 3.7.1 In Example 3.2.3, we estimated the average selling price of 15-year-old houses with 2500 square feet of living space to be

$$\begin{aligned}\widehat{E[y]} = \mathbf{t}'\mathbf{b} = \mathbf{x}'\mathbf{b} &= [1 \quad 15 \quad 2.5]\mathbf{b} \\ &= 33.06 - 0.189(15) + 10.718(2.5) \\ &= 57.02\end{aligned}$$

To extend this to a 95% confidence interval, we need the matrix $(X'X)^{-1}$ and the estimated variance s^2. These are found by computer to be

$$(X'X)^{-1} = \begin{bmatrix} 2.307551 & 0.1565378 & -1.88398 \\ 0.1565378 & 0.02578269 & -0.20442 \\ -1.88398 & -0.20442 & 1.977901 \end{bmatrix}$$

and

$$s^2 = 48.837937 \quad \text{and} \quad s = \sqrt{48.837937} \doteq 6.99$$

For these data,

$$\begin{aligned} x'_*(X'X)^{-1}\mathbf{x}_* &= [1 \quad 15 \quad 2.5](X'X)^{-1}\begin{bmatrix} 1 \\ 15 \\ 2.5 \end{bmatrix} \\ &\doteq 0.415 \end{aligned}$$

The bounds for a 95% confidence interval are found via the t distribution with $n - p = 5 - 3 = 2$ degrees of freedom. They are given by

$$\mathbf{x}'_*\mathbf{b} \pm t_{0.025}s\sqrt{\mathbf{x}'_*(X'X)^{-1}\mathbf{x}_*}$$
$$57.02 \pm 4.303(6.99)\sqrt{0.415}$$
$$57.02 \pm 19.376$$

We are 95% confident that the true average selling price of 15-year-old houses with 2500 square feet of living space lies between \$37,644 and \$76,396, but this interval is much too long to be very informative! This is because the sample is very small ($n = 5$). The purpose of this example is to show you the technique for creating confidence intervals on a mean response. We do not intend to imply that researchers routinely make inferences based on such small samples. In Computing Supplement B at the end of this chapter, the construction of this confidence interval via SAS is demonstrated.

In the case of simple linear regression, it is useful to construct a graphical display that lets the experimenter see at a glance the confidence bounds for the mean value of y at a given value of x over a wide range of x values. Such a display is often called a *confidence band.* To form such a band, simply pick several x values, find confidence bounds at each of these values, and then join the upper confidence bounds with a smooth curve and the lower confidence bounds with a second smooth curve to form the boundaries for the confidence band. Although no exact probability can be attached to the band as a whole, this procedure does allow a region in the xy plane that should contain the true regression line to be pinpointed. The next example illustrates this technique.

Example 3.7.2 A composite of 616 nylon and steel is being developed for use in cam gears. A study is conducted to investigate the relationship between the noise level in cast iron gears (x) and that in identical gears made from the new material (y). These data are available:

Noise Level in Decibels

Cast Iron (x)	Composite (y)
75	74
80	81
110	107
93	90
65	64

A simple linear regression model is assumed. For these data,

$$(X'X)^{-1} = \frac{1}{6066}\begin{bmatrix} 36999 & -423 \\ -423 & 5 \end{bmatrix}, \quad \mathbf{b} \doteq \begin{bmatrix} 3.95549 \\ 0.936696 \end{bmatrix}, \quad \text{and} \quad s \doteq 1.453$$

To form a 95% confidence band, arbitrarily select x values lying between 65 and 110, the range of the data. It is customary to choose $\bar{x}$ as one of these points. In Exercise 55 you are asked to verify that the confidence band is at its narrowest at this point. For these data $\bar{x} = 84.6$ and $\mathbf{x}'_* = [1 \quad 84.6]$. Here

$$\begin{aligned} \mathbf{x}'_*(X'X)^{-1}\mathbf{x}_* &= [1 \quad 84.6]\begin{bmatrix} 36999 & -423 \\ -423 & 5 \end{bmatrix}\begin{bmatrix} 1 \\ 84.6 \end{bmatrix}\frac{1}{6066} \\ &= [1213.2 \quad 0]\begin{bmatrix} 1 \\ 84.6 \end{bmatrix}\frac{1}{6066} \\ &= 0.2 \end{aligned}$$

The 95% confidence bounds on the mean noise level for gears made with the composite material when similar gears made from cast iron produce a noise level of 84.6 decibels are

$$\mathbf{x}'_*\mathbf{b} \pm t_{0.025}s\sqrt{\mathbf{x}'_*(X'X)^{-1}\mathbf{x}_*}$$

or

$$3.95549 + 0.936696(84.6) \pm 3.182(1.453)\sqrt{0.2}$$

This reduces to

$$83.199 \pm 2.067$$

This procedure is used repeatedly to find these bounds for selected values of x:

x	Lower bound (○)	Upper bound (*)
65	61.516	68.166
70	66.689	72.360
75	71.778	76.638
84.6 ($\bar{x}$)	81.132	85.266
90	86.069	90.448
105	98.900	105.717
110	103.035	110.949

These bounds are plotted in Figure 3.1. When the upper and lower bounds are connected via smooth curves, the 95% confidence band shown is obtained. Computing Supplement C at the end of this chapter shows how to use SAS to find these bounds.

FIGURE 3.1 **A 95% confidence band on the mean noise level for gears made from a composite**

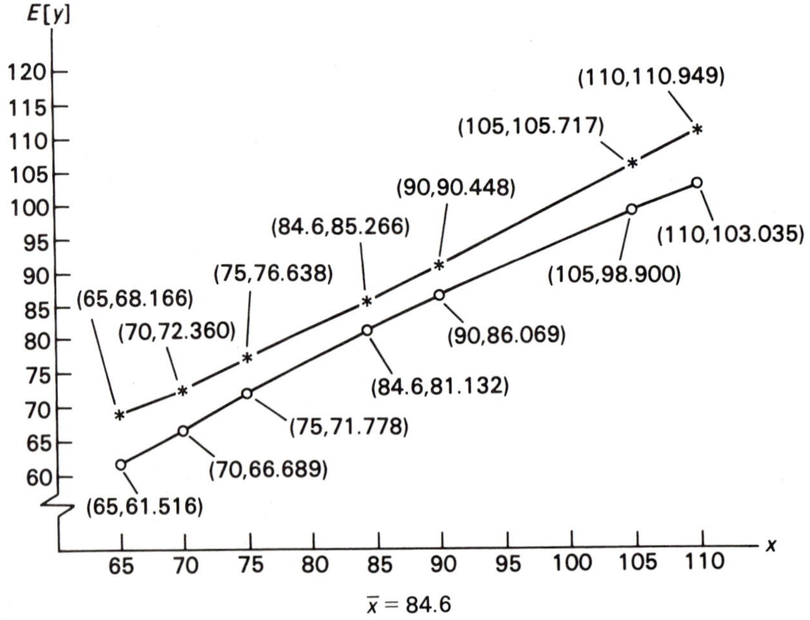

When there are two x variables in the model, the graph of the mean response as a function of x_1 and x_2 is a surface in three-dimensional space. Although it is a bit difficult to visualize, the procedure just demonstrated can be extended to create a confidence region about this theoretical surface. If the model contains three or more x variables, then there is no longer a convenient geometrical interpretation of the confidence region.

PREDICTION INTERVALS

One other interval estimator should be considered. We already know that the estimated model

$$b_0 + b_1x_1 + b_2x_2 + \cdots + b_kx_k$$

provides an unbiased estimate for the average value of y for given values of the x's. If asked to estimate the value that y itself will assume for some specific values of $x_1, x_2, \ldots, x_k$, not all of which were used to develop the estimated model, the only logical choice is the average value of y predicted by the model. That is,

$$\widehat{E[y]} = \hat{y}$$

Suppose we want to develop a bound on the value that y will assume in some future run of the experiment. To create such an interval, called a *prediction interval*, consider the random variable

$$y_* - \hat{y}_*$$

where y_* denotes the true value of the response when $x_1 = x_{*1}, x_2 = x_{*2}, \ldots, x_k = x_{*k}$ and $\hat{y}_*$ denotes the predicted response at this point. Note that

$$y_* = \mathbf{x}'_*\boldsymbol{\beta} + \varepsilon_*$$

and

$$\hat{y}_* = \mathbf{x}'_*\mathbf{b}$$

By substitution,

$$y_* - \hat{y}_* = \mathbf{x}'_*\boldsymbol{\beta} + \varepsilon_* - \mathbf{x}'_*\mathbf{b}$$

Since $\mathbf{b}$ depends only on the observed data and ε_* is a random error associated with a data point to be observed in the future, $\mathbf{b}$ and ε_* are assumed to be independent. Hence

$$\text{var}(y_* - \hat{y}_*) = \text{var}\ \varepsilon_* + \text{var}\ \mathbf{x}'_*\mathbf{b}$$

We have assumed that var $\varepsilon_* = \sigma^2$, and it has been shown that var $\mathbf{b} = (X'X)^{-1}\sigma^2$. Applying the rules for variance,

$$\begin{aligned}\text{var}\,(y_* - \hat{y}_*) &= \sigma^2 + \mathbf{x}'_*(X'X)^{-1}\mathbf{x}_*\sigma^2 \\ &= [1 + \mathbf{x}'_*(X'X)^{-1}\mathbf{x}_*]\sigma^2\end{aligned}$$

Following the same type of argument as that presented earlier, the random variable

$$\frac{y_* - \hat{y}_*}{s\sqrt{1 + \mathbf{x}'_*(X'X)^{-1}\mathbf{x}'_*}}$$

follows a t distribution with $n-p$ degrees of freedom. This random variable is used to create the following prediction interval:

Prediction interval on a future response for a particular set of x values ($n-p$ df): $\mathbf{x}_*'\mathbf{b} \pm t_{\alpha/2} s\sqrt{1+\mathbf{x}_*'(X'X)^{-1}\mathbf{x}_*}$

There is a difference in this interval and the one presented earlier. A prediction interval is an interval that is expected to contain a *single* future response for a specified set of x values; a confidence interval is expected to contain the *average* value of the response for these values. As you can see, the intervals are similar. However, since it is harder to anticipate an individual response than it is a group response, the prediction interval at $\mathbf{x}_*$ is always wider than the corresponding confidence interval. To illustrate, let us reconsider the data of Example 3.7.1.

Example 3.7.3 In Example 3.7.1 we estimated the average selling price of 15-year-old houses with 2500 square feet of living space to be

$$\mathbf{x}_*'\mathbf{b} \pm t_{0.025} s\sqrt{\mathbf{x}_*'(X'X)^{-1}\mathbf{x}_*}$$
$$57.02 \pm 19.376$$

Suppose we are now asked to estimate the selling price of a randomly selected house with these characteristics. Here interest centers on a single house rather than on an average for a group of houses. The prediction interval is given by

$$\mathbf{x}_*'\mathbf{b} \pm t_{0.025} s\sqrt{1+\mathbf{x}_*'(X'X)^{-1}\mathbf{x}_*}$$

From Example 3.7.1,

$$t_{0.025} = 4.303, \quad s = 6.99, \quad \text{and} \quad \mathbf{x}'(X'X)^{-1}\mathbf{x} = 0.415$$

By substitution, the desired interval is

$$57.02 \pm 4.303(6.99)\sqrt{1+0.415}$$

or 57.02 ± 35.78.

We are 95% confident that the selling price of an individual randomly selected 15-year-old house with 2500 square feet of living space lies between \$21,240 and \$92,800. As expected, this interval is wider than that obtained earlier. Computing Supplement D at the end of this chapter illustrates the use of SAS in constructing this prediction interval.

3.8

JOINT CONFIDENCE REGION ON THE REGRESSION COEFFICIENTS — THE *F* DISTRIBUTION

We have already derived techniques that can be used to find confidence bounds on each of the parameters $\beta_0, \beta_1, \beta_2, \ldots, \beta_k$ individually. There are times when it is useful to make a statement concerning the simultaneous behavior of these parameters. That is, we would like to develop a method by which we can claim that the vector $\boldsymbol{\beta}$ lies in some specified region in $k+1$ dimensional space with a stated confidence. This is a little harder to do than it might seem. One possibility is to apply the usual procedure to form confidence intervals on each of the parameters $\beta_0, \beta_1, \ldots, \beta_k$ and then use these $k+1$ intervals to form a $k+1$ dimensional "rectangle". However, this procedure is misleading. Since $b_0, b_1, \ldots, b_k$ are not independent, it is not possible to attach an exact confidence to such a rectangle. Another method must be found for making simultaneous inferences on the elements of $\boldsymbol{\beta}$. The method we shall derive is based on the F distribution. We begin by reviewing some of the more important properties of this distribution.

The family of F random variables is a family of continuous distributions. Each member of the family is identified by two parameters, γ_1 and γ_2, that assume only positive integer values and that are called *degrees of freedom.* An F random variable assumes only positive values, and the general shape of the graph of its density, which is skewed, is shown in Figure 3.2. It is possible to define an F random variable by stating the equation for its density. However, just as with the t distribution, it is more helpful to define it as the ratio of two random variables whose distributions are already familiar.

FIGURE 3.2 General form for the graph of the density of an F random variable

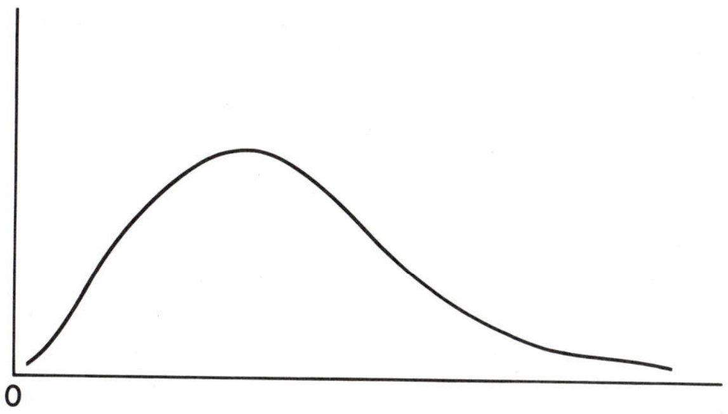

Definition 3.8.1

Let $X^2_{\gamma 1}$ and $X^2_{\gamma 2}$ be independent chi-squared random variables with γ_1 and γ_2 degrees of freedom, respectively. Then the random variable

$$F_{\gamma 1,\gamma 2} = \frac{X^2_{\gamma 1}/\gamma_1}{X^2_{\gamma 2}/\gamma_2}$$

follows an F distribution with γ_1 and γ_2 degrees of freedom.

Note that γ_1 is the degrees of freedom associated with the chi-squared random variable of the numerator; γ_2 is that associated with the denominator. The jist of this definition is that to show that a particular random variable follows the F distribution it is sufficient to show that it can be written as the ratio of two independent chi-squared random variables divided by their respective degrees of freedom. This idea is used in deriving a confidence region for $\boldsymbol{\beta}$.

Consider the least squares estimators

$$\mathbf{b} = (X'X)^{-1}X'\mathbf{y}$$

under the assumption that $\boldsymbol{\varepsilon}$ is a normally distributed random vector with mean $\mathbf{0}$ and variance $\sigma^2 I$. Let q denote the quadratic form

$$\frac{(\mathbf{b}-\boldsymbol{\beta})'X'X(\mathbf{b}-\boldsymbol{\beta})}{\sigma^2}$$

It can be shown that this quadratic form follows a chi-squared distribution with p degrees of freedom where p denotes the number of parameters in the model (see Exercise 61). In Section 3.6 we saw that the random variable

$$\frac{(n-p)s^2}{\sigma^2}$$

follows a chi-squared distribution with $n-p$ degrees of freedom and that $\mathbf{b}$ and s^2 are independent. Applying Definition 3.8.1, the quadratic form

$$\begin{aligned} q_2 &= \frac{(\mathbf{b}-\boldsymbol{\beta})'X'X(\mathbf{b}-\boldsymbol{\beta})}{\sigma^2 p} \Bigg/ \frac{(n-p)s^2}{\sigma^2(n-p)} \\ &= \frac{(\mathbf{b}-\boldsymbol{\beta})'X'X(\mathbf{b}-\boldsymbol{\beta})}{ps^2} \end{aligned}$$

follows an F distribution with p and $n-p$ degrees of freedom. This random variable is used to derive a confidence region on $\boldsymbol{\beta}$.

Let f_α denote the point associated with the $F_{p,n-p}$ distribution with α area to its right. Then

$$P[q_2 \leq f_\alpha] = 1-\alpha$$

Substituting,

$$P[(\mathbf{b}-\boldsymbol{\beta})'X'X(\mathbf{b}-\boldsymbol{\beta})/ps^2 \leqslant f_\alpha] = 1-\alpha$$

The confidence region desired is the set of all points $\boldsymbol{\beta}$ in p dimensional space such that

$$(\mathbf{b}-\boldsymbol{\beta})'X'X(\mathbf{b}-\boldsymbol{\beta}) \leqslant ps^2 f_\alpha$$

To create such a region, we determine the numerical values of $\mathbf{b}$, $X'X$, and s^2 from experimental data and then locate the set of all vectors satisfying the above inequality. The idea is illustrated in Example 3.8.1.

Example 3.8.1 It is known that humidity influences evaporation in water-reducible paints. These data are obtained on x, the relative humidity, and y, the extent of solvent evaporation in the paint during sprayout. (Adapted from "Evaporation During Sprayout of a Typical Water-Reducible Paint at Various Humidities," *Journal of Coating Technology* 65, 1983.)

Relative Humidity (x)	Solvent Evaporation % wt. (y)
35.3	11.0
29.7	11.1
30.8	12.5
58.8	8.4
61.4	9.3
71.3	8.7
74.4	6.4
76.7	8.5
70.7	7.8
57.5	9.1
46.4	8.2
28.9	12.2
28.1	11.9
39.1	9.6
46.8	10.9
48.5	9.6
59.3	10.1
70.0	8.1
70.0	6.8
74.4	8.9
72.1	7.7
58.1	8.5
44.6	8.9
33.4	10.4
28.6	11.1

Using SAS, it is found that

$$\mathbf{b} = \begin{bmatrix} 13.64 \\ -0.08 \end{bmatrix}, \quad s^2 = 0.785, \quad s = 0.886,$$

$$X'X = \begin{bmatrix} 25 & 1314.9 \\ 1314.9 & 76308.53 \end{bmatrix}, \quad \text{and} \quad (X'X)^{-1} = \begin{bmatrix} 0.4269 & -0.0074 \\ -0.0074 & 0.00014 \end{bmatrix}$$

Assuming simple linear regression, the matrix $(X'X)^{-1}$ can be used to find 95% confidence intervals on β_0 and β_1 individually with the t distribution with $n - p = 23$ degrees of freedom. In Section 3.6, the bounds for such intervals were shown to be

$$b_j \pm t_{\alpha/2} s \sqrt{c_{jj}}$$

where c_{jj} is the jth diagonal entry of $(X'X)^{-1}$. Here the 95% confidence bounds on β_0 are

$$13.64 \pm 2.069(0.886)\sqrt{0.4269}$$

or 13.64 ± 1.198; those for β_1 are

$$-0.08 \pm 2.069(0.886)\sqrt{0.00014}$$

or -0.08 ± 0.02.

Remember that these intervals pertain to β_0 and β_1 individually. We can say that we are 95% confident that β_0 lies between 12.442 and 14.838 and that we are 95% confident that β_1 lies between -0.10 and -0.06. We cannot say that we are 95% confident that both of these statements hold simultaneously. To obtain a statement concerning the joint behavior of β_0 and β_1, we use the procedure just developed. The required point from the $F_{2,23}$ distribution is $f_{0.05} = 3.42$. To find the confidence region, we need to find the set of points in the plane satisfying the property that

$$(\mathbf{b} - \boldsymbol{\beta})' X'X (\mathbf{b} - \boldsymbol{\beta}) \leqslant ps^2 f_\alpha$$

Here we need to find the region in two-dimensional space such that

$$[13.64 - \beta_0 \quad -0.08 - \beta_1] \begin{bmatrix} 25 & 1314.9 \\ 1314.9 & 76308.53 \end{bmatrix} \begin{bmatrix} 13.64 - \beta_0 \\ -0.08 - \beta_1 \end{bmatrix} \leqslant 2(0.785)(3.42)$$

Multiplying and simplifying, the desired region is the set of points such that

$$25b_0^2 - 471.616b_0 + 76308.53b_1^2 - 23661.11b_1 + 2629.8b_0b_1 + 2269.977 \leqslant 5.3694$$

or

$$25b_0^2 - 471.616b_0 + 76308.53b_1^2 - 23661.11b_1 + 2629.8b_0b_1 + 2264.6076 \leqslant 0$$

This inequality identifies an ellipse, the interior of which is the desired confidence region. Figure 3.3 illustrates the joint confidence region (β_0, β_1) for the data in this example.

FIGURE 3.3 Joint confidence region on β_0, β_1 for the data of Example 3.8.1

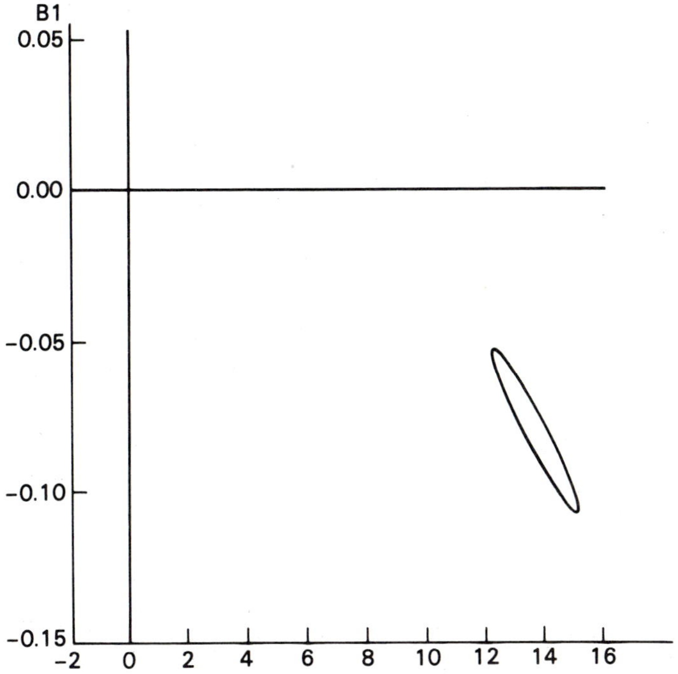

3.9

GENERALIZED LEAST SQUARES (ADVANCED)

Until now we have assumed ideal conditions relative to the vector of random errors. In particular, we have assumed that $\boldsymbol{\varepsilon}$ is a random vector with mean $\mathbf{0}$ and variance $\sigma^2 I$ or that it is normally distributed with these properties. That is, we have assumed that $\varepsilon_1, \varepsilon_2, \ldots, \varepsilon_n$ are uncorrelated with common variance. In many situations one or both of these assumptions is unrealistic. Here we consider techniques for estimating $\boldsymbol{\beta}$ when we assume only that the variance of $\boldsymbol{\varepsilon}$ is an $n \times n$ positive definite matrix V.

To begin, assume that $\boldsymbol{\varepsilon}$ follows a multivariate normal distribution. The joint density for $\varepsilon_1, \varepsilon_2, \ldots, \varepsilon_n$ is given by

$$f(\boldsymbol{\varepsilon}; \boldsymbol{\beta}, V) = \frac{1}{(2\pi)^{n/2}|V|^{1/2}} \exp\,[(-1/2)\boldsymbol{\varepsilon}'V^{-1}\boldsymbol{\varepsilon}]$$

To find the maximum likelihood estimator for $\boldsymbol{\beta}$ we must find the vector $\boldsymbol{\beta}^*$ that maximizes this function. It can be shown (see Exercise 60) that since V is positive definite so is V^{-1}. Hence by definition $\boldsymbol{\varepsilon}'V^{-1}\boldsymbol{\varepsilon}$ is positive, and therefore to maximize f we must minimize the term $\boldsymbol{\varepsilon}'V^{-1}\boldsymbol{\varepsilon}$. A vector $\boldsymbol{\beta}^*$ must be found such that

$$\boldsymbol{\varepsilon}'V^{-1}\boldsymbol{\varepsilon} = (\mathbf{y} - X\boldsymbol{\beta}^*)'V^{-1}(\mathbf{y} - X\boldsymbol{\beta}^*)$$

is at a minimum. This term is similar to the residual sum of squares that is minimized in ordinary least squares. For this reason, we shall denote it by $SS_{\text{Res},V}$. To minimize this "residual" sum of squares, we first differentiate with respect to $\boldsymbol{\beta}^*$ as follows:

$$\begin{aligned}\frac{\partial SS_{\text{Res},V}}{\partial \boldsymbol{\beta}^*} &= \frac{\partial(\mathbf{y} - X\boldsymbol{\beta}^*)'V^{-1}(\mathbf{y} - X\boldsymbol{\beta}^*)}{\partial \boldsymbol{\beta}^*} \\ &= \frac{\partial}{\partial \boldsymbol{\beta}^*}[\mathbf{y}'V^{-1}\mathbf{y} - 2\boldsymbol{\beta}^{*\prime}X'V^{-1}\mathbf{y} + \boldsymbol{\beta}^{*\prime}X'V^{-1}X\boldsymbol{\beta}^*]\end{aligned}$$

Applying the rules for differentiation of Section 2.2, it can be shown that this derivative is given by

$$\frac{SS_{\text{Res},V}}{\partial \boldsymbol{\beta}^*} = -2X'V^{-1}\mathbf{y} + 2X'V^{-1}X\boldsymbol{\beta}^*$$

Setting this derivative equal to zero, these generalized "normal" equations are obtained:

$$X'V^{-1}X\boldsymbol{\beta}^* = X'V^{-1}\mathbf{y}$$

The solution to this equation is

$$\mathbf{b}^* = (X'V^{-1}X)^{-1}X'V^{-1}\mathbf{y}$$

The estimator $\mathbf{b}^*$ is called the *generalized least squares* estimator. It is similar to the ordinary least squares solution $\mathbf{b}$ and reduces to $\mathbf{b}$ when $V = \sigma^2 I$. The matrix appearing in the generalized solution provides a type of weighting that takes into account the special variance-covariance structure of $\boldsymbol{\varepsilon}$.

The rules for variance and expectation can be applied to show that

$$E[\mathbf{b}^*] = \boldsymbol{\beta}$$

and that

$$\text{var } \mathbf{b}^* = (X'V^{-1}X)^{-1}$$

As you will see, the parallel between ordinary and generalized least squares continues nicely. In particular, it can be shown that $\mathbf{b}^*$ is the best linear unbiased estimator for $\boldsymbol{\beta}$. To do so, note that since V and V^{-1} are positive definite, the result stated in Exercise 64 can be applied to write V^{-1} and V as

$$V^{-1} = R'R \quad \text{and} \quad V = R^{-1}(R')^{-1}$$

where R is a nonsingular matrix. The original model is

$$\mathbf{y} = X\boldsymbol{\beta} + \boldsymbol{\varepsilon}$$

This model can be rewritten by multiplying by R to obtain the restructured model

$$R\mathbf{y} = RX\boldsymbol{\beta} + R\boldsymbol{\varepsilon}$$

or

$$\mathbf{v} = Z\boldsymbol{\beta} + \boldsymbol{\varepsilon}^*$$

where $\mathbf{v} = R\mathbf{y}$, $Z = RX$, and $\boldsymbol{\varepsilon}^* = R\boldsymbol{\varepsilon}$. The vector $\boldsymbol{\varepsilon}^*$ maintains the status of a random error vector and $E[\boldsymbol{\varepsilon}^*] = \mathbf{0}$. Consider now the variance-covariance structure of $\boldsymbol{\varepsilon}^*$:

$$\begin{aligned} \text{var } \boldsymbol{\varepsilon}^* &= \text{var } R\boldsymbol{\varepsilon} \\ &= RVR' \\ &= RR^{-1}(R')^{-1}R' \\ &= I \end{aligned}$$

As a result, the restructured model allows us to estimate $\boldsymbol{\beta}$ in a context in which the errors are uncorrelated. Applying the Gauss–Markoff theorem,

$$\mathbf{b} = (Z'Z)^{-1}Z'\mathbf{v}$$

is a BLUE estimator of $\boldsymbol{\beta}$. However,

$$\begin{aligned} \mathbf{b} &= (Z'Z)^{-1}Z'\mathbf{v} \\ &= [(RX)'RX]^{-1}(RX)'R\mathbf{y} \\ &= (X'R'RX)^{-1}X'R'R\mathbf{y} \\ &= (X'V^{-1}X)^{-1}X'V^{-1}\mathbf{y} \\ &= \mathbf{b}^* \end{aligned}$$

Thus, the generalized least squares estimator is a BLUE estimator as claimed.

WEIGHTED LEAST SQUARES

An important special case of generalized least squares occurs when the variance-covariance matrix of $\boldsymbol{\varepsilon}$ is diagonal and assumes the form

$$\text{var } \boldsymbol{\varepsilon} = \begin{bmatrix} \sigma_1^2 & & & \\ & \sigma_2^2 & 0 & \\ & 0 & \ddots & \\ & & & \sigma_n^2 \end{bmatrix}$$

Here we are assuming that $\varepsilon_1, \varepsilon_2, \ldots, \varepsilon_n$ are uncorrelated but that they do not have common variance. This situation occurs frequently in scientific studies when the variation in response is dependent upon the magnitude of the x variables. In this case, we seek an estimator that takes into account the heterogeneous nature of the variances by weighting the observations in some manner. In this way, an observation drawn from a population with a smaller variance is given more weight or importance than one from a population with more variability. When generalized least squares is used in this context, the technique is referred to as *weighted regression* or *weighted least squares.* It is easy to see that in weighted least squares, we are minimizing

$$\begin{aligned} SS_{\text{Res},V} &= (\mathbf{y} - X\mathbf{b}^*)'V^{-1}(\mathbf{y} - X\mathbf{b}^*) \\ &= \sum_{i=1}^{n} \left(\frac{y_i - \hat{y}_i}{\sigma_i} \right)^2 \end{aligned}$$

(See Exercise 69.)

Note that each residual $y_i - \hat{y}_i$ is weighted by $1/\sigma_i$, the reciprocal of the standard deviation of the population from which the data point is drawn. In this way, a point drawn from a population with small variability receives a heavy weight; one drawn from a population with a large degree of variability receives less importance in estimating $\boldsymbol{\beta}$.

The formula for **b** was developed assuming normality. However, normality is not essential. The generalized least squares estimators **b*** is BLUE whether normality is assumed or not. When we do assume normality, it can be shown that **b*** is UMVUE as is **b** in the case of ordinary least squares.

EXERCISES

Section 3.1

1. In analyzing performance on a production line, y, the total elapsed time from starting the first job on the first machine to finishing the last job on the last machine, is used. This variable is called make-span. The make-span is thought to depend primarily on x, the number of jobs to be done. A simple linear regression model is assumed.

a. Write the expression for the model.

b. These data are obtained:

Number of Jobs (x)	Make-span (y)
4	3.8
5	4.9
6	4.9
7	7.2
8	7.3
9	9.0

Find the system of linear equations generated by this model.

c. Find the response, parameter, and random error vectors.
d. Find the X matrix and express the model in matrix form.
(Based on a study described in *Engineering Costs and Production Economics* 1983, p. 137.)

2. It is assumed that the gasoline mileage of an automobile is dependent linearly upon its weight and the speed at which it is driven.
a. Write the model.
b. These data are obtained:

Miles per Gallon (y)	Weight in Tons (x_1)	Miles per Hour (x_2)
17.0	1.35	55
16.5	1.33	58
16.0	2.00	60
20.0	1.40	55
22.0	1.40	50
30.0	1.20	50
27.0	1.30	53
19.0	1.28	65

Find the system of linear equations generated by the model.
c. Find the response, parameter, and random error vectors.
d. Find the X matrix and express the model in matrix form.

3. Suppose that psychologists think that the attention span of a small child when exposed to a new task involving the manipulation of building blocks depends only on his or her IQ. The proposed model is

$$y = \beta_0 + \beta_1 x^2 + \varepsilon$$

a. Is this a linear model? Explain.
b. Assume that these data are obtained:

Attention Span in Minutes (y)	IQ (x)
0.5	70
1.0	77
2.0	85
2.5	92
2.6	98
3.3	105
2.1	112
1.5	120
1.0	130
0.5	150

Find the response vector, the parameter vector, and the X matrix for these data.

Section 3.2

4. Let $\mathbf{y} = X\boldsymbol{\beta} + \boldsymbol{\varepsilon}$ where X is an $n \times (k+1)$ full rank matrix of real numbers, $\boldsymbol{\beta}$ is a $(k+1) \times 1$ vector of parameters, and $\boldsymbol{\varepsilon}$ is an $n \times 1$ random vector with mean $\mathbf{0}$ and variance $\sigma^2 I$. Let $\mathbf{b}$ be the least squares estimator for $\boldsymbol{\beta}$.
a. Show that $E[\mathbf{y}] = X\boldsymbol{\beta}$.
b. Use the rules of expectation to show that $\mathbf{b}$ is unbiased for $\boldsymbol{\beta}$.
c. Use the rules for variance to show that $\text{var}\,\mathbf{b} = (X'X)^{-1}\sigma^2$.

5. Let $\mathbf{e}$ be the vector of residuals as defined in the proof of Theorem 3.2.1. Let $\mathbf{b}$ be the least squares estimator for $\boldsymbol{\beta}$.
a. Show that $\Sigma e_i = 0$. (*Hint*: Note that $\mathbf{e} = \mathbf{y} - X\mathbf{b}$ and hence $\Sigma e_i = [1 \;\; 1 \;\; \cdots \;\; 1]$ $[\mathbf{y} - X\mathbf{b}]$. Show that

$$\sum e_i = \sum y_i - nb_0 - b_1 \sum x_{i1} - b_2 \sum x_{i2} - \cdots - b_k \sum x_{ik}$$

Now expand the first row of the normal equation $X'X\mathbf{b} = X'\mathbf{y}$.)
b. Show that $E[\mathbf{e}] = \mathbf{0}$.
c. Let $H = X(X'X)^{-1}X'$. What are the dimensions of H? Is H symmetric? Let h_{ij} denote the element in the ith row and jth column of H. Write the matrix H in expanded form.
d. Use the rules for variance to show that $\text{var}\,\mathbf{e} = \sigma^2[I - H]$.
e. Let e_i and e_j denote the ith and jth residuals, respectively. Find an expression for $\text{var}\,e_i$ in terms of σ^2 and elements of H. Find an expression for $\text{cov}\,(e_i, e_j)$ in terms of σ^2 and elements of H. Is $\text{cov}\,(e_i, e_j)$ necessarily equal to 0?

6. Show that in simple linear regression, b_0, the estimator for the intercept found in Example 3.2.2, can be expressed as $\bar{y} - b_1\bar{x}$ as claimed.

7. Let x denote the number of years of formal education and let y denote an individual's income at age 30. Assume that simple linear regression is applicable and consider these data:

Years of Formal Education (x)	Income in Thousands of Dollars (y)
8	8
12	15
14	16
16	20
16	25
20	40

a. Find $\mathbf{y}$ and X.
b. Find $X'X$, $X'\mathbf{y}$, and $(X'X)^{-1}$.
c. Find the least squares estimates for β_0 and β_1 by calculating $(X'X)^{-1}X'\mathbf{y}$.
d. Verify your calculation in part c by finding b_0 and b_1 using the formulas derived in Example 3.2.2.

8. These data are obtained on x, household income, and y, energy consumption in units of 10 Btu/year.

Income in Thousands of Dollars (x)	Energy Consumption (y)
20	1.8
30	3.0
40	4.8
55	5.0
60	6.5

Assume that simple linear regression is appropriate.

a. Find $\mathbf{y}$ and X.

b. Find $X'X$, $X'\mathbf{y}$, and $(X'X)^{-1}$.

c. Find the least squares estimates for β_0 and β_1.

(Based on a study reported in *Energy Policy*, September 1982.)

***9.** Let $\mathbf{l}'$ and $\mathbf{t}'$ be $1 \times n$ and $1 \times (k+1)$ vectors of real numbers, respectively. Assume that $\mathbf{t}' \neq \mathbf{0}'$. It is evident that $\mathbf{l}'$ can be expressed in the form

$$\mathbf{l}' = \mathbf{t}'(X'X)^{-1}X' + \mathbf{z}'$$

where X is an $n \times (k+1)$ matrix of full rank and $\mathbf{z}'$ is an appropriately chosen vector of real numbers.

a. Argue that $\mathbf{z}'$ can be expressed in the form $\mathbf{z}' = \mathbf{t}'B$ where B is a $(k+1) \times n$ matrix. (*Hint*: Argue that k entries in each column of B are arbitrary and that the remaining entry of each column can be chosen in such a way that the equality holds.)

b. Use the result of part a to prove Theorem 3.2.4. (*Hint*: Assume that $\mathbf{l}'\mathbf{y}$ is another unbiased estimator for $\mathbf{t}'\boldsymbol{\beta}$. Write $\mathbf{l}'\mathbf{y}$ in the form

$$\mathbf{l}'\mathbf{y} = [\mathbf{t}'(X'X)^{-1}X' + \mathbf{t}'B]\mathbf{y}$$

and parallel the argument given in the proof of Theorem 3.2.3.)

c. Consider a system of m linear functions of $\beta_0, \beta_1, \ldots, \beta_k$. This system can be expressed in the form $T\boldsymbol{\beta}$ where T is an $m \times (k+1)$ matrix. Argue that the best linear unbiased estimator of $T\boldsymbol{\beta}$ is $T\mathbf{b}$. (*Hint*: Think of T as being expressed in the form

$$T = \begin{bmatrix} \mathbf{t}'_1 \\ \mathbf{t}'_2 \\ \vdots \\ \mathbf{t}'_m \end{bmatrix}$$

where $\mathbf{t}'_i$ is a $1 \times (k+1)$ vector of real numbers for $i = 1, 2, \ldots, m$ and apply part b.)

10. Consider the data of Exercise 7. Estimate the average salary of individuals who have had 15 years of formal education.

11. Consider the data of Exercise 8. Estimate the average energy consumption of households in which the yearly income is $50,000.

12. Consider the data of Example 3.2.1. Suppose we want to estimate the average selling prices for houses with these values of x_1 and x_2:

x_1	x_2
7	2
15	1.5
15	2.8

a. Set up the matrix T required to estimate these expectations.
b. Calculate $T\mathbf{b}$, thus finding estimates for these theoretical average selling prices.

Section 3.3

***13.** Let X be an $n \times (k+1)$ matrix of full rank. Show that $I - X(X'X)^{-1}X'$ is idempotent.

***14.** Show that $E[s^2] = \sigma^2$, thus verifying that s^2 is an unbiased estimator for σ^2.

***15.** Use the results of Exercise 4 and Example 3.2.2 to find expressions for var b_0, var b_1, and cov (b_0, b_1) in simple linear regression. Find unbiased estimators for var b_0 and var b_1.

***16.** The sample variance s^2 has been defined by

$$s^2 = \frac{SS_{\text{Res}}}{n-p} = \frac{(\mathbf{y} - X\mathbf{b})'(\mathbf{y} - X\mathbf{b})}{n-p}$$

Alternative forms for this estimator can be derived by expressing SS_{Res} in different ways. Show that each of the following expressions is equal to s^2.
a. $\mathbf{y}'[I - X(X'X)^{-1}X']\mathbf{y}/(n-p)$
b. $[\mathbf{y}'\mathbf{y} - \mathbf{y}'X(X'X)^{-1}X'\mathbf{y}]/(n-p)$
c. $(\mathbf{y}'\mathbf{y} - \mathbf{y}'X\mathbf{b})/(n-p)$
d. $(\mathbf{y}'\mathbf{y} - \mathbf{b}'X'\mathbf{y})/(n-p)$
e. $(\mathbf{y}'\mathbf{y} - \mathbf{b}'(X'X)\mathbf{b})/(n-p)$

17. Use the data of Exercise 7 and the definition of s^2 to find the least squares estimate for σ^2. Use the results of Exercise 15 to estimate var b_0 and var b_1 based on these data.

18. Use the data of Exercise 8 and part d of Exercise 16 to find the least squares estimate for σ^2. Use Exercise 15 to estimate var b_0 and var b_1 based on these data.

19. We have already argued that the estimated model $b_0 + b_1x_1 + b_2x_2 + \cdots + b_kx_k$ provides an unbiased estimate for the *average* value of y for given values of the x's. If asked to estimate the value that y itself will assume for some specific values of $x_1, x_2, \ldots, x_k$ not all of which were used to develop the estimated model, the only logical choice is the average value of y predicted by the model. That is, $\widehat{E[y]} = \hat{y}$.
a. Use the data of Exercise 7 to estimate the income of a randomly selected individual with 15 years of formal education; with 10 years of formal education.
b. Use the data of Exercise 8 to estimate the energy consumption in a randomly selected household in which the income is $50,000; in which the income is $25,000.
c. Use the data of Example 3.2.1 to estimate the selling price of a randomly selected

house that is 15 years old with 2500 square feet of living space; that is 9 years old with 2000 square feet of living space.

d. Show that in general $s^2 = \Sigma_{i=1}^{n}(y_i - \hat{y}_i)^2/(n-p)$ where y_i denotes the ith response and $\hat{y}_i = \widehat{E[y_i]}$.

20. Consider the simple linear regression model

$$y = X\boldsymbol{\beta} + \boldsymbol{\varepsilon} = \beta_0 + \beta_1 x + \boldsymbol{\varepsilon}$$

and assume that these data are available:

$y_1 = 4 \quad x_1 = 3$

$y_2 = 3 \quad x_2 = 2$

$y_3 = 5 \quad x_3 = 3$

$y_4 = 6 \quad x_4 = 5$

a. Find X, $X'X$, $(X'X)^{-1}$, and $X'\mathbf{y}$.
b. Find $\mathbf{b}$.
c. Find s^2 via any part of Exercise 16.
d. Find $\hat{y}_i = \widehat{E[y_i]}$ for $i = 1, 2, 3, 4$. Calculate s^2 via part d of Exercise 19. Verify that your answer matches the one found earlier.

***21.** The simplest linear model is called the location model. This model states that the response y varies randomly about its mean value μ. Mathematically, the model assumes the form

$$\mathbf{y} = \boldsymbol{\mu} + \boldsymbol{\varepsilon}$$

where

$$\boldsymbol{\mu} = \begin{bmatrix} \mu \\ \mu \\ \vdots \\ \mu \end{bmatrix}$$

a. The general form for a linear model is

$$\mathbf{y} = X\boldsymbol{\beta} + \boldsymbol{\varepsilon}$$

What is the X matrix for the location model? What is $\boldsymbol{\beta}$ for this model?

b. Find $X'X$, $(X'X)^{-1}$, and $X'\mathbf{y}$.
c. Find the least squares estimator for the mean value of the response. Is this estimator unbiased? In this exercise you have shown that the sample mean, $\bar{y}$, is an unbiased estimator for the true mean, μ.
d. Note that $y_i = \mu + \varepsilon_i$ implies that $E[y_i] = \mu$. Hence $\widehat{E[y_i]} = \hat{\mu} = \bar{y}$. Write the expression for the ith residual e_i. Use this to find the expression for s^2. Is this estimator unbiased for σ^2? In this exercise you have shown that the sample variance introduced in elementary courses is an unbiased estimator for the population variance.

22. A classic problem in operations research is the vehicle routing problem. In such a problem, a system consisting of a given number of customers with known locations and known demand for a product is being serviced from a single depot by a fixed number of vehicles. The purpose of the study is to route the vehicles so that the distance traveled is minimized. These data represent the cpu time in seconds required to solve such a problem (taken from a study reported in *European Journal of Operational Research*, April 1983, pp. 388–393):

2.0	2.4	3.6
3.5	2.8	2.0
4.9	3.3	2.5

The location model is assumed (see Exercise 21).

a. Use the results of Exercise 21 to estimate μ.
b. Find each of the nine residuals. Use these to compute SS_{Res} and s^2.
c. Calculate s^2 using any procedure learned in an earlier statistics class or by using your calculator and compare your answers to that found in part b.

***23.** (Regression through the origin) Thus far we have assumed that an intercept, β_0, must be estimated from the data. This results in an X matrix whose first column is a column of ones. Since β_0 gives the average response when $x_1, x_2, \ldots, x_k$ are all zero, it is not always necessary to estimate its value; sometimes it is known to be zero. For example, in a simple linear regression we might want to model the amount of diesel fuel used by trucks as a function only of the time during which their motors are running. It is evident that if $x = 0$, the motor does not run at all, then $y = 0$, no fuel is used. The desired model should have intercept 0 and hence should take the form

$$\mathbf{y} = \beta_1 \mathbf{x} + \boldsymbol{\varepsilon}$$

Consider this model.

a. What is the X matrix for such a model?
b. Find $X'X$, $(X'X)^{-1}$, and $X'\mathbf{y}$.
c. Find the least squares estimator for β_1 and compare it to that of the usual simple linear regression model given in Example 3.2.2.
d. Find an expression for s^2 based on a sample of size n.
e. Find an estimator for var b_1.
f. Assume that these data are available:

Amount of Fuel in Gallons (y)	Time Motor Runs in Hours (x)
3	0.6
5	2.0
7	2.1
9	2.0
10	2.4

Use these data to estimate β_1, σ^2, and var b_1.

24. In a study of staffing needs at naval installations, the data below are obtained on x, the number of items processed by a clerical staff, and y, the monthly work-hours required to do the work. Theory suggests that as the workload approaches zero, the personnel required to do the work as measured by monthly work-hours should approach zero. Thus the model to be fitted is the simple linear regression model through the origin developed in Exercise 23. (*Procedures and Analyses for Staffing Standards Development:*

Data/Regression Analysis Handbook, San Diego, California: Navy Manpower and Material Analysis Center, 1979.)

Items Processed (x)	Monthly Work-hours (y)
15	85
25	125
57	203
67	293
197	763
166	639
162	673
131	499
158	657
241	939
399	1546
527	2158
533	2182
563	2302
563	2202
932	3678
986	3894
1021	4034
1643	6622
1985	7890
1640	6610
2143	8522

a. Estimate β_1, σ^2, and var b_1.
b. Estimate the average number of monthly work-hours required to process 2000 items.
Note: To use SAS to fit a model through the origin, follow the procedure given in Computing Supplement A with one important change. In particular, the model statement is changed to read

```
MODEL Y = X/XPX I P NOINT;
```

Section 3.4

25. Prove Theorem 3.4.2. (*Hint*: Differentiate ln (L) with respect to σ^2. Treat all other terms in the expression as constants. Replace $\boldsymbol{\beta}$ by $\tilde{\boldsymbol{\beta}}$ when solving for $\tilde{\sigma}^2$.)

26. In the location model of Exercise 21, what is the maximum likelihood estimator for μ? for σ^2?

27. Find the maximum likelihood estimate for σ^2 based on the data of Exercise 22.

28. In the simple linear regression model, what is the denominator for s^2? for $\tilde{\sigma}^2$?

29. Find the maximum likelihood estimate for σ^2 based on the data of Exercise 7.

30. Find the maximum likelihood estimate for σ^2 based on the data of Exercise 8.

31. Argue that if $n = p$ then s^2 does not exist.

32. Argue that if $n = p$, then $SS_{\text{Res}} = 0$ and hence $\tilde{\sigma}^2 = 0$. (*Hint*: If $n = p$, then X is $p \times p$ of rank p and hence is nonsingular. Use this fact to show that $X\mathbf{b} = X(X'X)^{-1}X'\mathbf{y} = \mathbf{y}$ and hence that $\mathbf{e} = \mathbf{0}$.)

Section 3.5

33. Let $X_1, X_2, \ldots, X_n$ be a random sample from a distribution with density

$$f(x; \theta) = \theta x^{\theta - 1} \qquad 0 < x < 1 \quad \text{and} \quad \theta > 0$$

Show that the statistic $Y = \prod_{i=1}^{n} X_i$ is a sufficient statistic for θ.

34. Let $X_1, X_2, \ldots, X_n$ be a random sample from a normal distribution with mean 0 and variance σ^2. Show that $Y = \Sigma_{i=1}^{n} X_i^2$ is a sufficient statistic for σ^2. *Hint*: The function g required in the Fisher–Neyman theorem can be a constant function.

35. Let $X_1, X_2, \ldots, X_n$ be a random sample from a distribution with density

$$f(x; \theta) = (1 - \theta)^x \theta \qquad x = 0, 1, 2, \ldots \quad 0 < \theta < 1$$

Show that $Y = \sum_{i=1}^{n} X_i$ is a sufficient statistic for θ.

36. Show that $(\mathbf{y} - X\boldsymbol{\beta})'(\mathbf{y} - X\boldsymbol{\beta})$ can be expressed as $(\mathbf{y} - X\mathbf{b})'(\mathbf{y} - X\mathbf{b}) + (\mathbf{b} - \boldsymbol{\beta})'X'X(\mathbf{b} - \boldsymbol{\beta})$ as claimed in the proof of Theorem 3.5.3. (*Hint*: Expand the latter term keeping in mind that $\mathbf{b} = (X'X)^{-1}X'\mathbf{y}$. Show that the resulting expression can be reduced to the form

$$\mathbf{y}'\mathbf{y} - 2\boldsymbol{\beta}'X'\mathbf{y} + \boldsymbol{\beta}'X'X\boldsymbol{\beta} = (\mathbf{y} - X\boldsymbol{\beta})'(\mathbf{y} - X\boldsymbol{\beta})$$

Section 3.6

37. Let X be an $n \times p$ matrix of full rank.

a. Let $H = X(X'X)^{-1}X'$. Show that H is symmetric and idempotent.
b. Find tr (H) and use Theorem 1.5.2 to find $r(H)$.
c. Consider the matrix $I - H$. Show that $I - H$ is symmetric and idempotent.
d. Find tr $(I - H)$ and use Theorem 1.5.2 to find $r(I - H)$.

38. Show that in simple linear regression the bounds for confidence intervals on β_0 and β_1, respectively, are given by

$$b_0 + t_{\alpha/2} s \sqrt{\frac{\sum_{i=1}^{n} x_i^2}{n s_{xx}}}$$

and

$$b_1 \pm t_{\alpha/2} s / \sqrt{s_{xx}}$$

where

$$s_{xx} = \left[n \sum_{i=1}^{n} x_i^2 - \left(\sum_{i=1}^{n} x_i \right)^2 \right] \Big/ n$$

39. Use the data of Example 3.3.1 to find a 95% confidence interval on β_1, the slope of the line of regression. Can we conclude with 95% confidence that the estimated average cracking rate in the field increases as x increases? Explain this based on the geometric interpretation of the slope of a straight line. (*Hint*: For these data, $s^2 = 0.2771$.)

40. Consider the data of Exercise 7. Find and interpret the 90% confidence interval on β_1.

41. Consider the data of Exercise 8. Find and interpret the 99% confidence interval on β_1.

42. The following is the $(X'X)^{-1}$ matrix for the data of Example 3.2.1:

$$(X'X)^{-1} = \begin{bmatrix} 2.307551 & 0.1565378 & -1.88398 \\ 0.1565378 & 0.02578269 & -0.20442 \\ -1.88398 & -0.20442 & 1.977901 \end{bmatrix}$$

Here the model is

$$y = \beta_0 + \beta_1 x_1 + \beta_2 x_2 + \varepsilon$$

where y denotes the selling price of a house, x_1 is its age in years, and x_2 is its livable square footage. For the data given, s^2 is found via computer to be 47.837937 and $n = 5$.

a. Estimate var b_0; var b_1; var b_2.
b. Estimate cov (b_0, b_1); cov (b_0, b_2); cov (b_1, b_2).
c. The least squares estimates for $\boldsymbol{\beta}$ are

$$\mathbf{b} = \begin{bmatrix} 33.06 \\ -0.189 \\ 10.718 \end{bmatrix}$$

Find and interpret the 95% confidence interval on β_1; on β_2.

43. An experiment is conducted to estimate the demand for automobiles (y) based on cost (x_1), the current unemployment rate (x_2), and the current interest rate (x_3). The proposed linear model is

$$y = \beta_0 + \beta_1 x_1 + \beta_2 x_2 + \beta_3 x_3 + \varepsilon$$

These data are obtained:

Units Sold in Thousands (y)	Cost in Thousands (x_1)	Unemployment Rate % (x_2)	Interest Rate % (x_3)
6.5	9.0	10.0	4.0
5.9	5.5	9.0	7.0
8.0	9.0	12.0	5.0
9.0	9.8	11.0	6.2
10.0	14.5	12.0	5.8
10.5	8.0	14.0	3.9

a. Find $\mathbf{y}$ and X.
b. Find $X'X$ and $X'\mathbf{y}$.
c. For these data $(X'X)^{-1}$ is found via computer to be

$$(X'X)^{-1} = \begin{bmatrix} 27.17251 & 0.0506362 & -1.53063 & -1.90526 \\ 0.0506362 & 0.02681862 & -0.0205869 & -0.0125512 \\ -1.53063 & -0.0205869 & 0.1115842 & 0.08604356 \\ -1.90526 & -0.0125512 & 0.08604356 & 0.196896 \end{bmatrix}$$

Use this information to estimate $\boldsymbol{\beta}$.

d. The estimated value of σ^2 is 0.549038. Use this information to estimate var b_0, var b_1, var b_2, and var b_3.

44. Consider the location model of Exercise 21. It has been shown that the least squares estimator for μ, the mean value of the response, is $\bar{y}$, the sample mean. Use the methods in Section 3.6 to show that the general form for a confidence interval on μ is

$$\bar{y} \pm t_{\alpha/2} s/\sqrt{n}$$

45. Consider the model for regression through the origin given in Exercise 23. Find the general form for a $100(1-\alpha)\%$ confidence interval on β_1. How many degrees of freedom are involved here? Use the data given to find and interpret the 95% confidence interval on β_1.

46. Use the data of Exercise 24 to find a 90% confidence interval on β_1.

Section 3.7

47. Prove that $\mathbf{t}'\mathbf{b}$ is independent of SS_{Res}/σ^2. (*Hint*: Parallel the proof of Theorem 3.6.3.)

48. Show that the confidence bounds for an individual coefficient β_j derived in Section 3.6 are a special case of those derived here for $\mathbf{t}'\boldsymbol{\beta}$.

49. Use the data of Exercise 7 to find a 95% confidence interval on the average salary of individuals who have had 15 years of formal education.

50. Use the data of Exercise 8 to find a 90% confidence interval on the average energy consumption in households in which the yearly income is $50,000.

51. Use the data of Exercise 43 to find a 95% confidence interval on the average demand for cars costing $7000 when the unemployment rate is 10% and the current interest rate is 6.5%.

52. Verify the confidence band given in Example 3.7.2 for $x = 70$.

***53.** Most elementary statistics texts give these bounds for a confidence interval on the mean response when $x = x_*$ in simple linear regression:

$$(b_0 + b_1 x_*) \pm t_{\alpha/2} s \sqrt{\frac{1}{n} + \frac{(x_* - \bar{x})^2}{s_{xx}}}$$

where

$$s_{xx} = \left[n \sum_{i=1}^{n} x_i^2 - \left(\sum_{i=1}^{n} x_i \right)^2 \right] \Big/ n$$

Derive this expression.

54. Use the expression in Exercise 53 to verify your answer to Exercise 49.

55. Show that in simple linear regression the confidence band on the mean response is narrowest when $x_* = \bar{x}$.

56. Use the data of Exercise 43 to find a 95% prediction interval on the demand for a particular model of automobile that costs $7000 when the unemployment rate is 10% and the current interest rate is 6.5%. Is this interval wider or narrower than that found in Exercise 51? If a 99% prediction interval were required would it be wider or narrower than the 95% interval? Explain.

***57.** Find a nonmatrix expression for the prediction bounds in simple linear regression.

58. Use the data of Exercise 7 to find a 95% prediction interval on the salary of a randomly selected individual with 15 years of formal education.

59. Use the data of Exercise 8 to find a 90% prediction interval on the energy consumption of a randomly selected household in which the yearly income is $50,000.

60. Use the data of Exercise 1 to
a. Find $X'X$, $(X'X)^{-1}$, and $X'\mathbf{y}$.
b. Estimate β_0 and β_1.
c. Estimate σ^2.
d. Estimate var b_0 and var b_1.
e. Estimate the average make-span required when six jobs are run.
f. Find a 95% confidence interval on the average make-span when six jobs are run.
g. Estimate the make-span required when a future group of six jobs is run.
h. Find a 95% prediction interval on the make-span required for a future run of six jobs.

Section 3.8

61. Use Corollary 2.3.4 to prove that the quadratic form

$$\frac{(\mathbf{b}-\boldsymbol{\beta})'X'X(\mathbf{b}-\boldsymbol{\beta})}{\sigma^2}$$

follows a chi-squared distribution with p degrees of freedom.

62. In the manufacture of commercial wood products, it is important to estimate the relationship between the density of a wood product and its stiffness. A new type of particleboard is being considered. Thirty of these boards are produced at densities ranging roughly from 8 to 26 pounds per cubic foot, and the stiffness is measured in pounds per square inch. The data below are obtained. (Based on data presented in "Investigation of Certain Mechanical Properties of a Wood-Foam Composite," Terrance E. Conners, M.S. thesis, Department of Forestry and Wildlife Management, University of Massachusetts, Amherst, Massachusetts, 1979.)

Density (x) lb/ft^3	Stiffness (y) lb/in^2	Density (x) lb/ft^3	Stiffness (y) lb/in^2
15.0	2,622	8.4	17,502
14.5	22,148	9.8	14,007
14.8	26,751	11.0	19,443
13.6	18,036	8.3	7,573
25.6	96,305	9.9	14,191
23.4	104,170	8.6	9,714
24.4	72,594	6.4	8,076
23.3	49,512	7.0	5,304
19.5	32,207	8.2	10,728
21.2	48,218	17.4	43,243
22.8	70,453	15.0	25,319
21.7	47,661	15.2	28,028
19.8	38,138	16.4	41,792
21.3	54,045	16.7	49,499
9.5	14,814	15.4	25,312

For these data,

$$X'X = \begin{bmatrix} 30 & 464.1 \\ 464.1 & 8166.29 \end{bmatrix}, \quad (X'X)^{-1} = \begin{bmatrix} 0.2758892 & -0.0156791 \\ -0.0156791 & 0.001013517 \end{bmatrix},$$

$$X'\mathbf{y} = \begin{bmatrix} 1017405 \\ 19589339 \end{bmatrix}, \quad \text{and} \quad s^2 = 165242295.59$$

Use the simple linear regression model.

a. Estimate $\boldsymbol{\beta}$.
b. Find a point estimate for the mean stiffness reading when the density of the particleboard is 10 lb/ft^3. Find a 95% confidence interval on this mean reading.
c. Find a point estimate for the stiffness reading of an individual particleboard whose density is 10 lb/ft^3. Find a 95% prediction interval on the stiffness reading for such a board.
d. Find a 95% confidence interval on the slope of the regression line.
e. Find a 95% joint confidence region on the pair of parameters (β_0, β_1).

63. Consider the information given in Exercise 42. Use this to find a 95% joint confidence region on $(\beta_0, \beta_1, \beta_2)$.

Section 3.9

***64.** A theorem from the field of linear algebra states that "A necessary and sufficient condition for the symmetric matrix A to be positive definite is that there exist a nonsingular matrix P such that $A = PP'$" [17]. Use this result to prove that if V is symmetric and positive definite, then V^{-1} is also positive definite.

65. Verify that $\partial SS_{\text{Res},V}/\partial\boldsymbol{\beta}^* = -2X'V^{-1}\mathbf{y} + 2X'V^{-1}X\boldsymbol{\beta}^*$ as claimed. (*Hint*: Remember that since V is a variance-covariance matrix it is symmetric.)

66. Verify that when $V = \sigma^2 I$, $\mathbf{b}^* = \mathbf{b}$.

67. Use the rules of variance and expectation given in Section 2.2 to show that $\mathbf{b}^*$ is an unbiased estimator for $\boldsymbol{\beta}$ and that var $\mathbf{b}^* = (X'V^{-1}X)^{-1}\sigma^2$.

68. Prove that the matrix $Z = RX$ defined in Section 3.9 is an $n \times p$ matrix of full rank and hence that $\mathbf{b} = (Z'Z)^{-1}Z'\mathbf{v}$ is the least squares estimator for $\boldsymbol{\beta}$ in the restructured model.

69. Verify that in weighted least squares the residual sum of squares can be expressed as

$$\sum_{i=1}^{n} \left(\frac{y_i - \hat{y}_i}{\sigma_i}\right)^2$$

as claimed.

70. In solving the normal equations in generalized least squares it is found that

$$\mathbf{b}^* = (X'V^{-1}X)^{-1}X'V^{-1}\mathbf{y}$$

This exercise shows that $X'V^{-1}X$ is a nonsingular matrix as implied in the derivation of $\mathbf{b}^*$.

a. Argue that if V is an $n \times n$ positive definite symmetric matrix then V^{-1} is positive definite.

b. Argue that V^{-1} can be expressed in the form

$$V^{-1} = Q'Q$$

where Q is a nonsingular matrix.

c. Argue that $X'V^{-1}X$ can be expressed in the form

$$X'V^{-1}X = R'R$$

where R is an $n \times p$ matrix of rank p. (*Hint*: See Properties of Rank, Section 1.4.)

d. Argue that $R'R$ is $p \times p$ of rank p and hence is nonsingular.

COMPUTING SUPPLEMENT A (Introduction to SAS and PROC REG)

As you can see from the simple examples given in this chapter, finding the least squares estimates for $\beta_0, \beta_1, \ldots, \beta_k$ is easy conceptually but not so easy computationally. Because the matrix $(X'X)^{-1}$ is usually fairly difficult to compute, estimates are rarely found by hand. Rather, they are obtained from one of the many statistical packages currently on the market. In this section we present a very brief introduction to one of the most popular packages available, SAS. If this package is available at your installation, you should try these programs. A consultant at your installation can show you how to access the package.

We begin by writing a SAS program to analyze the data of Example 3.2.1 and answer the question posed in Example 3.2.3.

The first step in writing a SAS program is the data step. In this step a name is chosen for the data and the data are entered. Three statements are required: the DATA statement, the INPUT statement, and the CARDS statement. The DATA statement names the data set, the INPUT statement names the variable and describes the arrangement of the data lines, and the CARDS statement signals that the data lines follow. SAS statements may begin in any column and always end with a semicolon.

The name chosen for a data set should be a one-word name from one to eight characters in length. It is helpful to choose a name indicative of the data being analyzed. In this example, let us call the data set "house". We inform the computer of this choice by typing the statement

```
DATA HOUSE;
```

on a single line with at least one blank space separating these two words.

The INPUT statement names the variables and describes the order in which they will appear on the data lines. The programmer chooses these names following the naming convention mentioned earlier. In our example, there are three variables: y, the selling price of the house; x_1, its age; and x_2, its livable square footage. In this case, logical forms for the input statement are

```
INPUT Y X1 X2;
```

or

```
INPUT PRICE AGE FOOTAGE;
```

Either is acceptable.

The input statement is followed by the CARDS statement. This statement signals the fact that the data follow immediately and takes the form

```
CARDS;
```

The data lines follow with one y value and its corresponding x values listed in order per line with at least one blank space between each entry. Data lines do not end with a semicolon. SAS recognizes the end of the data when it sees a semicolon, so enter a single line containing only a semicolon after the last data entry. Thus far, the program looks like this:

```
DATA HOUSE;
INPUT     PRICE     AGE    FOOTAGE;
CARDS;
50     1     1
40     5     1
52     5     2
47    10     2
65    20     3
;
```

We must now tell the computer what to do with the data. This is done via what are called procedure statements. A *procedure* statement begins with the SAS keyword PROC followed by the name of the procedure desired. Here we want to use the regression procedure whose keyword is REG. Hence the next line of input takes the form

```
PROC REG;
```

Since many possible models are available, we must specify the form of the assumed model via a *model* statement. To specify a linear model with y (price) as the response, we write

```
MODEL PRICE = AGE FOOTAGE;
```

This statement will produce estimates for β_0, β_1, and β_2. To obtain other information such as the $X'X$ matrix and its inverse, residuals, and predicted values, some options must be specified. To ask for these, the model statement is rewritten as

```
MODEL PRICE = AGE FOOTAGE/XPX  I  P;
```

Finally a title statement can be added to identify the output. This statement assumes the form TITLE*n* where *n* is an integer that gives the line where the title is to be printed. Here we might use the title

```
TITLE1    ANALYSIS;
TITLE2    OF;
TITLE3    REAL ESTATE DATA;
```

Title lines can contain up to 132 characters.
The entire program follows:

Statement	Purpose
`DATA HOUSE;`	Names the data set
`INPUT PRICE AGE FOOTAGE;`	Names the variables
`CARDS;`	Signals that the data follow

`50 1 1`
`40 5 1`
`52 5 2`
`47 10 2`
`65 20 3`
	Observed data

`. 15 2.5`	Allows SAS to estimate the mean cost of 15-year-old houses with 2500 square feet of living space
`;`	Signals the end of the data
`PROC REG;`	Asks for the regression procedure
`MODEL PRICE = AGE FOOTAGE/XPX I P;`	Specifies the model and asks for $X'X$, $(X'X)^{-1}$, and predicted values to be printed

`TITLE1 ANALYSIS;`
`TITLE2 OF;`
`TITLE3 REAL ESTATE DATA;`
	Titles the output

The output of this program is shown in Figure 3.4. The matrices $X'X$, $(X'X)^{-1}$ and the vector $X'\mathbf{y}$ are given by (1), (2), and (3), respectively. The estimates for $\beta_0, \beta_1, \beta_2$, and σ^2 are given by (4), (5), (6), and (7), respectively. The residuals are shown in column (8). Notice that if the observations in this column are squared and the sum of these squares is divided by $n - p = 5 - 3 = 2$, we obtain the value for s^2 given in (7) apart from roundoff error, as expected. The estimated value of y and $E[y]$ when $x_1 = 15$ and $x_2 = 2.5$ is given by (9). (*Note*: The first line of the program should be OPTIONS LINESIZE = 72; if you want the output to appear on the screen or be printed on $8\frac{1}{2} \times 11''$ paper.)

FIGURE 3.4

```
                              ANALYSIS                              1
                                 OF
                          REAL ESTATE DATA

                  MODEL CROSSPRODUCTS X'X X'Y Y'Y

X'X              INTERCEP             AGE          FOOTAGE        PRICE

INTERCEP              5               41                9          254
AGE           (1)    41              551               96         2280 (3)
FOOTAGE               9               96               19          483
PRICE               254             2280              483        13238
```

FIGURE 3.4 Continued

FIGURE 3.4 Continued

X'X INVERSE, B, SSE

INVERSE		INTERCEP	AGE	FOOTAGE	PRICE
INTERCEP		2.307551	0.1565378	-1.88398	33.06262
AGE	②	0.1565378	0.02578269	-0.20442	-0.189687
FOOTAGE		-1.88398	-0.20442	1.977901	10.71823
PRICE		33.06262	-0.189687	10.71823	95.67587

DEP VARIABLE: PRICE

ANALYSIS OF VARIANCE

SOURCE	DF	SUM OF SQUARES	MEAN SQUARE	F VALUE	PROB>F
MODEL	2	239.12412523	119.56206262	2.499	0.2858
ERROR	2	95.67587477	47.83793738 ⑦		
C TOTAL	4	334.80000000			

ROOT MSE	6.916497	R-SQUARE	0.7142
DEP MEAN	50.8	ADJ R-SQ	0.4285
C.V.	13.61515		

PARAMETER ESTIMATES

VARIABLE	DF	PARAMETER ESTIMATE	STANDARD ERROR	T FOR H0: PARAMETER=0	PROB > |T|
INTERCEP	1	33.06261510 ④	10.50659142	3.147	0.0879
AGE	1	-0.189687 ⑤	1.11058122	-0.171	0.8801
FOOTAGE	1	10.71823204 ⑥	9.72721352	1.102	0.3854

ANALYSIS
OF
REAL ESTATE DATA

2

OBS	ACTUAL	PREDICT VALUE	RESIDUAL ⑧
1	50.0000	43.5912	6.4088
2	40.0000	42.8324	-2.8324
3	52.0000	53.5506	-1.5506
4	47.0000	52.6022	-5.6022
5	65.0000	61.4236	3.5764
6	.	57.0129 ⑨	.

SUM OF RESIDUALS	0
SUM OF SQUARED RESIDUALS	95.67587

COMPUTING SUPPLEMENT B (Finding Confidence Intervals on $\mu_{Y|x}$)

To create a 95% confidence interval on the average price of 15-year-old houses with 2500 square feet of living space, the code given in Computing Supplement A requires only a small change. The model statement is changed to

```
MODEL PRICE = AGE FOOTAGE/P CLM;
```

The statement CLM asks for "Confidence Limits on the Mean". The output is shown in Figure 3.5. The point estimate for the mean response is given by (1). The

FIGURE 3.5

```
                                   ANALYSIS                                     1
                                      OF
                               REAL ESTATE DATA

DEP VARIABLE: PRICE
                             ANALYSIS OF VARIANCE

                          SUM OF          MEAN
     SOURCE      DF       SQUARES        SQUARE          F VALUE          PROB>F

     MODEL        2 239.12412523 119.56206262             2.499           0.2858
     ERROR        2  95.67587477  47.83793738
     C TOTAL      4 334.80000000

         ROOT MSE       6.916497      R-SQUARE           0.7142
         DEP MEAN           50.8      ADJ R-SQ           0.4285
         C.V.           13.61515

                             PARAMETER ESTIMATES

                    PARAMETER           STANDARD      T FOR H0:
  VARIABLE  DF       ESTIMATE              ERROR   PARAMETER=0        PROB > |T|

  INTERCEP   1    33.06261510        10.50659142         3.147            0.0879
  AGE        1      -0.189687         1.11058122        -0.171            0.8801
  FOOTAGE    1    10.71823204         9.72721352         1.102            0.3854


                         PREDICT        STD ERR      LOWER95%      UPPER95%
  OBS          ACTUAL      VALUE        PREDICT          MEAN          MEAN     RESIDUAL

       1      50.0000    43.5912         4.6269       23.6830       63.4993       6.4088
       2      40.0000    42.8324         5.7171       18.2336       67.4312      -2.8324
       3      52.0000    53.5506         6.2048       26.8533       80.2480      -1.5506
       4      47.0000    52.6022         3.2105       38.7882       66.4162      -5.6022
       5      65.0000    61.4236         6.3729       34.0029       88.8442       3.5764
       6            .    57.0129 (1)     4.4572       37.8350 (2)   76.1908 (3)        .

 SUM OF RESIDUALS                          0
 SUM OF SQUARED RESIDUALS           95.67587
 PREDICTED RESID SS (PRESS)         889.4871
```

lower and upper 95% confidence bounds are shown in (2) and (3), respectively. Notice that these values agree with those found by hand apart from roundoff error (see Example 3.7.1).

COMPUTING SUPPLEMENT C (Finding a Confidence Band)

In this supplement the code needed to create the confidence bounds used to find the confidence band shown in Figure 3.1 is given. The XPX and I options are used so that you can check the work done in Exercise 52.

Statement	Purpose
`DATA NOISE;`	Names the data set
`INPUT IRON COMPOS;`	Names the variables
`CARDS;`	Signals that the data follow

`75 74`
`80 81`
`110 107`
`93 90`
`65 64`
	Observed data

`70 .`
`84.6 .`
`90 .`
`105 .`
	Lines added to enable SAS to compute confidence bounds for x values not in the original data set

`;`	Signals the end of the data
`PROC REG;`	Asks for the regression procedure
`MODEL COMPOS=IRON/P XPX I CLM;`	Identifies the model as simple linear regression, asks for predicted values, $(X'X)$, $(X'X)^{-1}$, and confidence intervals on the mean response for each x value given

In the output, shown in Figure 3.6, these quantities are given:

(1) $(X'X)^{-1}$
(2) $\mathbf{b}$
(3) $X'\mathbf{y}$
(4) $\mathbf{y}'\mathbf{y}$
(5) s^2
(6) s
(7)–(13) confidence intervals $x = 65, 70, 75, 84.6, 90, 105$, and 110, respectively

FIGURE 3.6

SAS 1

MODEL CROSSPRODUCTS X'X X'Y Y'Y

X'X	INTERCEP	IRON	COMPOS
INTERCEP	5	423	416 ③
IRON	423	36999	36330
COMPOS	416	36330	35682 ④

X'X INVERSE, B, SSE

INVERSE	INTERCEP	IRON	COMPOS
INTERCEP	6.099407	-0.0697329 ①	3.95549 ②
IRON	-0.0697329	0.0008242664	0.9366963
COMPOS	3.95549	0.9366963	6.338279

DEP VARIABLE: COMPOS

ANALYSIS OF VARIANCE

SOURCE	DF	SUM OF SQUARES	MEAN SQUARE	F VALUE	PROB>F
MODEL	1	1064.46172	1064.46172	503.825	0.0002
ERROR	3	6.33827893	2.11275964 ⑤		
C TOTAL	4	1070.80000			

ROOT MSE	1.453534 ⑥	R-SQUARE	0.9941
DEP MEAN	83.2	ADJ R-SQ	0.9921
C.V.	1.747035		

PARAMETER ESTIMATES

VARIABLE	DF	PARAMETER ESTIMATE	STANDARD ERROR	T FOR H0: PARAMETER=0	PROB > |T|
INTERCEP	1	3.95548961	3.58978829	1.102	0.3510
IRON	1	0.93669634	0.04173101	22.446	0.0002

OBS	ACTUAL	PREDICT VALUE	STD ERR PREDICT	LOWER95% MEAN	UPPER95% MEAN	RESIDUAL
1	74.0000	74.2077	0.7636	71.7776	76.6378	-.207715 ⑨
2	81.0000	78.8912	0.6778	76.7341	81.0483	2.1088
3	107.0000	106.9921	1.2434	103.0349	110.9493	0.007913 ⑬
4	90.0000	91.0682	0.7385	88.7179	93.4186	-1.0682
5	64.0000	64.8408	1.0448	61.5158	68.1658	-.840752 ⑦

FIGURE 3.6 Continued

FIGURE 3.6 Continued

SAS 2

OBS	ACTUAL	PREDICT VALUE	STD ERR PREDICT	LOWER95% MEAN	UPPER95% MEAN	RESIDUAL
6	.	69.5242	0.8909	66.6888	72.3596	. (8)
7	.	83.2000	0.6500	81.1312	85.2688	. (10)
8	.	88.2582	0.6880	86.0686	90.4477	. (11)
9	.	102.3086	1.0711	98.8998	105.7174	. (12)

SUM OF RESIDUALS	7.10543E-15
SUM OF SQUARED RESIDUALS	6.338279
PREDICTED RESID SS (PRESS)	12.44408

COMPUTING SUPPLEMENT D (Finding a Prediction Interval)

To create the 95% prediction interval of Example 3.7.3, the code given in Computing Supplement A requires one change. The model statement is changed to

```
MODEL PRICE = AGE FOOTAGE/P CLI;
```

In Figure 3.7 the lower and upper bounds for the interval are given by (1) and (2), respectively. These agree closely with those found by hand.

FIGURE 3.7

ANALYSIS
OF
REAL ESTATE DATA 1

DEP VARIABLE: PRICE

ANALYSIS OF VARIANCE

SOURCE	DF	SUM OF SQUARES	MEAN SQUARE	F VALUE	PROB>F
MODEL	2	239.12412523	119.56206262	2.499	0.2858
ERROR	2	95.67587477	47.83793738		
C TOTAL	4	334.80000000			

ROOT MSE	6.916497	R-SQUARE	0.7142
DEP MEAN	50.8	ADJ R-SQ	0.4285
C.V.	13.61515		

PARAMETER ESTIMATES

VARIABLE	DF	PARAMETER ESTIMATE	STANDARD ERROR	T FOR H0: PARAMETER=0	PROB > |T|
INTERCEP	1	33.06261510	10.50659142	3.147	0.0879
AGE	1	-0.189687	1.11058122	-0.171	0.8801
FOOTAGE	1	10.71823204	9.72721352	1.102	0.3854

OBS	ACTUAL	PREDICT VALUE	STD ERR PREDICT	LOWER95% PREDICT	UPPER95% PREDICT	RESIDUAL
1	50.0000	43.5912	4.6269	7.7866	79.3957	6.4088
2	40.0000	42.8324	5.7171	4.2224	81.4425	-2.8324
3	52.0000	53.5506	6.2048	13.5709	93.5304	-1.5506
4	47.0000	52.6022	3.2105	19.7927	85.4117	-5.6022
5	65.0000	61.4236	6.3729	20.9572	101.8899	3.5764
6	.	57.0129	4.4572	21.6091 (1)	92.4166 (2)	.

SUM OF RESIDUALS	0
SUM OF SQUARED RESIDUALS	95.67587
PREDICTED RESID SS (PRESS)	889.4871

CHAPTER 4

HYPOTHESIS TESTING IN THE FULL RANK MODEL

In Chapter 3 we developed methods for estimating the coefficients $\beta_0, \beta_1, \beta_2, \ldots, \beta_k$ and linear combinations of these coefficients in the full rank model. In this chapter we consider the problem of testing various hypotheses concerning these coefficients. In particular, we use hypothesis-testing techniques to answer these and other questions:

1. Does the proposed model explain a large proportion of the observed variation in response?
2. Is there a subset of variables that adequately explains the observed variation in response, or are all the proposed variables necessary to do so?
3. Is a particular variable in the model useful in helping to predict the response?

Although each of these questions can be answered via the *general linear hypothesis*, it is instructive to proceed at a slower pace. For this reason, we consider each question separately and then generalize the results once the ideas are fixed in these important specific settings.

4.1

TESTING FOR MODEL ADEQUACY

Recall that our basic model assumes the form

$$y_i = \beta_0 + \beta_1 x_{i1} + \beta_2 x_{i2} + \cdots + \beta_k x_{ik} + \varepsilon_i \qquad i = 1, 2, \ldots, n.$$

The first question to be asked is, Is the proposed model adequate? That is, does the linear model really help explain the observed variability in response? If not, then all coefficients $\beta_0, \beta_1, \ldots, \beta_k$ will have value 0; otherwise, at least one of these coefficients will be nonzero. Hence to test for model adequacy, we consider

$$H_0 : \boldsymbol{\beta} = \mathbf{0} \quad \text{versus} \quad H_1 : \boldsymbol{\beta} \neq \mathbf{0} \quad \text{(Model is useful)}$$

To perform the test, we make the same model assumptions that we made in Section 3.6. In particular, we assume throughout this chapter that the random errors are each normally distributed with $E[\boldsymbol{\varepsilon}] = \mathbf{0}$ and var $\boldsymbol{\varepsilon} = \sigma^2 I$. This in turn implies that $\mathbf{y}$ is an $n \times 1$ normally distributed random vector with mean $X\boldsymbol{\beta}$ and variance $\sigma^2 I$.

The method used to test H_0 is *analysis of variance* (ANOVA). You have probably used this general procedure in earlier courses in applied statistics. Briefly, ANOVA is an analytic technique in which a sum of squares is subdivided into components that can be attributed to important sources. These components can then be used to test useful hypotheses. Here we want to subdivide $\mathbf{y}'\mathbf{y}$, the sum of the squares of the responses, in a meaningful way. This is easy to do. In Exercise 16 in Chapter 3, we saw that the residual sum of squares, which reflects random or unexplained variations in response, can be expressed as

$$SS_{\text{Res}} = \mathbf{y}'\mathbf{y} - \mathbf{y}'X(X'X)^{-1}X'\mathbf{y}$$

Solving this equation for $\mathbf{y}'\mathbf{y}$, we obtain

$$\mathbf{y}'\mathbf{y} = \mathbf{y}'X(X'X)^{-1}X'\mathbf{y} + SS_{\text{Res}}$$

The term $\mathbf{y}'X(X'X)^{-1}X'\mathbf{y}$ reflects the variation in response that is not random. That is, it reflects the variation in response that is explained by the linear regression model. For this reason, we call $\mathbf{y}'X(X'X)^{-1}X'\mathbf{y}$ the *model or regression sum of squares* and denote it by SS_{Model} or SS_{Reg}. Letting $\mathbf{y}'\mathbf{y} = SS_{\text{Total}}$, we have subdivided the total sum of squares as follows:

$$SS_{\text{Total}} = SS_{\text{Reg}} + SS_{\text{Res}}$$

If the model is appropriate, then SS_{Reg} should be large relative to SS_{Res}; most of the variability in response should be attributed to the regression model rather than to random sources. Thus to develop a test statistic for testing H_0 it is natural to compare these two components in some way. We begin by considering the probability distributions of SS_{Reg}/σ^2 and SS_{Res}/σ^2, respectively, and the relationship between them.

Theorem 4.1.1 Let SS_{Reg} denote the regression sum of squares in the full rank linear model. Then SS_{Reg}/σ^2 follows a noncentral chi-squared distribution with $p = k + 1$ degrees of freedom and noncentrality parameter

$$\lambda = \frac{1}{2\sigma^2}\boldsymbol{\beta}'(X'X)\boldsymbol{\beta}$$

Proof

By definition,

$$SS_{\text{Reg}}/\sigma^2 = \mathbf{y}'X(X'X)^{-1}X'\mathbf{y}/\sigma^2$$

Note that $\mathbf{y}$ is an $n \times 1$ normally distributed random vector with mean $X\boldsymbol{\beta}$ and variance $\sigma^2 I$. Note also that $X(X'X)^{-1}X'$ is an $n \times n$ idempotent and symmetric matrix. Furthermore, by Theorem 1.5.2, the rank of $X(X'X)^{-1}X'$ is equal to its trace. Now

$$\text{tr}[X(X'X)^{-1}X'] = \text{tr}[(X'X)^{-1}X'X] = \text{tr}[I_{k+1}] = k + 1$$

By Corollary 2.3.2,

$$\frac{1}{\sigma^2}\mathbf{y}'X(X'X)^{-1}X'\mathbf{y} = SS_{\text{Reg}}/\sigma^2$$

follows a noncentral chi-squared distribution with $k + 1$ degrees of freedom and noncentrality parameter

$$\lambda = \frac{1}{2\sigma^2}(X\boldsymbol{\beta})'X(X'X^{-1})X'(X\boldsymbol{\beta})$$

$$\lambda = \frac{1}{2\sigma^2}\boldsymbol{\beta}'X'X\boldsymbol{\beta}$$

as claimed. ■

In Section 3.4 the distribution of SS_{Res}/σ^2 was derived. The results of this derivation are summarized in Theorem 4.1.2 for easy reference.

Theorem 4.1.2 Let SS_{Res} denote the residual sum of squares in the full rank linear model. Then SS_{Res}/σ^2 follows a chi-squared distribution with $n - p$ degrees of freedom. ■

Theorem 4.1.3 gives the relationship between the quadratic forms SS_{Reg}/σ^2 and SS_{Res}/σ^2. Its proof is left as an exercise (see Exercise 1).

Theorem 4.1.3 SS_{Reg}/σ^2 and SS_{Res}/σ^2 are independent quadratic forms. ■

To understand the logic behind a statistical test, it is helpful to know what to expect if H_1 is true as well as the distribution of the test statistic when H_0 is true. Theorem 4.1.4 will help justify intuitively the test statistic used to test for model adequacy.

Theorem 4.1.4 If X is $n \times p$ of full rank, then $X'X$ is positive definite.

Proof

Consider any $1 \times p$ row vector $\mathbf{y}'$. Note that $\mathbf{y}'X'X\mathbf{y} = (X\mathbf{y})'(X\mathbf{y})$ is a sum of squares and hence is nonnegative. It must now be shown that if $\mathbf{y}'X'X\mathbf{y} = 0$, then $\mathbf{y} = \mathbf{0}$. To do this, note that if $\mathbf{y}'X'X\mathbf{y} = 0$, then $X\mathbf{y} = \mathbf{0}$ and $X'X\mathbf{y} = X'\mathbf{0} = \mathbf{0}$. Since $X'X$ is nonsingular, it can be concluded that $\mathbf{y} = \mathbf{0}$ as desired. ■

With these results available, it is easy to derive a test statistic for testing the null hypothesis that $\boldsymbol{\beta} = \mathbf{0}$. If this hypothesis is true, then the noncentrality parameter, λ, associated with SS_{Reg}/σ^2 has value 0, implying that this quadratic form follows a chi-squared distribution with p degrees of freedom. Since SS_{Reg}/σ^2 and SS_{Res}/σ^2 are independent, if H_0 is true the ratio

$$\frac{SS_{\text{Reg}}/p\sigma^2}{SS_{\text{Res}}/(n-p)\sigma^2} = \frac{SS_{\text{Reg}}/p}{SS_{\text{Res}}/(n-p)}$$

follows an F distribution with p and $n-p$ degrees of freedom. It is customary to call SS_{Reg}/p the *regression or model mean square* and denote it by MS_{Reg}. Similarly, $SS_{\text{Res}}/(n-p) = s^2$ is called the *error or residual mean square* and is denoted by MS_{Res}. Since it has already been shown that s^2 is an unbiased estimator for σ^2, $E[MS_{\text{Res}}] = \sigma^2$. What is the expected value of the regression mean square? To answer this question, apply Theorem 2.2.1 as follows:

$$\begin{aligned}
E[MS_{\text{Reg}}] &= E[(1/p)\mathbf{y}'X(X'X)^{-1}X'\mathbf{y}] \\
&= (1/p)\{\text{tr}[X(X'X)^{-1}X']\sigma^2 I + (X\boldsymbol{\beta})'X(X'X)^{-1}X'X\boldsymbol{\beta}] \\
&= (1/p)[p\sigma^2 + \boldsymbol{\beta}'X'X\boldsymbol{\beta}] \\
&= \sigma^2 + (1/p)\boldsymbol{\beta}'X'X\boldsymbol{\beta}
\end{aligned}$$

As you can see, if $H_0{:}\boldsymbol{\beta} = \mathbf{0}$ is true, then $E[MS_{\text{Reg}}] = \sigma^2$ and MS_{Res} and MS_{Reg} are each unbiased estimators for σ^2. Intuitively, we expect that if H_0 is true, then the ratio $MS_{\text{Reg}}/MS_{\text{Res}}$ should have a value close to 1. However, if the null hypothesis is not true, then $\boldsymbol{\beta} \neq \mathbf{0}$. Since $X'X$ is positive definite, $\boldsymbol{\beta}'X'X\boldsymbol{\beta} > 0$ and $E[MS_{\text{Reg}}] > \sigma^2$. In this case, one would expect the ratio $MS_{\text{Reg}}/MS_{\text{Res}}$ to exceed 1. Our test then is to reject the null hypothesis $\boldsymbol{\beta} = \mathbf{0}$ in favor of the alternative that the model is useful in explaining the variation in response, $\boldsymbol{\beta} \neq \mathbf{0}$, for large values of $MS_{\text{Reg}}/MS_{\text{Res}}$, with the proper distribution being the F distribution with p and $n-p$ degrees of freedom.

As indicated in Section 3.3, the number of model parameters is $p=k+1$ in a model that contains an intercept term. If no intercept is included, then $p=k$. Regardless of the situation, the adequacy of the model is tested as described above. The general results for both models are summarized in Table 4.1.

TABLE 4.1 **ANOVA table for testing for model adequacy. Here p = number of model parameters**

Source of Variation	Sum of Squares	Degrees of Freedom	Mean Square	F Ratio
Regression or model	$\mathbf{y}'X(X'X)^{-1}X'\mathbf{y}$	p	SS_{Reg}/p	$MS_{\text{Reg}}/MS_{\text{Res}}$
Residual or error	$\mathbf{y}'\mathbf{y} - \mathbf{y}'X(X'X)^{-1}X'\mathbf{y}$	$n-p$	$SS_{\text{Res}}/(n-p)$	
Total	$\mathbf{y}'\mathbf{y}$	n		

An example will illustrate the ideas presented here.

Example 4.1.1 A data processing system entails three basic structural elements: files (x_1), flows (x_2), and processes (x_3). Files are permanent records, flows are data interfaces, and processes are functionally defined logical manipulations of the data. An investigation of the cost of developing software was reported in "A Software Matrix for Cost Estimation and Efficiency Measurement in Data Processing System Development," *Journal of Systems Software* 3, 1983. These data are based on that study:

Cost (in units of 1000) (y)	Files (x_1)	Flows (x_2)	Processes (x_3)
22.6	4	44	18
15.0	2	33	15
78.1	20	80	80
28.0	6	24	21
80.5	6	227	50
24.5	3	20	18
20.5	4	41	13
147.6	16	187	137
4.2	4	19	15
48.2	6	50	21
20.5	5	48	17

The assumed linear regression model is

$$y_i = \beta_0 + \beta_1 x_{i1} + \beta_2 x_{i2} + \beta_3 x_{i3} + \varepsilon_i \qquad i = 1, 2, \ldots, 11$$

Let us test

$$H_0: \boldsymbol{\beta} = \mathbf{0} \quad \text{versus} \quad H_1: \boldsymbol{\beta} \neq \mathbf{0} \quad \text{(Model is useful)}$$

For these data,

$$X = \begin{bmatrix} 1 & 4 & 44 & 18 \\ 1 & 2 & 33 & 15 \\ \vdots & \vdots & & \\ 1 & 5 & 48 & 17 \end{bmatrix} \quad \text{and} \quad \mathbf{y} = \begin{bmatrix} 22.6 \\ 15.0 \\ \vdots \\ 20.5 \end{bmatrix}$$

SAS is used to find that

$$(X'X)^{-1} = \begin{bmatrix} 0.3197263 & -0.0408268 & -0.00202208 & 0.005305965 \\ -0.0408268 & 0.0140738 & 0.0003717104 & -0.00224159 \\ -0.00202208 & 0.0003717104 & 0.00005188447 & 0.000113861 \\ 0.005305965 & -0.00224159 & -0.000113861 & 0.0004938527 \end{bmatrix}$$

$$SS_{\text{Reg}} = \mathbf{y}'X(X'X)^{-1}X'\mathbf{y} = 38978.38$$

$$\mathbf{y}'\mathbf{y} = 39667.01$$

$$SS_{\text{Res}} = \mathbf{y}'\mathbf{y} - SS_{\text{Reg}} = 688.63$$

$$MS_{\text{Reg}} = SS_{\text{Reg}}/p = SS_{\text{Reg}}/4 = 38978.38/4 = 9744.595$$

$$MS_{\text{Res}} = SS_{\text{Res}}/(n-p) = SS_{\text{Res}}/7 = 688.63/7 = 98.375$$

$$F_{4,7} = MS_{\text{Reg}}/MS_{\text{Res}} = 9744.595/98.375 = 99.055$$

Since the F ratio far exceeds 1, it is expected that H_0 will be rejected based on the $F_{4,7}$ distribution. Since the critical point for an $\alpha = 0.01$ level test is 7.85, the true P value is much less than 0.01. There is strong evidence that $\boldsymbol{\beta} \neq \mathbf{0}$. That is, at least one of the parameters β_0, β_1, β_2, or β_3 is not zero. Our task eventually is to discover exactly which of these parameters is nonzero. The results of this analysis are summarized in Table 4.2.

TABLE 4.2 **ANOVA for cost data of Example 4.1.1**

Source of Variation	Sum of Squares	Degrees of Freedom	Mean Square	F Ratio
Regression	38978.38	4	9744.595	99.055
Residual	688.63	7	98.375	
Total	39667.01	11		

4.2

HYPOTHESIS TESTS ON A SUBVECTOR OF $\boldsymbol{\beta}$

In Section 4.1 we tested

$$H_0:\boldsymbol{\beta}=\mathbf{0} \quad \text{versus} \quad H_1:\boldsymbol{\beta}\neq\mathbf{0}$$

Recall that the linear regression model can be expressed as

$$\mathbf{y}=X\boldsymbol{\beta}+\boldsymbol{\varepsilon}$$

where $E[\boldsymbol{\varepsilon}]=\mathbf{0}$ and var $\boldsymbol{\varepsilon}=\sigma^2 I$. If H_0 is true, then $E[\mathbf{y}]=E[\boldsymbol{\varepsilon}]=\mathbf{0}$ and var $\mathbf{y}=\sigma^2 I$. That is, the null hypothesis can be viewed as stating that the variability in response is random about a *mean of 0*. If $\boldsymbol{\beta}\neq\mathbf{0}$, then $E[\mathbf{y}]=X\boldsymbol{\beta}\neq\mathbf{0}$ but var $\mathbf{y}$ is still $\sigma^2 I$ (see Exercise 4). Hence the alternative hypothesis states that the variation in response is random about a *nonzero mean.* To say that $\boldsymbol{\beta}\neq\mathbf{0}$ implies that at least one of the parameters $\beta_0, \beta_1, \ldots, \beta_k$ is not zero. In practice most physical responses such as lengths, heights, weights, speeds, prices, and other measurement variables are not centered at 0. For this reason, β_0 is not usually zero in a realistic setting. Often when we reject $H_0:\boldsymbol{\beta}=\mathbf{0}$ via the method introduced in Section 4.1, we are in fact primarily detecting this shift away from 0. That is, rejection of $H_0:\boldsymbol{\beta}=\mathbf{0}$ might only be detecting the fact that $\beta_0\neq 0$. It should be evident that the rejection of H_0: $\boldsymbol{\beta}=\mathbf{0}$ does not conclude a regression study. The real questions to be answered are yet to come. In particular, we want to know whether or not there is evidence that the regressors $x_1, x_2, \ldots, x_k$ are useful in explaining the variation in response, and, if so, which of these regressors are most important. To determine this, we need to develop a method for testing hypotheses concerning arbitrary subsets of the set of parameters $\{\beta_0, \beta_1, \ldots, \beta_k\}$. This topic is discussed in this section.

To begin, recall that in expanded form

$$X=\begin{bmatrix} 1 & x_{11} & x_{12} & \cdots & x_{1k} \\ 1 & x_{21} & x_{22} & & x_{2k} \\ \vdots & \vdots & \vdots & & \vdots \\ 1 & x_{n1} & x_{n2} & & x_{nk} \end{bmatrix} \quad \text{and} \quad \boldsymbol{\beta}=\begin{bmatrix} \beta_0 \\ \beta_1 \\ \beta_2 \\ \vdots \\ \beta_k \end{bmatrix}$$

where X is $n\times p$ of rank p. Consider any subset of r parameters chosen from $\{\beta_0, \beta_1, \beta_2, \ldots, \beta_k\}$. Without loss of generality, it can be assumed that the first r parameters are selected. Now partition $\boldsymbol{\beta}$ as

$$\boldsymbol{\beta} = \begin{bmatrix} \beta_0 \\ \beta_1 \\ \vdots \\ \beta_{r-1} \\ --- \\ \beta_r \\ \beta_{r+1} \\ \vdots \\ \beta_k \end{bmatrix} = \begin{bmatrix} \boldsymbol{\gamma}_1 \\ -- \\ \boldsymbol{\gamma}_2 \end{bmatrix}$$

Note that $\boldsymbol{\gamma}_1$ is an $r \times 1$ column vector in its own right and that $\boldsymbol{\gamma}_2$ is a $(p-r) \times 1$ column vector. The matrix X can be partitioned as $[X_1 \mid X_2]$ where X_1 consists of the first r columns of X, and X consists of the remaining $p-r$ columns. We want to test

$$H_0: \boldsymbol{\gamma}_1 = \mathbf{0} \quad \text{versus} \quad H_1: \boldsymbol{\gamma}_1 \neq \mathbf{0}$$

Practically speaking, we are testing the null hypothesis that the first r parameters are not needed to explain the variation in response satisfactorily versus the alternative that they are needed. Mathematically, two models are being compared. The model under H_0 contains only the last $p-r$ parameters and is called a *reduced* model. In matrix form it is written as

$$\mathbf{y} = X_2 \boldsymbol{\gamma}_2 + \boldsymbol{\varepsilon}^*$$

The model indicated in H_1 contains all the original parameters and is called the *full* model. It is written in the usual form:

$$\mathbf{y} = X\boldsymbol{\beta} + \boldsymbol{\varepsilon}$$

In essence, we are choosing between a reduced model and a full model with the philosophy that the reduced model will be retained unless it is shown to be inadequate.

To decide between H_0 and H_1, a test statistic must be developed. The logic behind the test is easy to understand. Recall that the regression sum of squares for the full model is

$$SS_{\text{Reg}} = \mathbf{y}'X(X'X)^{-1}X'\mathbf{y}$$

In this context it is useful to denote this quadratic form by $R(\boldsymbol{\beta})$. This sum of squares measures the variation in response explained by the model that contains all the parameters $\beta_0, \beta_1, \beta_2, \ldots, \beta_k$. The regression sum of squares for the reduced model is denoted by $R(\boldsymbol{\gamma}_2)$ and is given by

$$R(\boldsymbol{\gamma}_2) = \mathbf{y}'X_2(X_2'X_2)^{-1}X_2'\mathbf{y}$$

(see Exercise 5). The difference between $R(\boldsymbol{\beta})$ and $R(\boldsymbol{\gamma}_2)$ is the amount of variation in response that is not random but that cannot be accounted for by the reduced model alone. This difference is called the *sum of squares for regression on* γ_1 *in the presence of* γ_2 and is denoted by $R(\gamma_1|\gamma_2)$. Thus

$$R(\gamma_1|\gamma_2) = R(\beta) - R(\gamma_2)$$

Logic dictates that when H_0 is true most of the variability in response should be explained by the reduced model and hence $R(\beta)$ and $R(\gamma_2)$ should be close in value, forcing $R(\gamma_1|\gamma_2)$ to be small. On the other hand, if H_1 is true, the parameters $\beta_r, \beta_{r+1}, \ldots, \beta_k$ alone will not adequately explain the observed variability. This should be reflected by $R(\gamma_1|\gamma_2)$ assuming a relatively large value. Hence it is reasonable to expect that the quadratic form $R(\gamma_1|\gamma_2)$ will play a major role in the test statistic used to choose between H_0 and H_1.

To develop a test statistic mathematically, consider the identity

$$\begin{aligned} \mathbf{y}'\mathbf{y} = \mathbf{y}'[X_2(X_2'X_2)^{-1}X_2']\mathbf{y} + \mathbf{y}'[X(X'X)^{-1}X' - X_2(X_2'X_2)^{-1}X_2']\mathbf{y} \\ + \mathbf{y}[I - X(X'X)^{-1}X']\mathbf{y} \end{aligned}$$

Rewriting this identity using the notation just presented, it is easy to see that

$$\mathbf{y}'\mathbf{y} = R(\gamma_2) + R(\gamma_1|\gamma_2) + SS_{\text{Res}}$$

and that

$$\mathbf{y}'\mathbf{y}/\sigma^2 = R(\gamma_2)/\sigma^2 + R(\gamma_1|\gamma_2)/\sigma^2 + SS_{\text{Res}}/\sigma^2$$

To determine the distribution of each component of this identity, the following four lemmas are needed.

Lemma 4.2.1 The rank of $X_2(X_2'X_2)^{-1}X_2'$ is $p - r$ (see Exercise 6). ■

Lemma 4.2.2 The matrix $A = X(X'X)^{-1}X' - X_2(X_2'X_2)^{-1}X_2'$ is idempotent.

Proof

Note that $X'[I - X(X'X)^{-1}X'] = 0$. In partitioned form

$$\begin{bmatrix} X_1' \\ --- \\ X_2' \end{bmatrix} [I - X(X'X)^{-1}X'] = 0$$

This implies that

$$X_1'[I - X(X'X)^{-1}X'] = 0$$

and

$$X_2'[I - X(X'X)^{-1}X'] = 0$$

The latter equation guarantees that

$$X_2' = X_2'X(X'X)^{-1}X'$$

and

$$X_2 = X(X'X)^{-1}X'X_2$$

To show that A is idempotent, consider its square. Keep in mind the fact that $X(X'X)^{-1}X'$ and $X_2(X_2'X_2)^{-1}X_2'$ are idempotent.

$$\begin{aligned} A^2 &= [X(X'X)^{-1}X' - X_2(X_2'X_2)^{-1}X_2']^2 \\ &= X(X'X)^{-1}X' - X(X'X)^{-1}X'X_2(X_2'X_2)^{-1}X_2' \\ &\quad - X_2(X_2'X_2)^{-1}X_2'X(X'X)^{-1}X' + X_2(X_2'X_2)^{-1}X_2' \end{aligned}$$

Substituting X_2 for $X(X'X)^{-1}X'X_2$ in the second term on the right and X_2' for $X_2'X(X'X)^{-1}X'$ in the third term,

$$\begin{aligned} A^2 &= X(X'X)^{-1}X' - X_2(X_2'X_2)^{-1}X_2' - X_2(X_2'X_2)^{-1}X_2' + X_2(X_2'X_2)^{-1}X_2' \\ &= X(X'X)^{-1}X' - X_2(X_2'X_2)^{-1}X_2' = A \end{aligned}$$

Thus A is idempotent as claimed. ■

Lemma 4.2.3 The matrix $A = X(X'X)^{-1}X' - X_2(X_2'X_2)^{-1}X_2'$ is of rank r (see Exercise 7). ■

Lemma 4.2.4 The matrix $[I - X(X'X)^{-1}X']$ is of rank $n - p$ (see Exercise 37 in Chapter 3). ■

The following theorem, the *Cochran–Fisher theorem*, was partially proved in Section 2.4. It provides the key for determining the distribution of each of the quadratic forms under study. In particular, it focuses on the distribution of $R(\gamma_1|\gamma_2)/\sigma^2$.

Theorem 4.2.1 (Cochran–Fisher) Let $\mathbf{z}$ be an $n \times 1$ multivariate normal random variable with mean $\boldsymbol{\mu}$ and variance I. Let

$$\mathbf{z}'\mathbf{z} = \sum_{i=1}^{m} \mathbf{y}'A_i\mathbf{y}$$

A necessary and sufficient condition for the quadratic forms to be independent and distributed as noncentral chi-squared random variables with parameters r_i and λ_i, respectively, where $r_i = r(A_i)$ and $\lambda_i = (\frac{1}{2})\boldsymbol{\mu}'A_i\boldsymbol{\mu}$ is that $\Sigma_{i=1}^{m} r_i = n$. ■

To apply this theorem here, let $\mathbf{z} = \mathbf{y}/\sigma$. Note that under the current model assumptions,

$$E[\mathbf{z}] = \boldsymbol{\mu} = X\boldsymbol{\beta}/\sigma$$

and

$$\text{var } \mathbf{z} = \text{var } \mathbf{y}/\sigma = (1/\sigma^2) \text{ var } \mathbf{y} = I$$

Now reconsider the basic identity

$$\mathbf{y}'\mathbf{y}/\sigma^2 = \mathbf{z}'\mathbf{z} = \frac{\mathbf{y}'[X_2(X_2'X_2)^{-1}X_2']\mathbf{y}}{\sigma^2} + \frac{\mathbf{y}'[X(X'X)^{-1}X' - X_2(X_2'X_2)^{-1}X_2']\mathbf{y}}{\sigma^2} + \frac{\mathbf{y}'[I - X(X'X)^{-1}X']\mathbf{y}}{\sigma^2}$$

Since the rank of a matrix is not affected by division by a constant, Lemmas 4.2.1, 4.2.3, and 4.2.4 can be applied to determine that the sum of the ranks of the matrices on the right-hand side of the identity is

$$(p - r) + r + (n - p) = n$$

By the Cochran–Fisher theorem, it can be concluded that the quadratic forms involved are independent noncentral chi-squared random variables. In particular, the quadratic form

$$\mathbf{y}'[X(X'X)^{-1}X' - X_2(X_2'X_2)^{-1}X_2']\mathbf{y}/\sigma^2 = R(\boldsymbol{\gamma}_1|\boldsymbol{\gamma}_2)/\sigma^2$$

follows a noncentral chi-squared distribution with rank r and noncentrality parameter

$$\lambda = (1/2\sigma^2)(X\boldsymbol{\beta})'[X(X'X)^{-1}X' - X_2(X_2'X_2)^{-1}X_2'](X\boldsymbol{\beta})$$

It has been argued intuitively that the magnitude of $R(\boldsymbol{\gamma}_1|\boldsymbol{\gamma}_2)$ is indicative of whether or not $H_0\!:\boldsymbol{\gamma}_1 = \mathbf{0}$ should be rejected. As you know, a test statistic must be one whose distribution is known under the assumption that the null hypothesis is true. To develop such a statistic in this case, consider the ratio

$$\frac{R(\boldsymbol{\gamma}_1|\boldsymbol{\gamma}_2)/\sigma^2 r}{SS_{\text{Res}}/\sigma^2(n-p)}$$

At this point it is known that the numerator is a noncentral chi-squared random variable divided by its degrees of freedom; the denominator is a chi-squared random variable divided by its degrees of freedom and is independent of the numerator. If it can be shown that the noncentrality parameter associated with the numerator has value 0 whenever the null hypothesis is true, then the above statistic will follow an F distribution with r and $n-p$ degrees of freedom. It will be a suitable test statistic for testing the null hypothesis that a reduced model adequately explains the variability in response. Theorem 4.2.2 establishes this result.

Theorem 4.2.2 If $H_0: \gamma_1 = \mathbf{0}$ is true, then the statistic

$$\frac{R(\gamma_1 | \gamma_2)/r}{SS_{\text{Res}}/(n-p)}$$

follows an F distribution with r and $n-p$ degrees of freedom.

Proof

Via the Cochran–Fisher theorem, it is known that

$$\begin{aligned}
\lambda &= (1/2\sigma^2)(X\boldsymbol{\beta})'[X(X'X)^{-1}X' - X_2(X_2'X_2)^{-1}X_2']X\boldsymbol{\beta} \\
&= (1/2\sigma^2)[\boldsymbol{\beta}'X'X(X'X)^{-1}X'X\boldsymbol{\beta} - \boldsymbol{\beta}'X'X_2(X_2'X_2)^{-1}X_2'X\boldsymbol{\beta}] \\
&= (1/2\sigma^2)\left\{\boldsymbol{\beta}'X'X\boldsymbol{\beta} - [\gamma_1 \mid \gamma_2]\begin{bmatrix} X_1' \\ -- \\ X_2' \end{bmatrix} X_2(X_2'X_2)^{-1}X_2'[X_1 \mid X_2]\begin{bmatrix} \gamma_1 \\ -- \\ \gamma_2 \end{bmatrix}\right\} \\
&= (1/2\sigma^2)\{\boldsymbol{\beta}'X'X\boldsymbol{\beta} - [\gamma_1'X_1' + \gamma_2'X_2']X_2(X_2'X_2)^{-1}X_2'[X_1\gamma_1 + X_2\gamma_2]\} \\
&= (1/2\sigma^2)\left\{[\gamma_1' \mid \gamma_2']\begin{bmatrix} X_1' \\ -- \\ X_2' \end{bmatrix}[X_1 \mid X_2]\begin{bmatrix} \gamma_1 \\ -- \\ \gamma_2 \end{bmatrix}\right. \\
&\qquad \left. - [\gamma_1'X_1' + \gamma_2'X_2']X_2(X_2'X_2)^{-1}X_2'[X_1\gamma_1 + X_2\gamma_2]\right\} \\
&= (1/2\sigma^2)\{[\gamma_1'X_1' + \gamma_2'X_2'][X_1\gamma_1 + X_2\gamma_2] \\
&\qquad - [\gamma_1'X_1' + \gamma_2'X_2']X_2(X_2'X_2)^{-1}X_2'[X_1\gamma_1 + X_2\gamma_2]\}
\end{aligned}$$

If $H_0: \gamma_1 = \mathbf{0}$ is true, then λ reduces to

$$\lambda = (1/2\sigma^2)\{\gamma_2'X_2'X_2\gamma_2 - \gamma_2'X_2'X_2(X_2'X_2)^{-1}X_2'X_2\gamma_2\}$$
$$\lambda = 0$$

as claimed. ■

Should the test statistic just developed be used to conduct a right, left, or two-tailed test? Our intuitive arguement points to a right-tailed test since if H_1 is true, the numerator of the F ratio should be rather large. A mathematical argument that this is the case is somewhat involved and is presented in Section 4.4.

In summary, the ANOVA table for testing a hypothesis on a subvector of $\boldsymbol{\beta}$ is as shown in Table 4.3.

TABLE 4.3
ANOVA table for testing a hypothesis on a subvector of β. Here p = number of parameters in the full model, r = number of parameters whose presence in the model is questioned

Source of Variation	Sum of Squares	Degrees of Freedom	Mean Square	F Ratio
Regression				
Full model	$R(\boldsymbol{\beta})$	p		
Reduced model	$R(\boldsymbol{\gamma}_2)$	$p-r$		
$\boldsymbol{\gamma}_1$ in presence of $\boldsymbol{\gamma}_2$	$R(\boldsymbol{\beta})-R(\boldsymbol{\gamma}_2)=R(\boldsymbol{\gamma}_1\mid\boldsymbol{\gamma}_2)$	r	$R(\boldsymbol{\gamma}_1\mid\boldsymbol{\gamma}_2)/r$	$\dfrac{R(\boldsymbol{\gamma}_1\mid\boldsymbol{\gamma}_2)/r}{SS_{\text{Res}}/(n-p)}$
Residual	$\mathbf{y}'\mathbf{y}-R(\boldsymbol{\beta})=SS_{\text{Res}}$	$n-p$	$SS_{\text{Res}}/(n-p)$	
Total	$\mathbf{y}'\mathbf{y}$	n		

As an illustration, let us continue to analyze the data of Example 4.1.1.

Example 4.2.1 The null hypothesis of random variation of cost about zero, $H_0{:}\boldsymbol{\beta}=\mathbf{0}$, was rejected in favor of a model containing an intercept term and three regressors, x_1 (files), x_2 (flows), and x_3 (processes). Since it is almost certainly true that $\beta_0 \neq 0$, let us continue the investigation by testing to see if the regressors x_1, x_2, and x_3 are needed in addition to the intercept. To do so, the columns of the X matrix are rearranged and partitioned as follows:

$$X=\left[\begin{array}{ccc|c} x_{11} & x_{12} & x_{13} & 1 \\ x_{21} & x_{22} & x_{23} & 1 \\ x_{31} & x_{32} & x_{33} & 1 \\ \vdots & \vdots & \vdots & \vdots \\ x_{11\,1} & x_{11\,2} & x_{11\,3} & 1 \end{array}\right]=[X_1 \mid X_2]$$

The parameter vector is written as

$$\boldsymbol{\beta}=\begin{bmatrix} \beta_1 \\ \beta_2 \\ \beta_3 \\ -- \\ \beta_0 \end{bmatrix}=\begin{bmatrix} \boldsymbol{\gamma}_1 \\ -- \\ \boldsymbol{\gamma}_2 \end{bmatrix}$$

We want to test $H_0{:}\boldsymbol{\gamma}_1=\mathbf{0}$ (the reduced model is adequate) versus $H_1{:}\boldsymbol{\gamma}_1 \neq \mathbf{0}$ (the full model is preferable). We are choosing between the reduced model

$$\mathbf{y}=\beta_0+\boldsymbol{\varepsilon}^*$$

and the full model

$$\mathbf{y} = X\boldsymbol{\beta} + \boldsymbol{\varepsilon}$$

Here

$$R(\boldsymbol{\gamma}_2) = \mathbf{y}'X_2(X_2'X_2)^{-1}X_2'\mathbf{y}$$

where

$$X_2 = \begin{bmatrix} 1 \\ 1 \\ 1 \\ \vdots \\ 1 \end{bmatrix}$$

It can be verified easily that

$$\begin{aligned} R(\boldsymbol{\gamma}_2) &= \left(\sum_{i=1}^{11} y_i\right)^2 \Big/ 11 \\ &= (22.6 + 15.0 + \cdots + 20.5)^2/11 \\ &= (489.7)^2/11 \\ &= 21{,}800.55 \end{aligned}$$

From previous work, it is known that

$$R(\boldsymbol{\beta}) = 38978.38$$

Hence

$$\begin{aligned} R(\boldsymbol{\gamma}_1 | \boldsymbol{\gamma}_2) &= R(\boldsymbol{\beta}) - R(\boldsymbol{\gamma}_2) \\ &= 38978.38 - 21{,}800.55 \\ &= 17{,}177.83 \end{aligned}$$

Is this value large enough to cause the rejection of H_0? From Table 4.2 it can be seen that the residual mean square is given by

$$SS_{\text{Res}}/(n-p) = SS_{\text{Res}}/(11-4) = 98.375$$

The F ratio used to test H_0 is

$$\frac{R(\boldsymbol{\gamma}_1 | \boldsymbol{\gamma}_2)/r}{SS_{\text{Res}}/(n-p)} = \frac{17{,}177.83/3}{98.375} = 58.2$$

Since $\boldsymbol{\gamma}_1$ contains three parameters, $r = 3$ and $n - p = 11 - 4 = 7$. Thus the test is based on an F ratio with 3 and 7 degrees of freedom. Based on the $F_{3,7}$ distribution, H_0 can be rejected with $P < 0.01$ (critical point, 8.45). We conclude that the model containing only an intercept term does not adequately explain the observed variability in response. The complete analysis is summarized in Table 4.4.

TABLE 4.4 ANOVA for cost data of Example 4.2.1

Source of Variation	Sum of Squares	Degrees of Freedom	Mean Square	F Ratio
Regression				
Full model	38,978.38	4		
Reduced model	21,800.55	1		
γ_1 in presence of γ_2	17,177.83	3	5725.9	58.2
Residual	688.63	7	98.375	
Total	39,667.01	11		

Since the full parameter set $\{\beta_0, \beta_1, \beta_2, \ldots, \beta_k\}$ contains $k+1$ elements, it has 2^{k+1} subsets. Of these, $2^{k+1}-2$ lead to subvectors that are viable candidates for $\boldsymbol{\gamma}_1$. The two subsets that are not of use are the full parameter set and the empty set. For example, in a model containing β_0, β_1, and β_2, there are $2^3 = 8$ subsets that generate these six partitions of $\boldsymbol{\beta}$:

$$\begin{bmatrix}\beta_0\\ --\\ \beta_1\\ \beta_2\end{bmatrix} \begin{bmatrix}\beta_1\\ --\\ \beta_0\\ \beta_2\end{bmatrix} \begin{bmatrix}\beta_2\\ --\\ \beta_0\\ \beta_1\end{bmatrix} \begin{bmatrix}\beta_0\\ \beta_1\\ --\\ \beta_2\end{bmatrix} \begin{bmatrix}\beta_0\\ \beta_2\\ --\\ \beta_1\end{bmatrix} \begin{bmatrix}\beta_1\\ \beta_2\\ --\\ \beta_0\end{bmatrix}$$

To decide which of these partitions is useful in practice, the statistician often relies on the expert opinion of the researcher, who is usually well-versed in the subject matter under consideration. He or she can usually suggest which parameters are essential to the model and which are questionable.

With the advent and widespread availability of sophisticated statistical software packages, it is not hard to examine all possible partitions. However, certain partitions have received special attention and have become part of the standard output of these packages. They are discussed in Section 4.3.

TEST BASED ON "CORRECTED" SUMS OF SQUARES

The ANOVA tables presented so far are based on what is called an "uncorrected" total sum of squares. That is, the total sum of squares is defined to be $\mathbf{y}'\mathbf{y}$, the variability of the response about zero. This allows us to begin the analysis by testing the most general of all hypotheses, namely $H_0\!:\boldsymbol{\beta} = \mathbf{0}$. Since it is often known at the outset that $\beta_0 \neq 0$, this hypothesis is not usually of primary interest to the researcher. Rather, the real question to be answered first is, Are the regressors important in the presence of an intercept? This hypothesis can be tested by following the procedure outlined in Example 4.2.1 and noting that the sum of squares for the reduced model, the model containing only an intercept, is given by

$$R(\gamma_2) = \left(\sum_{i=1}^{n} y_i\right)^2 \Big/ n$$

The ANOVA for the general test is shown in Table 4.5.

TABLE 4.5
ANOVA used to test for the importance of the regressors in the presence of an intercept based on the "uncorrected" total sum of squares, y′y

Source of Variation	Sum of Squares	Degrees of Freedom	Mean Square	F Ratio
Regression				
Full model	$R(\boldsymbol{\beta})$	$p = k + 1$		
Reduced model (intercept)	$R(\gamma_2) = \left(\sum_{i=1}^{n} y_i\right)^2 \Big/ n$	1		
γ_1 in presence of γ_2 (regressors in presence of intercept)	$R(\boldsymbol{\beta}) - R(\gamma_2) = R(\gamma_1 \mid \gamma_2)$	$p - 1 = k$	$R(\gamma_1 \mid \gamma_2)/k$	$\dfrac{R(\gamma_1 \mid \gamma_2)/k}{SS_{\text{Res}}/(n-p)}$
Residual	$\mathbf{y}'\mathbf{y} - R(\boldsymbol{\beta}) = SS_{\text{Res}}$	$n - p = n - k - 1$	$SS_{\text{Res}}/(n-p)$	
Total (uncorrected)	$\mathbf{y}'\mathbf{y}$	n		

If it is assumed from scientific consideration that $\beta_0 \neq 0$ and hence that there is no interest in testing $H_0{:}\boldsymbol{\beta} = \mathbf{0}$, then another procedure can be used to test for the importance of the regressors in the presence of an intercept. In particular, rather than measure the total variation in response by $\mathbf{y}'\mathbf{y}$, the variability about zero, we assume that even in the absence of regressors, the responses are varying about some nonzero mean. This mean is estimated by $\bar{y}$, and the measure of total variation in response is taken to be

$$\sum_{i=1}^{n} (y_i - \bar{y})^2 = \sum_{i=1}^{n} y_i^2 - \left(\sum_{i=1}^{n} y_i\right)^2 \Big/ n = \mathbf{y}'\mathbf{y} - R(\gamma_2)$$

This sum of squares is called the "corrected" total sum of squares, or the total sum of squares adjusted for the mean or intercept. The term

$$R(\gamma_2) = \left(\sum_{i=1}^{n} y_i\right)^2 \Big/ n$$

is called the *correction factor*. A test for the importance of the regressors in the presence of an intercept can be developed based on this corrected total sum of squares. The theoretical development of the test is outlined in Exercise 9. The end result of this exercise is shown in Table 4.6. Although the formats of Tables 4.5 and 4.6 differ a little, the F tests given are identical. Either version can be used to test for the importance of the regressors in the presence of an intercept. Both are presented here because both are widely used. In most elementary courses in applied statistics and on SAS printouts, tables based on corrected totals are given; however, in advanced texts in the theory of linear models, tables are often based on uncorrected totals because of their more general nature. You should be familiar with both. Since one of the primary purposes of this text is to prepare you for more advanced theoretical work, we emphasize the use of uncorrected totals.

TABLE 4.6 **ANOVA used to test for the importance of the regressors in the presence of an intercept based on the "corrected" total sum of squares, $y'y - (\sum_{i=1}^{n} y_i)^2/n$. Here $\gamma_1' = [\beta_1, \beta_2, \ldots, \beta_k]$ and $\gamma_2' = [\beta_0]$**

Source of Variation	Sum of Squares	Degrees of Freedom	Mean Square	F Ratio
Regression	$R(\gamma_1 \mid \gamma_2)$	k	$R(\gamma_1 \mid \gamma_2)/k$	$\dfrac{R(\gamma_1 \mid \gamma_2)/k}{SS_{\text{Res}}/(n-p)}$
Residual	SS_{Res}	$n-p=n-k-1$	$SS_{\text{Res}}/(n-k-1)$	
Total (corrected)	$\mathbf{y}'\mathbf{y} - \left(\sum_{i=1}^{n} y_i\right)^2 \Big/ n$	$n-1$		

4.3

PARTIAL AND SEQUENTIAL TESTS

In the previous section a method for testing any subset of parameters was developed. When $\boldsymbol{\gamma}_1$ is 1×1, that is, when a single parameter is being tested for inclusion in the model in the presence of all the others, then the F test based on 1 and $n-p$ degrees of freedom is called a *partial* F *test*. In a model with p parameters, p partial tests can be conducted. These tests are not based on sums of squares that add to the full model regression sum of squares, $R(\boldsymbol{\beta})$. That is, in general

$$R(\boldsymbol{\beta}) \neq R(\beta_0 \mid \beta_1, \beta_2, \ldots, \beta_k) + R(\beta_1 \mid \beta_0, \beta_2, \beta_3, \ldots, \beta_k) + R(\beta_2 \mid \beta_0, \beta_1, \beta_3, \ldots, \beta_k) + \cdots + R(\beta_k \mid \beta_0, \beta_1, \beta_2, \ldots, \beta_{k-1})$$

As a result, the sums of squares and the resulting F tests are *not independent.* This makes interpreting these F tests difficult. Each gives information regarding the importance of a single parameter in a model that contains *all* the others. If a parameter is important in the presence of all the others, it does not mean that it is important in every subset. That is, rejection of

$$H_0{:}\beta_j = 0 \quad \text{versus} \quad H_1{:}\beta_j \neq 0$$

via the partial F test does not guarantee that the best model for explaining variability in response necessarily contains the parameter β_j. Conversely, failure to reject H_0 does not guarantee that the best model does not include β_j. Clearly, it is very difficult to draw meaningful conclusions concerning the best model via these partial F tests. However, because these tests are included as standard output in many software packages, familiarity with them is a necessity.

It is possible to generate a series of F tests that are based on sums of squares that do sum to $R(\boldsymbol{\beta})$. To do so, it is necessary to clarify the notation used. We are dealing with a series of models that progress from the simplest possible to the most complex:

$$\begin{aligned}
\mathbf{y} &= \beta_0 + \boldsymbol{\varepsilon}^{(0)} \\
\mathbf{y} &= \beta_0 + \beta_1 x_1 + \boldsymbol{\varepsilon}^{(1)} \\
\mathbf{y} &= \beta_0 + \beta_1 x_1 + \beta_2 x_2 + \boldsymbol{\varepsilon}^{(2)} \\
&\vdots \\
\mathbf{y} &= \beta_0 + \beta_1 x_1 + \beta_2 x_2 + \cdots + \beta_k x_k + \boldsymbol{\varepsilon}^{(k)}
\end{aligned}$$

where $\boldsymbol{\varepsilon}^{(j)}$ denotes the vector of residuals for a model containing the first j regressors. The X matrices associated with each of these models differ and are denoted by $X^{(j)}$ where

$$X^{(j)} = \begin{bmatrix} 1 & x_{11} & x_{12} & \cdots & x_{1j} \\ 1 & x_{21} & x_{22} & & x_{2j} \\ \vdots & \vdots & \vdots & & \vdots \\ 1 & x_{n1} & x_{n2} & & x_{nj} \end{bmatrix}$$

for $j = 1, 2, \ldots, k$ and

$$X^{(0)} = \begin{bmatrix} 1 \\ 1 \\ \vdots \\ 1 \end{bmatrix}$$

The regression sum of squares for each of these models is denoted by $R(\beta_0, \beta_1, \beta_2, \ldots, \beta_j)$ and is given by

$$R(\beta_0, \beta_1, \beta_2, \ldots, \beta_j) = \mathbf{y}'X^{(j)}(X^{(j)\prime}X^{(j)})^{-1}X^{(j)\prime}\mathbf{y}$$

Hence we are actually dealing with a series of essentially "full" model regression sums of squares, namely,

$$\begin{aligned} &R(\beta_0) \\ &R(\beta_0, \beta_1) \\ &R(\beta_0, \beta_1, \beta_2) \\ &\vdots \\ &R(\beta_0, \beta_1, \beta_2, \ldots, \beta_k) = R(\boldsymbol{\beta}) \end{aligned}$$

These can be used to determine a series of "extra sums of squares for regression" by finding the differences between these "full" model regression sums of squares as more and more parameters are added to the model. These are given by

$$\begin{aligned} &R(\beta_1 | \beta_0) = R(\beta_0, \beta_1) - R(\beta_0) \\ &R(\beta_2 | \beta_0, \beta_1) = R(\beta_0, \beta_1, \beta_2) - R(\beta_0, \beta_1) \\ &R(\beta_3 | \beta_0, \beta_1, \beta_2) = R(\beta_0, \beta_1, \beta_2, \beta_3) - R(\beta_0, \beta_1, \beta_2) \\ &\vdots \\ &R(\beta_k | \beta_0, \beta_1, \ldots, \beta_{k-1}) = R(\beta_0, \beta_1, \ldots, \beta_k) - R(\beta_0, \beta_1, \beta_2, \ldots, \beta_{k-1}) \end{aligned}$$

It is easy to see that

$$\begin{aligned} R(\boldsymbol{\beta}) = R(\beta_0) + R(\beta_1 | \beta_0) + R(\beta_2 | \beta_0, \beta_1) + R(\beta_3 | \beta_0, \beta_1, \beta_2) \\ + \cdots + R(\beta_k | \beta_0, \beta_1, \beta_2, \ldots, \beta_{k-1}) \end{aligned}$$

These extra sums of squares each have 1 degree of freedom associated with them. The F ratio

$$\frac{R(\beta_j | \beta_0, \beta_1, \ldots, \beta_{j-1})}{SS_{\text{Res}}/(n-p)} = F_{1,n-p}$$

is used to test the null hypothesis that β_j is not needed in a model that already contains all of the *preceding* parameters $\beta_0, \beta_1, \ldots, \beta_{j-1}$. These tests are called *sequential* F tests. It should be evident that the order in which the parameters are listed can have a profound effect on the conclusions drawn. For example, suppose that the regressors prime rate (x_1), estimated inventory (x_2), quarter of the year (x_3), and local demand (x_4) are variables that are thought to affect the interest rate charged on new-car loans. Since these variables have no natural order, any imposed order is arbitrary. It is possible that information concerning the quarter of the year is not important in a model that already contains information on the prime rate and the estimated inventory, whereas it would be important if the first two regressors entered into the model were prime rate and local demand. In other words, the appropriateness of a regressor variable often depends on what regressors are already in the model. Thus a full-scale model-building process cannot be accomplished effectively using sequential F tests unless they are used in harmony with an appropriate selection of order based on subject matter expertise.

Sequential and partial F tests are illustrated in Example 4.3.1.

Example 4.3.1 An experiment was conducted to study the size of squid eaten by sharks and tuna. The regressors are characteristics of the beak or mouth of the squid. They are

x_1: Beak length in inches
x_2: Wing length in inches
x_3: Beak to notch length
x_4: Notch to wing length
x_5: Width in inches

The response is the weight of the squid in pounds. These data are obtained:

y	x_1	x_2	x_3	x_4	x_5
1.95	1.31	1.07	0.44	0.75	0.35
2.90	1.55	1.49	0.53	0.90	0.47
0.72	0.99	0.84	0.34	0.57	0.32
0.81	0.99	0.83	0.34	0.54	0.27
1.09	1.05	0.90	0.36	0.64	0.30
1.22	1.09	0.93	0.42	0.61	0.31
1.02	1.08	0.90	0.40	0.51	0.31
1.93	1.27	1.08	0.44	0.77	0.34
0.64	0.99	0.85	0.36	0.56	0.29
2.08	1.34	1.13	0.45	0.77	0.37
1.98	1.30	1.10	0.45	0.76	0.38
1.90	1.33	1.10	0.48	0.77	0.38
8.56	1.86	1.47	0.60	1.01	0.65
4.49	1.58	1.34	0.52	0.95	0.50
8.49	1.97	1.59	0.67	1.20	0.59
6.17	1.80	1.56	0.66	1.02	0.59
7.54	1.75	1.58	0.63	1.09	0.59
6.36	1.72	1.43	0.64	1.02	0.63
7.63	1.68	1.57	0.72	0.96	0.68
7.78	1.75	1.59	0.68	1.08	0.62
10.15	2.19	1.86	0.75	1.24	0.72
6.88	1.73	1.67	0.64	1.14	0.55

The ANOVA table for these data is given in Table 4.7.

As expected, the null hypothesis that $\boldsymbol{\beta} = \mathbf{0}$ is rejected soundly. The sequential and partial sums of squares are found using SAS to be as follows (see Computing Supplement E at the end of this chapter):

Sequential

$$R(\beta_0) = 387.156$$
$$R(\beta_0|\beta_1) = 199.145$$
$$R(\beta_2|\beta_0, \beta_1) = 0.126664$$
$$R(\beta_3|\beta_0, \beta_1, \beta_2) = 4.119539$$
$$R(\beta_4|\beta_0, \beta_1, \beta_2, \beta_3) = 0.263496$$
$$R(\beta_5|\beta_0, \beta_1, \beta_2, \beta_3, \beta_4) = 4.352193$$

Partial

$$R(\beta_0|\beta_1, \beta_2, \beta_3, \beta_4, \beta_5) = 24.0792$$
$$R(\beta_1|\beta_0, \beta_2, \beta_3, \beta_4, \beta_5) = 0.298731$$
$$R(\beta_2|\beta_0, \beta_1, \beta_3, \beta_4, \beta_5) = 0.868761$$
$$R(\beta_3|\beta_0, \beta_1, \beta_2, \beta_4, \beta_5) = 0.078273$$
$$R(\beta_4|\beta_0, \beta_1, \beta_2, \beta_3, \beta_5) = 0.982690$$
$$R(\beta_5|\beta_0, \beta_1, \beta_2, \beta_3, \beta_4) = 4.352193$$

Notice that, apart from roundoff error, the sequential sums of squares add to the regression sum of squares (595.164) as expected; the partial sums of squares do not add to anything meaningful.

The sequential sums of squares can be used to illustrate the test on a general subset presented in Section 4.2. For example, suppose we want to test

$$H_0\colon \boldsymbol{\gamma}_1 = \mathbf{0}$$

where

$$\boldsymbol{\gamma}_1 = \begin{bmatrix} \beta_4 \\ \beta_5 \end{bmatrix}$$

The proper numerator for the F test based on 2 and 16 degrees of freedom is

$$R(\beta_4, \beta_5|\beta_0, \beta_1, \beta_2, \beta_3)/2$$

The desired sum of squares can be obtained from the sequential sums of squares by noting that

$$\begin{aligned}
R(\beta_4, \beta_5|\beta_0, \beta_1, \beta_2, \beta_3) &= R(\boldsymbol{\beta}) - R(\beta_0, \beta_1, \beta_2, \beta_3) \\
&= [R(\beta_0) + R(\beta_1|\beta_0) + R(\beta_2|\beta_0, \beta_1) \\
&\quad + R(\beta_3|\beta_0, \beta_1, \beta_2) + R(\beta_4|\beta_0, \beta_1, \beta_2, \beta_3) \\
&\quad + R(\beta_5|\beta_0, \beta_1, \beta_2, \beta_3, \beta_4)] \\
&\quad - [R(\beta_0) + R(\beta_1|\beta_0) \\
&\quad + R(\beta_2|\beta_0, \beta_1) + R(\beta_3|\beta_0, \beta_1, \beta_2)] \\
&= R(\beta_4|\beta_0, \beta_1, \beta_2, \beta_3) + R(\beta_5|\beta_0, \beta_1, \beta_2, \beta_3, \beta_4) \\
&= 0.263496 + 4.352193 \\
&= 4.6157
\end{aligned}$$

The desired F ratio is

$$F = \frac{(4.6157)/2}{0.4948} = 4.6642$$

The null hypothesis can be rejected with $P \doteq 0.025$ (critical point, 4.69). The implication is that the information on notch to wing length and width in inches is useful *in addition* to the other available information.

TABLE 4.7 **ANOVA for squid data of Example 4.3.1. The null hypothesis $\beta = 0$ is rejected**

Source of Variation	Sum of Squares	Degrees of Freedom	Mean Square	F Ratio
Regression	595.164	6	99.194	200.4729
Residual	7.917	16	0.4948	
Total	603.081	22		

4.4

AN ALTERNATIVE APPROACH TO HYPOTHESIS TESTS ON SUBVECTORS OF β

In Section 4.2 the statistic

$$\frac{R(\gamma_1|\gamma_2)/r}{SS_{\text{Res}}/(n-p)} = F_{r,n-p}$$

was developed for testing $H_0{:}\gamma_1 = \mathbf{0}$. However, as was admitted at the time, the argument that the test is right-tailed is only an intuitive one. In this section an alternative method for computing $R(\gamma_1|\gamma_2)$ is derived. In the alternative form, it will be evident that the F test developed is, in fact, right-tailed. The derivation of the alternative is based on the ability to write $X'X$ and its inverse in partitioned form.

In particular, we are interested in using Theorem 2.1.3 to pinpoint the form of a crucial portion of the matrix $(X'X)^{-1}$. The proof of Theorem 4.4.1 is left as an exercise (see Exercise 18).

Theorem 4.4.1 Let X be $n \times p$ of rank p expressed in partitioned form as

$$X = [X_1 \mid X_2]$$

where X_1 is $n \times r$ of rank r and X_2 is $n \times (p-r)$ of rank $p-r$. Then $X'X$ can be expressed as

$$X'X = \left[\begin{array}{c|c} X_1'X_1 & X_1'X_2 \\ \hline X_2'X_1 & X_2'X_2 \end{array}\right]$$

Furthermore, if $(X'X)^{-1}$ is expressed as

$$(X'X)^{-1} = \left[\begin{array}{c|c} A_{11} & A_{12} \\ \hline A_{21} & A_{22} \end{array}\right]$$

then

$$A_{11}^{-1} = X_1'X_1 - X_1'X_2(X_2'X_2)^{-1}X_2'X_1$$ ■

With the help of this theorem, an alternative method for writing $R(\gamma_1|\gamma_2)$ can be derived. Before attempting to follow the derivation presented in Theorem 4.4.2, you should review the algebraic material in Exercise 17.

Theorem 4.4.2 Let X be $n \times p$ of rank p expressed in partitioned form as

$$X = [X_1 \mid X_2]$$

where X_1 is $n \times r$ of rank r and X_2 is $n \times (p-r)$ of rank $p-r$. Let $\boldsymbol{\beta}$ be partitioned as

$$\boldsymbol{\beta} = \left[\begin{array}{c} \boldsymbol{\gamma}_1 \\ \hline \boldsymbol{\gamma}_2 \end{array}\right]$$

where $\boldsymbol{\gamma}_1$ is an $r \times 1$ vector and $\boldsymbol{\gamma}_2$ is a $(p-r) \times 1$ vector. Then

$$R(\boldsymbol{\gamma}_1|\boldsymbol{\gamma}_2) = \hat{\boldsymbol{\gamma}}_1' A_{11}^{-1} \hat{\boldsymbol{\gamma}}_1$$

where $\hat{\boldsymbol{\gamma}}_1$ is the least squares estimator for $\boldsymbol{\gamma}_1$ and

$$A_{11}^{-1} = X_1'X_1 - X_1'X_2(X_2'X_2)^{-1}X_2'X_1$$

Proof

We know that the least squares estimator for $\boldsymbol{\beta}$ is given by

$$\hat{\boldsymbol{\beta}} = \left[\begin{array}{c} \hat{\boldsymbol{\gamma}}_1 \\ \hline \hat{\boldsymbol{\gamma}}_2 \end{array}\right] = (X'X)^{-1}X'\mathbf{y}$$

Writing $(X'X)^{-1}$ in partitioned form and expanding,

$$\hat{\boldsymbol{\beta}} = \begin{bmatrix} A_{11} & A_{12} \\ A_{21} & A_{22} \end{bmatrix} \begin{bmatrix} X'_1 \\ X'_2 \end{bmatrix} \mathbf{y} = \begin{bmatrix} A_{11}X'_1\mathbf{y} + A_{12}X'_2\mathbf{y} \\ A_{21}X'_1\mathbf{y} + A_{22}X'_2\mathbf{y} \end{bmatrix}$$

From this, it is easy to see that

$$\hat{\boldsymbol{\gamma}}_1 = A_{11}X'_1\mathbf{y} + A_{12}X'_2\mathbf{y}$$

Since

$$X'[X(X'X)^{-1}X'] = X'$$

it can be concluded that

$$\begin{bmatrix} X'_1 \\ X'_2 \end{bmatrix} X(X'X)^{-1}X' = \begin{bmatrix} X'_1 \\ X'_2 \end{bmatrix}$$

which implies that

$$X'_1X(X'X)^{-1}X' = X'_1$$

and that

$$X_1 = X(X'X)^{-1}X'X_1$$

Now consider the expression $\hat{\boldsymbol{\gamma}}'_1 A_{11}^{-1}\hat{\boldsymbol{\gamma}}_1$. Using the above identities and Theorem 4.4.1,

$$\begin{aligned}
\hat{\boldsymbol{\gamma}}'_1 A_{11}^{-1}\hat{\boldsymbol{\gamma}}_1 &= \hat{\boldsymbol{\gamma}}'_1[X'_1X_1 - X'_1X_2(X'_2X_2)^{-1}X'_2X_1]\hat{\boldsymbol{\gamma}}_1 \\
&= \hat{\boldsymbol{\gamma}}'_1X'_1[X_1 - X_2(X'_2X_2)^{-1}X'_2X_1]\hat{\boldsymbol{\gamma}}_1 \\
&= \hat{\boldsymbol{\gamma}}'_1X'_1[X(X'X)^{-1}X'X_1 - X_2(X'_2X_2)^{-1}X'_2X_1]\hat{\boldsymbol{\gamma}}_1 \\
&= \hat{\boldsymbol{\gamma}}'_1X'_1[X(X'X)^{-1}X' - X_2(X'_2X_2)^{-1}X'_2]X_1\hat{\boldsymbol{\gamma}}_1
\end{aligned}$$

To show that the expression on the right is equal to $R(\boldsymbol{\gamma}_1|\boldsymbol{\gamma}_2)$, we need only show that $\hat{\boldsymbol{\gamma}}'_1X'_1 = \mathbf{y}'$. To do so, note that since $X'X$ is symmetric, so is $(X'X)^{-1}$. The implications of this are that

$$A'_{11} = A_{11}, \quad A'_{22} = A_{22}, \quad \text{and} \quad A'_{12} = A_{21}$$

Now

$$\begin{aligned}
\hat{\boldsymbol{\gamma}}'_1X'_1 &= [A_{11}X'_1\mathbf{y} + A_{12}X'_2\mathbf{y}]'X'_1 \\
&= [(A_{11}X'_1 + A_{12}X'_2)\mathbf{y}]'X'_1 \\
&= \mathbf{y}'[X_1A'_{11} + X_2A'_{12}]X'_1 \\
&= \mathbf{y}'[X_1A_{11} + X_2A_{21}]X'_1
\end{aligned}$$

Since $X'X$ and $(X'X)^{-1}$ are inverses of one another, Theorem 4.4.1 can be applied again to conclude that

$$\begin{bmatrix} X_1'X_1 & X_1'X_2 \\ X_2'X_1 & X_2'X_2 \end{bmatrix} \begin{bmatrix} A_{11} & A_{12} \\ A_{21} & A_{22} \end{bmatrix} = I_p$$

This, in turn, implies that

$$X_1'X_1A_{11} + X_1'X_2A_{21} = I_r$$

Post multiplication by X_1' yields

$$\begin{aligned} X_1' &= X_1'X_1A_{11}X_1' + X_1'X_2A_{21}X_1' \\ &= X_1'[X_1A_{11} + X_2A_{21}]X_1' \end{aligned}$$

implying that

$$[X_1A_{11} + X_2A_{21}]X_1' = I_n$$

Substituting, we see that

$$\begin{aligned} \hat{\boldsymbol{\gamma}}_1 X_1' &= \mathbf{y}'[X_1A_{11} + X_2A_{21}]X_1' \\ &= \mathbf{y}'I_n = \mathbf{y}' \end{aligned}$$

and the proof is complete. ■

Let us now reconsider the F statistic used to test $H_0{:}\boldsymbol{\gamma}_1 = \mathbf{0}$. This statistic can be expressed as

$$F_{r,n-p} = \frac{R(\boldsymbol{\gamma}_1|\boldsymbol{\gamma}_2)/r}{SS_{\text{Res}}/(n-p)} = \frac{(\hat{\boldsymbol{\gamma}}_1'A_{11}^{-1}\hat{\boldsymbol{\gamma}}_1)/r}{SS_{\text{Res}}/(n-p)}$$

It is known that $E[SS_{\text{Res}}/(n-p)] = E[s^2] = \sigma^2$. The least squares estimator for $\boldsymbol{\gamma}_1$, $\hat{\boldsymbol{\gamma}}_1$, is known to be normally distributed with mean $\boldsymbol{\gamma}_1$ and variance $A_{11}\sigma^2$. Applying Theorem 2.2.1,

$$\begin{aligned} E[\hat{\boldsymbol{\gamma}}_1'A_{11}^{-1}\hat{\boldsymbol{\gamma}}_1)/r] &= (1/r)[\text{tr}\,(A_{11}^{-1}A_{11}\sigma^2 + \boldsymbol{\gamma}_1'A_{11}^{-1}\boldsymbol{\gamma}_1)] \\ &= (1/r)[\text{tr}\,(I_r\sigma^2 + \boldsymbol{\gamma}_1'A_{11}^{-1}\boldsymbol{\gamma}_1)] \\ &= (1/r)[r\sigma^2 + \boldsymbol{\gamma}_1'A_{11}^{-1}\boldsymbol{\gamma}_1] \\ &= \sigma^2 + (1/r)\boldsymbol{\gamma}_1'A_{11}^{-1}\boldsymbol{\gamma}_1 \end{aligned}$$

It is easy to see that if H_0 is true, $E[(\hat{\boldsymbol{\gamma}}_1'A_{11}^{-1}\hat{\boldsymbol{\gamma}}_1)/r] = \sigma^2$. Thus if H_0 is true, the F ratio should assume a value near 1. Since A_{11} is a principal minor of the positive definite matrix $(X'X)^{-1}$, A_{11} is positive definite, as is its inverse. By definition,

$$\boldsymbol{\gamma}_1'A_{11}^{-1}\boldsymbol{\gamma}_1 > 0$$

for $\boldsymbol{\gamma}_1 \neq \mathbf{0}$. Hence if H_0 is not true, the numerator of the F statistic should be larger than σ^2, producing an F ratio that exceeds 1. Logic dictates that $H_0{:}\boldsymbol{\gamma}_1 = \mathbf{0}$ should be rejected for large values of the test statistic. That is, the F test developed for testing the null hypothesis that a subvector of zero is a right-tailed test as claimed.

t TESTS ON β_j

It is a special case when $\boldsymbol{\gamma}_1$ is a single parameter. Here we are testing

$$H_0:\beta_j=0 \quad \text{versus} \quad H_1:\beta_j\neq 0$$

As you know, this hypothesis can be tested via a partial F test of the form

$$F_{1,n-p}=\frac{R(\beta_j|\beta_0,\beta_1,\ldots,\beta_{j-1},\beta_{j+1},\ldots,\beta_k)}{SS_{\text{Res}}/(n-p)}$$

We now know that the numerator of this statistic can be expressed as $\hat{\boldsymbol{\gamma}}_1' A_{11}^{-1}\hat{\boldsymbol{\gamma}}_1$. In this case, $\hat{\boldsymbol{\gamma}}_1=b_j$, and A_{11} is a real number. In particular, $A_{11}=c_{jj}$, the jth diagonal element of $(X'X)^{-1}$. Substituting,

$$R(\beta_j|\beta_0,\beta_1,\ldots,\beta_{j-1},\beta_{j+1},\ldots,\beta_k)=b_j^2/c_{jj}$$

and the F statistic assumes the form

$$F_{1,n-p}=\frac{b_j^2}{c_{jj}SS_{\text{Res}}/(n-p)}=\frac{b_j^2}{c_{jj}s^2}$$

This should look familiar. It is the square of the t_{n-p} random variable used in Chapter 3 to develop a confidence interval on β_j. This should not be surprising since it is easy to show that in general the square of a t random variable follows an F distribution (see Exercise 19). As a result of this relationship, the null hypothesis that β_j has value 0 can be tested via a t test or a partial F test. Both of these tests are included as standard output of many statistical software packages. The t test has an advantage over the partial F test in that it can be interpreted in a directional sense; that is, the algebraic sign of the t statistic is indicative of the sign of β_j. The F ratio is always positive and gives an indication only of whether or not β_j differs from 0. These t tests must be interpreted with care. Remember that they indicate the need to include β_j in a model that already contains *all* the other parameters.

4.5

AN ALTERNATIVE APPROACH: THE NONCENTRALITY PARAMETER (ADVANCED)

In Section 4.4 it was argued that the F test used to test $H_0:\boldsymbol{\gamma}_1=\mathbf{0}$ is a right-tailed test. This was done by showing that if H_0 is true, both numerator and denominator of the test statistic are estimators of σ^2; otherwise, the numerator estimates a quantity that exceeds σ^2. Ideally then an F ratio near 1 suggests that H_0 is true; an F ratio somewhat larger than 1 is an indication that H_0 is probably not true. An alternative argument can be given to justify the fact that a right-tailed F test is appropriate. This argument is based on examining the role of the noncentrality parameter and its relationship to the test statistic already derived.

Recall that, in this case, the noncentrality parameter λ involved is that associated with the numerator of the ratio

$$F = \frac{R(\boldsymbol{\gamma}_1|\boldsymbol{\gamma}_2)/r}{SS_{\text{Res}}/(n-p)} = \frac{(\hat{\boldsymbol{\gamma}}_1' A_{11}^{-1}\hat{\boldsymbol{\gamma}}_1)/r}{SS_{\text{Res}}/(n-p)}$$

This nonnegative parameter has value 0 if H_0 is true; otherwise it is greater than 0. Logic dictates that if an estimator for λ can be found then this estimator can be used to help distinguish between H_0 and H_1 with large values of the estimator indicating that H_0 is probably not true.

We have already shown that

$$\lambda = \frac{1}{2\sigma^2}(X\boldsymbol{\beta})'[X(X'X)^{-1}X' - X_2(X_2'X_2)^{-1}X_2'](X\boldsymbol{\beta})$$

This expression can be rewritten using the matrix results derived in the last section,

$$X_1' = X_1'X(X'X)^{-1}X'$$
$$X_2' = X_2'X(X'X)^{-1}X'$$
$$A_{11}^{-1} = X_1'X_1 - X_1'X_2(X_2'X_2)^{-1}X_2'X_1$$

as follows:

$$\begin{aligned}
\lambda &= \frac{1}{2\sigma^2}[X_1\boldsymbol{\gamma}_1 + X_2\boldsymbol{\gamma}_2]'[X(X'X)^{-1}X' - X_2(X_2'X_2)^{-1}X_2'][X_1\boldsymbol{\gamma}_1 + X_2\boldsymbol{\gamma}_2]\\
&= \frac{1}{2\sigma^2}[\boldsymbol{\gamma}_1'X_1' + \boldsymbol{\gamma}_2'X_2'][X(X'X)^{-1}X' - X_2(X_2'X_2)^{-1}X_2'][X_1\boldsymbol{\gamma}_1 + X_2\boldsymbol{\gamma}_2]\\
&= \frac{1}{2\sigma^2}[\boldsymbol{\gamma}_1'X_1' - \boldsymbol{\gamma}_1'X_1'X_2(X_2'X_2)^{-1}X_2' + \boldsymbol{\gamma}_2'X_2' - \boldsymbol{\gamma}_2'X_2'][X_1\boldsymbol{\gamma}_1 + X_2\boldsymbol{\gamma}_2]\\
&= \frac{1}{2\sigma^2}[\boldsymbol{\gamma}_1'X_1'X_1\boldsymbol{\gamma}_1 - \boldsymbol{\gamma}_1'X_1'X_2(X_2'X_2)^{-1}X_2'X_1\boldsymbol{\gamma}_1 + \boldsymbol{\gamma}_1'X_1'X_2\boldsymbol{\gamma}_2 - \boldsymbol{\gamma}_1'X_1'X_2\boldsymbol{\gamma}_2]\\
&= \frac{1}{2\sigma^2}\boldsymbol{\gamma}_1'[X_1'X_1 - X_1'X_2(X_2'X_2)^{-1}X_1]\boldsymbol{\gamma}_1\\
&= \frac{1}{2\sigma^2}\boldsymbol{\gamma}_1'A_{11}^{-1}\boldsymbol{\gamma}_1
\end{aligned}$$

To estimate this parameter, replace $\boldsymbol{\gamma}_1$ by its estimator, $\hat{\boldsymbol{\gamma}}_1$, and σ^2 by its unbiased estimator, $SS_{\text{Res}}/(n-p)$, to obtain

$$\lambda = \frac{1}{2}\frac{\hat{\boldsymbol{\gamma}}_1'A_{11}^{-1}\hat{\boldsymbol{\gamma}}_1}{SS_{\text{Res}}/(n-p)}$$

Although this estimator is not exactly the F ratio derived in Section 4.4, they are functionally related. In particular, it is easy to see that

$$\lambda = \tfrac{1}{2} rF$$

Since r is constant, large values of the F ratio result in a *large* estimated noncentrality parameter. This, in turn, tends to support the contention that H_0 is not true. Thus H_0 is rejected for values of F that are considered to be too large to have occurred by chance. The test is right-tailed as claimed.

4.6

A MORE GENERAL HYPOTHESIS

Thus far we have seen how to test the null hypotheses that $\boldsymbol{\beta} = \mathbf{0}$ or that $\boldsymbol{\gamma}_1 = \mathbf{0}$. Suppose that it is hypothesized that one or the other of these vectors assumes some specified *nonzero* value. How can such a hypothesis be tested? To begin, let us derive a method for testing

$$H_0\colon \boldsymbol{\beta} = \boldsymbol{\beta}^* \quad \text{versus} \quad H_1\colon \boldsymbol{\beta} \neq \boldsymbol{\beta}^*$$

where $\boldsymbol{\beta}^* \neq \mathbf{0}$. The logical place to start is by considering $\mathbf{b}$, the least squares estimator for $\boldsymbol{\beta}$. It is known that under the usual model assumptions $\mathbf{b}$ is normally distributed with mean $\boldsymbol{\beta}$ and variance $(X'X)^{-1}\sigma^2$. Define the vector $\mathbf{z}$ by

$$\mathbf{z} = \mathbf{b} - \boldsymbol{\beta}^*$$

Using the rules for expectation and variance, it is easy to see that $\mathbf{z}$ is normally distributed with mean $\boldsymbol{\mu} = \boldsymbol{\beta} - \boldsymbol{\beta}^*$ and variance $V = (X'X)^{-1}\sigma^2$. Now consider the quadratic form

$$\mathbf{z}'A\mathbf{z} = \frac{(\mathbf{b} - \boldsymbol{\beta}^*)'X'X(\mathbf{b} - \boldsymbol{\beta}^*)}{\sigma^2}$$

Note that $A = (1/\sigma^2)X'X$ is a $p \times p$ symmetric matrix and that

$$AV = (1/\sigma^2)(X'X)(X'X)^{-1}\sigma^2 = I_p$$

Since I_p is idempotent of rank p, Theorem 2.3.3 can be applied to conclude that $\mathbf{z}'A\mathbf{z}$ follows a noncentral chi-squared distribution with p degrees of freedom and noncentrality parameter

$$\lambda = (1/2\sigma^2)(\boldsymbol{\beta} - \boldsymbol{\beta}^*)'(X'X)(\boldsymbol{\beta} - \boldsymbol{\beta}^*)$$

It is easy to see that if $\boldsymbol{\beta} = \boldsymbol{\beta}^*$ then $\lambda = 0$ and $\mathbf{z}'A\mathbf{z}$ follows a chi-squared distribution with p degrees of freedom.

From previous patterns it can be seen that the ratio

$$\frac{(\mathbf{b} - \boldsymbol{\beta}^*)'X'X(\mathbf{b} - \boldsymbol{\beta}^*)/p}{SS_{\mathrm{Res}}/(n-p)}$$

should serve as a test statistic for testing $H_0{:}\boldsymbol{\beta} = \boldsymbol{\beta}^*$. It is known that the expected value of the denominator is σ^2. The expected value of the numerator is given by

$$\begin{aligned}E[(\mathbf{b}-\boldsymbol{\beta}^*)'X'X(\mathbf{b}-\boldsymbol{\beta}^*)/p] &= (1/p)[\operatorname{tr}(X'X)(X'X)^{-1}\sigma^2 + (\boldsymbol{\beta}-\boldsymbol{\beta}^*)'X'X(\boldsymbol{\beta}-\boldsymbol{\beta}^*)]\\ &= (1/p)[p\sigma^2 + (\boldsymbol{\beta}-\boldsymbol{\beta}^*)'X'X(\boldsymbol{\beta}-\boldsymbol{\beta}^*)]\\ &= \sigma^2 + (\boldsymbol{\beta}-\boldsymbol{\beta}^*)'X'X(\boldsymbol{\beta}-\boldsymbol{\beta}^*)/p\end{aligned}$$

Since $X'X$ is positive definite, we can conclude that the expected value of the numerator is σ^2 when H_0 is true and that it exceeds σ^2 otherwise. Thus, the proposed test statistic ideally should assume a value near 1 when the null hypothesis is true; values much larger than 1 indicate that the null hypothesis is probably not true. The only question left to answer is, What is the distribution of this ratio? Is it an F ratio as other similar ones have been? It has already been shown that if $\boldsymbol{\beta} = \boldsymbol{\beta}^*$, then the above ratio is the quotient of two chi-squared random variables each divided by their respective degrees of freedom. To show that the statistic does indeed follow an $F_{p,n-p}$ distribution, we must show only that the numerator and denominator are independent. To do so requires a bit of a trick. In particular, we will write both the numerator and denominator as quadratic forms in $\mathbf{q}$ where

$$\mathbf{q} = \mathbf{y} - X\boldsymbol{\beta}^*$$

We will then apply Theorem 2.4.1 to show that the resulting quadratic forms are independent, thus completing the argument.

It is not difficult to verify that

$$(\mathbf{b}-\boldsymbol{\beta}^*)'X'X(\mathbf{b}-\boldsymbol{\beta}^*) = \mathbf{q}'X(X'X)^{-1}X'\mathbf{q} = \mathbf{q}'B\mathbf{q}$$

and that

$$SS_{\text{Res}} = \mathbf{q}'[I - X(X'X)^{-1}X']\mathbf{q} = \mathbf{q}'C\mathbf{q}$$

The verification involves straightforward matrix manipulations and is left as an exercise (see Exercises 21 and 22). Since the variance of $\mathbf{q}$ is the same as the variance of $\mathbf{y}$, namely $V = \sigma^2 I$, it is easy to see that

$$BVC = X(X'X)^{-1}X'(\sigma^2 I)[I - X(X'X)^{-1}X'] = 0$$

By Theorem 2.4.1, $\mathbf{q}'B\mathbf{q}$ and $\mathbf{q}'C\mathbf{q}$ are independent, and the statistic

$$\frac{(\mathbf{b}-\boldsymbol{\beta}^*)'X'X(\mathbf{b}-\boldsymbol{\beta}^*)/p}{SS_{\text{Res}}/(n-p)}$$

follows an F distribution with p and $n-p$ degrees of freedom as claimed.

Although we will not attempt the proof here, it can be shown that the appropriate statistic for testing $H_0{:}\boldsymbol{\gamma}_1 = \boldsymbol{\gamma}_1^*$ where $\boldsymbol{\gamma}_1^*$ is an $r \times 1$ vector not necessarily equal to $\mathbf{0}$ is

$$F_{r,n-p} = \frac{(\hat{\boldsymbol{\gamma}}_1 - \boldsymbol{\gamma}_1^*)'A_{11}^{-1}(\hat{\boldsymbol{\gamma}}_1 - \boldsymbol{\gamma}_1^*)/r}{SS_{\text{Res}}/(n-p)}$$

where

$$A_{11}^{-1} = X_1'X_1 - X_1'X_2(X_2'X_2)^{-1}X_2'X_1$$

4.7

THE GENERAL LINEAR HYPOTHESIS

In this section we consider the *general linear hypothesis.* As mentioned in the introduction, this hypothesis includes those presented earlier as special cases. The hypothesis assumes the form

$$H_0: C\boldsymbol{\beta} = \boldsymbol{\delta}^* \qquad H_1: C\boldsymbol{\beta} \neq \boldsymbol{\delta}^*$$

or

$$H_0: C\boldsymbol{\beta} - \boldsymbol{\delta}^* = \mathbf{0} \quad H_1: C\boldsymbol{\beta} - \boldsymbol{\delta}^* \neq \mathbf{0}$$

where C is an $r \times p$ matrix of rank $r \leqslant p$ and $\boldsymbol{\delta}^*$ is an $r \times 1$ vector. This hypothesis makes it possible to test for functional relationships among parameters as well as to test the standard hypotheses already discussed. Example 4.7.1 illustrates the idea.

Example 4.7.1 Consider the model

$$y_i = \beta_0 + \beta_1 x_{i1} + \beta_2 x_{i2} + \beta_3 x_{i3} + \varepsilon_i$$

Let

$$C = \begin{bmatrix} 0 & 1 & -1 & 0 \\ 0 & 0 & 1 & -1 \end{bmatrix}$$

Note that C is 2×4 of rank $3 \leqslant p$. Let

$$\boldsymbol{\delta}^* = \begin{bmatrix} 0 \\ 0 \end{bmatrix}$$

Suppose we test

$$H_0: C\boldsymbol{\beta} = \boldsymbol{\delta}^*$$

In order to determine what is really being tested, write the expression in matrix form and expand it to obtain the system

$$\beta_1 - \beta_2 = 0$$
$$\beta_2 - \beta_3 = 0$$

Solving this system, it is easy to see that the given hypothesis is equivalent to the null hypothesis that $\beta_1 = \beta_2 = \beta_3$. No restrictions are made concerning β_0.

The development of an appropriate test statistic for testing the general linear hypothesis follows the same lines used to develop the previous F ratios. The logical place to begin is with the statistic $C\mathbf{b} - \boldsymbol{\delta}^*$, the least squares estimator for $C\boldsymbol{\beta} - \boldsymbol{\delta}^*$. Since $\mathbf{b}$ is normally distributed with mean $\boldsymbol{\beta}$ and variance $(X'X)^{-1}\sigma^2$, the rules for expectation and variance can be applied to see that $C\mathbf{b} - \boldsymbol{\delta}^*$ is normally distributed with mean $\boldsymbol{\mu} = C\boldsymbol{\beta} - \boldsymbol{\delta}^*$ and variance $V = C(X'X)^{-1}\sigma^2 C'$. By Corollary 2.3.4, the quadratic form

$$\frac{(C\mathbf{b} - \boldsymbol{\delta}^*)'[C(X'X)^{-1}C']^{-1}(C\mathbf{b} - \boldsymbol{\delta}^*)}{\sigma^2}$$

follows a noncentral chi-squared distribution with r degrees of freedom and noncentrality parameter

$$= \frac{(\frac{1}{2})(C\boldsymbol{\beta} - \boldsymbol{\delta}^*)'[C(X'X)^{-1}C']^{-1}(C\boldsymbol{\beta} - \boldsymbol{\delta}^*)}{\sigma^2}$$

Note that if $H_0{:}C\boldsymbol{\beta} - \boldsymbol{\delta}^* = \mathbf{0}$ is true, then $\lambda = 0$ and the given quadratic form follows a chi-squared distribution with r degrees of freedom.

Following previous patterns, the ratio

$$\frac{(C\mathbf{b} - \boldsymbol{\delta}^*)'[C(X'X)^{-1}C']^{-1}(C\mathbf{b} - \boldsymbol{\delta}^*)/r}{SS_{\text{Res}}/(n-p)}$$

is a logical choice for a test statistic for testing the general linear hypothesis. It is left as an exercise (Exercise 23) to show that

$$E[(C\mathbf{b} - \boldsymbol{\delta}^*)'[C(X'X)^{-1}C']^{-1}(C\mathbf{b} - \boldsymbol{\delta}^*)/r] = \sigma^2 + (C\boldsymbol{\beta} - \boldsymbol{\delta}^*)'[C(X'X)^{-1}C']^{-1}(C\boldsymbol{\beta} - \boldsymbol{\delta}^*)/r$$

It is easy to see that if H_0 is true this expectation reduces to σ^2. Hence a ratio close to 1 is certainly evidence that the null hypothesis is true. It is not so easy to show that values that are much larger than 1 probably indicate that the null hypothesis is false. To do so, it is necessary to show that the matrix $[C(X'X)^{-1}C']^{-1}$ is positive definite. To do this it is sufficient to show that $C(X'X)^{-1}C'$ is positive definite. This is done in Lemma 4.7.1.

Lemma 4.7.1 Let C be an $r \times p$ matrix of rank $r \leqslant p$ and let X be an $n \times p$ full rank matrix of rank p. Then $C(X'X)^{-1}C'$ is positive definite.

Proof

By Theorem 4.1.4, $X'X$ is positive definite. Applying Exercise 17c, $(X'X)^{-1}$ is also positive definite. In Exercise 64 in Chapter 3 it was shown that there exists a $p \times p$ nonsingular matrix P such that

$$(X'X)^{-1} = PP'$$

Hence

$$C(X'X)^{-1}C' = CPP'C' = Z'Z$$

where $Z = P'C'$. Since the rank of a matrix is unaltered by multiplication by a nonsingular matrix,

$$r(Z) = r(P'C') = r(C') = r(C) = r$$

Note that Z is a $p \times r$ matrix of full rank. By Theorem 4.1.4, $Z'Z$ is positive definite as claimed. ■

It should now be evident that the proposed test statistic behaves in the same way as the F ratios developed in Sections 4.1, 4.2, and 4.4. Ideally, its values should be near 1 when H_0 is true; it should exceed 1 substantially otherwise. The only question to be answered is, Does the ratio follow an F distribution when H_0 is true? That is, are the chi-squared random variables in the numerator and denominator of the ratio independent as required? As you have probably noticed, independence arguments are not always simple! Unfortunately, that is the case here and no formal proof of independence will be attempted. However, independence is not too hard to see intuitively. Recall that it has already been shown that $\mathbf{b}$ and s^2 are independent. The numerator of the proposed test statistic is a function only of $\mathbf{b}$ and the denominator is s^2. Hence it is logical to conclude that the numerator and denominator are in fact independent. Thus, the test statistic used to test the general linear hypothesis is

$$F_{r,n-p} = \frac{(C\mathbf{b} - \boldsymbol{\delta}^*)'[C(X'X)^{-1}C']^{-1}(C\mathbf{b} - \boldsymbol{\delta}^*)/r}{SS_{\text{Res}}/(n-p)}$$

It was claimed earlier that the hypotheses considered in Sections 4.1 and 4.2 can be thought of as special cases of the general linear hypothesis. Example 4.7.2 illustrates the idea.

Example 4.7.2 Consider the null hypothesis $H_0\!:\boldsymbol{\beta} = \mathbf{0}$. To express this hypothesis in the form $C\boldsymbol{\beta} - \boldsymbol{\delta}^* = \mathbf{0}$, let C be the $p \times p$ identity matrix and let $\boldsymbol{\delta}^*$ be a zero vector. Then the proper F ratio for testing H_0 is

$$F_{p,n-p} = \frac{\mathbf{b}'(X'X)\mathbf{b}/p}{SS_{\text{Res}}/(n-p)} = \frac{[(X'X)^{-1}X'\mathbf{y}]'X'X[(X'X)^{-1}X'\mathbf{y}]/p}{SS_{\text{Res}}/(n-p)}$$

$$= \frac{\mathbf{y}'X(X'X)^{-1}X'\mathbf{y}/p}{SS_{\text{Res}}/(n-p)}$$

$$= \frac{SS_{\text{Reg}}/p}{SS_{\text{Res}}/(n-p)}$$

Notice that this is the same statistic as that derived in Section 4.1.

We leave it to you to show that the other hypotheses can be expressed in general linear form (see Exercise 24).

4.8

A SPECIAL CASE: ORTHOGONALITY

Consider the model

$$\mathbf{y} = X\boldsymbol{\beta} + \boldsymbol{\varepsilon}$$

with X and $\boldsymbol{\beta}$ partitioned as

$$X = [X_1 \mid X_2] \quad \text{and} \quad \boldsymbol{\beta} = \begin{bmatrix} \boldsymbol{\gamma}_1 \\ \hline \boldsymbol{\gamma}_2 \end{bmatrix}$$

where X_1 is an $n \times r$ matrix of rank r, X_2 is an $n \times (p-r)$ matrix of rank $p-r$, $\boldsymbol{\gamma}_1$ is an $r \times 1$ vector, and $\boldsymbol{\gamma}_2$ is an $n \times (p-r)$ vector. In Section 4.2 a method for testing hypotheses on subvectors of $\boldsymbol{\beta}$ was developed. The ANOVA is shown in Table 4.8.

TABLE 4.8 **ANOVA used to test $H_0:\boldsymbol{\gamma}_1 = \mathbf{0}$**

Source of Variation	Sum of Squares	Degrees of Freedom	Mean Square	F Ratio
Regression				
Full model	$R(\boldsymbol{\beta})$	p		$\frac{R(\boldsymbol{\beta})/p}{SS_{\text{Res}}/(n-p)}$
Reduced model	$R(\boldsymbol{\gamma}_2)$	$p-r$		
$\boldsymbol{\gamma}_1$ in presence of $\boldsymbol{\gamma}_2$	$R(\boldsymbol{\gamma}_1 \mid \boldsymbol{\gamma}_2)$	r		$\frac{R(\boldsymbol{\gamma}_1 \mid \boldsymbol{\gamma}_2)/r}{SS_{\text{Res}}/(n-p)}$
Residual	$\mathbf{y}'\mathbf{y} - R(\boldsymbol{\beta})$	$n-p$		
Total	$\mathbf{y}'\mathbf{y}$	n		

This analysis can be used to test $H_0:\boldsymbol{\gamma}_1 = \mathbf{0}$ via the F ratio

$$\frac{R(\boldsymbol{\gamma}_1 \mid \boldsymbol{\gamma}_2)/r}{SS_{\text{Res}}/(n-p)}$$

It can also be used to test $H_0:\boldsymbol{\beta} = \mathbf{0}$ via the ratio

$$\frac{R(\boldsymbol{\beta})/p}{SS_{\text{Res}}/(n-p)}$$

However, it does not provide a ratio that is appropriate for testing $H_0:\boldsymbol{\gamma}_2 = \mathbf{0}$. Using the technique developed earlier, it is known that this hypothesis is tested via the F ratio

$$\frac{R(\boldsymbol{\gamma}_2|\boldsymbol{\gamma}_1)/(p-r)}{SS_{\text{Res}}/(n-p)}$$

where

$$\begin{aligned} R(\boldsymbol{\gamma}_2|\boldsymbol{\gamma}_1) &= R(\boldsymbol{\beta}) - R(\boldsymbol{\gamma}_1) \\ &= \mathbf{y}'X(X'X)^{-1}X'\mathbf{y} - \mathbf{y}'X_1(X_1'X_1)^{-1}X_1'\mathbf{y} \end{aligned}$$

This sum of squares is not equal to $R(\boldsymbol{\gamma}_2)$ in general. Hence the regression sum of squares for the reduced model given in Table 4.8 cannot be used to test a null hypothesis concerning $\boldsymbol{\gamma}_2$ in all cases. However, if X_1 and X_2 are *orthogonal*, that is, if $X_1'X_2 = 0$, then it can be shown that $R(\boldsymbol{\gamma}_2|\boldsymbol{\gamma}_1) = R(\boldsymbol{\gamma}_2)$. Hence in this special case, $H_0:\boldsymbol{\gamma}_2 = \mathbf{0}$ can be tested by means of the ANOVA of Table 4.8. To show that this is true, first consider the form of $X'X$ and $(X'X)^{-1}$ when X_1 and X_2 are orthogonal. Note that

$$X'X = \begin{bmatrix} X_1' \\ \hline X_2' \end{bmatrix} [X_1 \mid X_2] = \left[\begin{array}{c|c} X_1'X_1 & X_1'X_2 \\ \hline X_2'X_1 & X_2'X_2 \end{array}\right] = \left[\begin{array}{c|c} X_1'X_1 & 0 \\ \hline 0 & X_2'X_2 \end{array}\right]$$

and

$$(X'X)^{-1} = \left[\begin{array}{c|c} (X_1'X_1)^{-1} & 0 \\ \hline 0 & (X_2'X_2)^{-1} \end{array}\right]$$

Substituting, we obtain

$$\begin{aligned} R(\boldsymbol{\gamma}_2|\boldsymbol{\gamma}_1) &= \mathbf{y}'X(X'X)^{-1}X'\mathbf{y} - \mathbf{y}'X_1(X_1'X_1)^{-1}X_1'\mathbf{y} \\ &= \mathbf{y}'[X_1 \mid X_2] \left[\begin{array}{c|c} (X_1'X_1)^{-1} & 0 \\ \hline 0 & (X_2'X_2)^{-1} \end{array}\right] \begin{bmatrix} X_1' \\ \hline X_2' \end{bmatrix} \mathbf{y} - \mathbf{y}'X_1(X_1'X_1)^{-1}X_1'\mathbf{y} \\ &= \mathbf{y}'[X_1(X_1'X_1)^{-1} + X_2(X_2'X_2)^{-1}] \begin{bmatrix} X_1' \\ \hline X_2' \end{bmatrix} \mathbf{y} - \mathbf{y}'X_1(X_1'X_1)^{-1}X_1'\mathbf{y} \\ &= \mathbf{y}'[X_1(X_1'X_1)^{-1}X_1' + X_2(X_2'X_2)^{-1}X_2']\mathbf{y} - \mathbf{y}'X_1(X_1'X_1)^{-1}X_1'\mathbf{y} \\ &= \mathbf{y}'X_2(X_2'X_2)^{-1}X_2'\mathbf{y} \\ &= R(\boldsymbol{\gamma}_2) \end{aligned}$$

From this it can be concluded that if X_1 and X_2 are orthogonal, then the null hypothesis that $\boldsymbol{\gamma}_2 = \mathbf{0}$ can be tested using the F ratio

$$F_{p-r,n-p} = \frac{R(\boldsymbol{\gamma}_2|\boldsymbol{\gamma}_1)/(p-r)}{SS_{\text{Res}}/(n-p)} = \frac{R(\boldsymbol{\gamma}_2)/(p-r)}{SS_{\text{Res}}/(n-p)}$$

The idea of orthogonality is illustrated in Example 4.8.1.

Example 4.8.1 Consider the model

$$y_i = \beta_0 + \beta_1 x_{i1} + \beta_2 x_{i2} + \cdots + \beta_k x_{ik} + \varepsilon_i$$

For this model,

$$X = \begin{bmatrix} 1 & x_{11} & x_{12} & \cdots & x_{1k} \\ 1 & x_{21} & x_{22} & & x_{2k} \\ \vdots & \vdots & \vdots & & \vdots \\ 1 & x_{n1} & x_{n2} & & x_{nk} \end{bmatrix}$$

If the columns of X are orthogonal (see Section 1.3) then the model is said to be *mutually orthogonal.* Suppose that we want to test a null hypothesis concerning a single parameter, say β_1. That is, we want to test

$$H_0:\beta_1 = 0 \quad \text{versus} \quad H_1:\beta_1 \neq 0$$

To do so, the X matrix is rewritten as

$$X = [X_1 \mid X_2] = \begin{bmatrix} x_{11} & 1 & x_{12} & x_{13} & \cdots & x_{1k} \\ x_{21} & 1 & x_{22} & x_{23} & & x_{2k} \\ \vdots & \vdots & \vdots & \vdots & & \vdots \\ x_{n1} & 1 & x_{n2} & x_{n3} & & x_{nk} \end{bmatrix}$$

Note that

$$X'_1 X_2 = [x_{11} \quad x_{21} \quad \cdots \quad x_{n1}] \begin{bmatrix} 1 & x_{12} & x_{13} & \cdots & x_{1k} \\ 1 & x_{22} & x_{23} & & x_{2k} \\ \vdots & \vdots & \vdots & & \vdots \\ 1 & x_{n2} & x_{n3} & & x_{nk} \end{bmatrix}$$

$$= \left[\sum_{i=1}^{n} x_{i1} \ \sum_{i=1}^{n} x_{i1} x_{i2} \ \sum_{i=1}^{n} x_{i1} x_{i3} \cdots \sum_{i=1}^{n} x_{i1} x_{ik} \right]$$

Since it is assumed that the columns of X are orthogonal, each entry in this vector is 0. The orthogonality assumption guarantees that $X'_1 X_2 = 0$. From this it can be concluded that

$$\begin{aligned} R(\beta_1 | \beta_0, \beta_2, \beta_3, \ldots \beta_k) &= R(\beta_1) \\ &= \mathbf{y}' X_1 (X'_1 X_1)^{-1} X'_1 \mathbf{y} \end{aligned}$$

A similar argument holds for any parameter β_j.

Example 4.8.2 Consider the model whose X matrix is

$$X = \begin{bmatrix} 1 & -1 & -1 & -1 \\ 1 & 1 & -1 & -1 \\ 1 & -1 & 1 & -1 \\ 1 & 1 & 1 & -1 \\ 1 & -1 & -1 & 1 \\ 1 & 1 & -1 & 1 \\ 1 & -1 & 1 & 1 \\ 1 & 1 & 1 & 1 \\ 1 & 0 & 0 & 0 \\ 1 & -2 & 0 & 0 \\ 1 & 2 & 0 & 0 \\ 1 & 0 & -2 & 0 \\ 1 & 0 & 2 & 0 \\ 1 & 0 & 0 & -2 \\ 1 & 0 & 0 & 2 \end{bmatrix} \begin{bmatrix} 57 \\ 40 \\ 19 \\ 40 \\ 54 \\ 41 \\ 21 \\ 43 \\ 63 \\ 28 \\ 11 \\ 2 \\ 18 \\ 56 \\ 46 \end{bmatrix}$$

Inspection will verify that the columns of X are orthogonal. Let us find the F ratio needed to test $H_0{:}\beta_1 = 0$. To do so, we must compute

$$R(\beta_1 | \beta_0, \beta_2, \beta_3) = R(\beta_1)$$

where

$$R(\beta_1) = \mathbf{y}' X_1 (X_1' X_1)^{-1} X_1' \mathbf{y}$$

In this case, $X_1 = \mathbf{x_1}$. Inspection shows that

$$X_1' X_1 = 16 \qquad (X_1' X_1)^{-1} = 1/16$$

and

$$\begin{aligned} X_1'\mathbf{y} &= (-57) + (40) + (-19) + (40) + (-54) + (41) + (-21) \\ &\quad + (43) + (-2)(28) + (2)(11) \\ &= -21 \end{aligned}$$

Hence

$$\begin{aligned} R(\beta_1 | \beta_0, \beta_2, \beta_3) &= (-21)^2/16 \\ &= 27.5625 \end{aligned}$$

The denominator of the required F ratio is the usual residual mean square for the full model, $SS_{\mathrm{Res}}/(n-p)$.

It is tempting to view the orthogonality concept as merely a computational simplification. In fact, it is considerably more. When an investigator has the ability to control the values of the regressors in a multiple linear regression setting, it becomes possible to make the columns of X mutually orthogonal. In the area of experimental design, such a model is referred to as an *orthogonal design.* The advantages of such designs are well-documented **[18]**, **[5]**, **[6]**. Recall that in Section 4.3 we saw that the importance of a regressor depends upon what other variables are already in the model. With an orthogonal design, the impact of each regressor can be measured without reference to the others. For this reason, there is certainly an intuitive appeal for orthogonality.

4.9

LIKELIHOOD RATIO CRITERION (ADVANCED)

The development of the test procedures introduced earlier in this chapter was based on the use of quadratic forms. Theorems dealing with distributions of quadratic forms and independence among them were prominent. For students who have been exposed to Neyman–Pearson concepts and, in particular, to the *likelihood ratio criterion,* it is of interest to determine whether these tests are indeed likelihood ratio tests. Recall that the likelihood ratio criterion is based on a ratio R of the form

$$R = \frac{L_\Omega}{L_\omega}$$

where L_ω is the maximum of the likelihood function with the parameters satisfying the restrictions set forth by the null hypothesis; L_Ω is the maximum of the likelihood function with the parameters unrestricted. It is evident that $L_\omega \leqslant L_\Omega$ and hence that $R \geqslant 1$. If L_ω lies close in value to L_Ω when these two functions are evaluated for a particular data set, then there is no reason to reject H_0; however, if they differ considerably, then it is suspected that H_0 is false. Thus, the likelihood ratio criterion requires rejection of H_0 for large values of R. A test based on R itself or a monotonic function of R is called a *likelihood ratio test.*

We begin by considering the most elementary of the hypotheses considered previously, namely

$$H_0:\boldsymbol{\beta} = \mathbf{0} \quad \text{versus} \quad H_1:\boldsymbol{\beta} \neq \mathbf{0}$$

This hypothesis is to be tested assuming, as usual, that $\boldsymbol{\varepsilon}$ is normally distributed with mean $\mathbf{0}$ and variance $\sigma^2 I$. In Section 3.4 the likelihood function was shown to be

$$\begin{aligned} L(\boldsymbol{\varepsilon}; \sigma^2) &= \left[\frac{1}{(2\pi)^{n/2}(\sigma^2)^{n/2}}\right] e^{(-1/2\sigma^2)\boldsymbol{\varepsilon}'\boldsymbol{\varepsilon}} \\ &= \left[\frac{1}{(2\pi)^{n/2}(\sigma^2)^{n/2}}\right] e^{(-1/2\sigma^2)[(\mathbf{y}-X\boldsymbol{\beta})'(\mathbf{y}-X\boldsymbol{\beta})]} \end{aligned}$$

To evaluate L_Ω for a particular data set, σ^2 and $\boldsymbol{\beta}$ are replaced by their maximum likelihood estimates. These were found earlier to be

$$\tilde{\boldsymbol{\beta}} = \mathbf{b} = (X'X)^{-1}X'\mathbf{y} \quad \text{and} \quad \tilde{\sigma}^2 = \frac{(\mathbf{y} - X\mathbf{b})'(\mathbf{y} - X\mathbf{b})}{n}$$

The likelihood function can now be expressed as a function of $\mathbf{y}$ as

$$\begin{aligned} L_\Omega(\mathbf{y}) &= \left[\frac{1}{(2\pi)^{n/2}(\tilde{\sigma}^2)^{n/2}}\right] e^{(-1/2\tilde{\sigma}^2)[(\mathbf{y}-X\mathbf{b})'(\mathbf{y}-X\mathbf{b})]} \\ &= \left[\frac{(n)^{n/2}}{(2\pi)^{n/2}[(\mathbf{y} - X\mathbf{b})'(\mathbf{y} - X\mathbf{b})]^{n/2}}\right] e^{-n/2} \end{aligned}$$

To evaluate L_ω, invoke the restriction that $\boldsymbol{\beta} = \mathbf{0}$ by setting $\mathbf{b} = \mathbf{0}$. In this case, the maximum likelihood estimate for σ^2 is $\tilde{\sigma}^2 = \mathbf{y}'\mathbf{y}/n$. Substituting,

$$\begin{aligned} L_\omega(\mathbf{y}) &= \frac{1}{(2\pi)^{n/2}\left(\dfrac{\mathbf{y}'\mathbf{y}}{n}\right)^{n/2}} \; e^{-\left(\frac{1}{2\mathbf{y}'\mathbf{y}/n}\mathbf{y}'\mathbf{y}\right)} \\ &= \left[\frac{(n)^{n/2}}{(2\pi)^{n/2}(\mathbf{y}'\mathbf{y})^{n/2}}\right] e^{-n/2} \end{aligned}$$

By definition,

$$R = \frac{L_\Omega(\mathbf{y})}{L_\omega(\mathbf{y})} = \left[\frac{\mathbf{y}'\mathbf{y}}{(\mathbf{y} - X\mathbf{b})'(\mathbf{y} - X\mathbf{b})}\right]^{n/2}$$

To show that the F test developed in Section 4.1 is a likelihood ratio test, it is sufficient to show that the F statistic used is a monotonic function of R. To do so, recall the basic sum of squares identity

$$\mathbf{y}'\mathbf{y} = \mathbf{y}'X(X'X)^{-1}X'\mathbf{y} + (\mathbf{y} - X\mathbf{b})'(\mathbf{y} - X\mathbf{b})$$

or

$$SS_{\text{Total}} = SS_{\text{Reg}} + SS_{\text{Res}}$$

Substituting,

$$\begin{aligned} R &= \left[\frac{SS_{\text{Reg}} + SS_{\text{Res}}}{SS_{\text{Res}}}\right]^{n/2} \\ &= \left[1 + \frac{SS_{\text{Reg}}}{SS_{\text{Res}}}\right]^{n/2} \\ &= \left[1 + \frac{pSS_{\text{Reg}}/p}{(n-p)SS_{\text{Res}}/(n-p)}\right]^{n/2} \end{aligned}$$

You should recognize

$$\frac{SS_{\text{Reg}}/p}{SS_{\text{Res}}/(n-p)}$$

as the F ratio used in Section 4.1 to test $H_0:\boldsymbol{\beta}=\mathbf{0}$. Hence it has been shown that

$$R=\left[1+\frac{p}{(n-p)}F\right]^{n/2}$$

This equation can be solved easily to see that

$$F=\frac{(n-p)}{p}[R^{2/n}-1]$$

Since the derivative of F with respect to R is

$$\left(\frac{n-p}{p}\right)\left(\frac{2}{n}\right)R^{2/n-1} \qquad R \geqslant 1$$

it can be seen that the derivative is always positive. This guarantees that F is a monotonic increasing function of R. Thus the rejection of H_0 on the basis of a large value of R is equivalent to the rejection of H_0 on the basis of a large value of F. The tests are equivalent, and the F test is a legitimate likelihood ratio test.

Using LaGrange multipliers, it can be shown that the F test used to test the general linear hypothesis

$$H_0:C\boldsymbol{\beta}=\boldsymbol{\delta}^* \quad \text{versus} \quad H_1:C\boldsymbol{\beta}\neq\boldsymbol{\delta}^*$$

is a likelihood ratio test. Since each test developed in this chapter is a special case of the general linear hypothesis, it can be concluded that each is a likelihood ratio test.

EXERCISES

Section 4.1

1. Show that under the conditions stated in Section 4.1, SS_{Res} and SS_{Reg} are independent. (*Hint*: See Theorem 2.4.1.)
2. Consider the data of Example 3.3.1. Construct the ANOVA table for these data and test $H_0:\boldsymbol{\beta}=\mathbf{0}$ versus $H_1:\boldsymbol{\beta}\neq\mathbf{0}$ (The simple linear regression model is useful in explaining variation in the field cracking rate.)
3. Consider the data of Example 3.7.1. Construct the ANOVA table for these data and test $H_0:\boldsymbol{\beta}=\mathbf{0}$ versus $H_1:\boldsymbol{\beta}\neq\mathbf{0}$ (The multiple regression model is useful in explaining the variation in the selling price of houses.)

Section 4.2

*4. A lemma from the field of linear algebra states that "A system of n homogeneous linear equations in p unknowns has a nontrivial solution if and only if the rank of the system is less than p" [11]. That is, if A is an $n \times p$ matrix, then the system

$$A\mathbf{x} = \mathbf{0}$$

has a nontrivial solution if and only if $r(A) < p$. Use an indirect proof to show the following:

Lemma: Let X be $n \times p$ of full rank where $n \geqslant p$. If $\boldsymbol{\beta} \neq \mathbf{0}$ then $X\boldsymbol{\beta} \neq \mathbf{0}$. ■

5. In writing $(X_2'X_2)^{-1}$ it has been assumed that this inverse exists. Show that this is true.

*6. Prove Lemma 4.2.1. (*Hint*: Recall the properties of the trace given in Section 1.5. Apply Theorem 1.5.2.)

*7. Prove Lemma 4.2.3. (*Hint*: Show that the trace of A is r. Apply Theorem 1.5.2.)

8. In Examples 4.1.1 and 4.2.1 assume that the researcher is fairly certain from practical considerations that both the intercept and x_1 should be in the model. However, the presence of x_2 and x_3 is in question. He wants to test $H_0{:}\boldsymbol{\gamma}_1 = \mathbf{0}$.
   a. What is $\boldsymbol{\gamma}_1$? $\boldsymbol{\gamma}_2$?
   b. Partition X appropriately.
   c. Use the data of Example 4.1.1 to test H_0. (*Hint*: If A is a 2×2 nonsingular matrix, then A^{-1} is given by

$$A^{-1} = \frac{1}{|A|}\begin{bmatrix} a_{22} & -a_{12} \\ -a_{21} & a_{11} \end{bmatrix}$$

   Are the variables x_2 and x_3 needed in a model that already includes x_1 and an intercept?)

*9. In courses in applied statistics, the presence of an intercept term is often assumed. Hence the first test that is conducted is usually one to distinguish between the full model and the reduced model that contains only β_0. Often in this case, the ANOVA table that is presented differs from the one developed in Table 4.5 (p. 159). In particular, it assumes the form given in Table 4.6 (on p. 160). The most obvious differences between the two tables appear in the Degrees of Freedom column. In parts a–c below you will justify Table 4.6 step by step.
   a. Define the "corrected" total sum of squares by

$$SS_{\text{Total(Corrected)}} = \mathbf{y}'\mathbf{y} - \mathbf{y}'[X_2(X_2'X_2)^{-1}X_2']\mathbf{y}$$

   where X_2 is a column vector of ones. Show that $SS_{\text{Total(Corrected)}}/\sigma^2$ follows a noncentral chi-squared distribution with $n-1$ degrees of freedom.
   b. Show that

$$\mathbf{y}'X_2(X_2'X_2)^{-1}X_2'\mathbf{y} = \left(\sum_{i=1}^{n} y_i\right)^2 \Big/ n$$

   and argue that

$$SS_{\text{Total(Corrected)}} = \sum_{i=1}^{n} y_i^2 - \frac{\left(\sum_{i=1}^{n} y_i\right)^2}{n}$$

c. Finally, argue that

$$SS_{\text{Total (Corrected)}} = R(\gamma_1 | \gamma_2) + SS_{\text{Res}}$$

where

$$\gamma_1 = \begin{bmatrix} \beta_1 \\ \beta_2 \\ \vdots \\ \beta_k \end{bmatrix} \quad \text{and} \quad \gamma_2 = [\beta_0]$$

d. Table 4.6 sometimes appears in expanded form. That is, an expanded formula for $R(\gamma_1 | \gamma_2)$ is given. Show that

$$R(\gamma_1 | \gamma_2) = \sum_{i=1}^{n} \hat{y}_i^2 - \frac{\left(\sum_{i=1}^{n} y_i\right)^2}{n}$$

(*Hint*: Note that

$$\mathbf{y}'X(X'X)^{-1}X'\mathbf{y} = \mathbf{b}'X'X\mathbf{b} = \hat{\mathbf{y}}'\hat{\mathbf{y}})$$

10. Consider the data of Exercises 1 and 60 in Chapter 3. Suppose we want to test $H_0: \beta_1 = 0$.

a. What is the full model in this case?
b. What is the reduced model?
c. Use the method presented in Section 4.2 to test H_0.
d. What is the practical interpretation of the test conducted?

11. Consider the information given in Exercise 43 in Chapter 3. Suppose we want to test $H_0: \beta_0 = \beta_1 = \beta_2 = 0$.

a. What is the full model in this case?
b. What is the reduced model?
c. Test H_0 and interpret the results in a practical sense.

Section 4.3

12. Explain how to compute $R(\beta_2 | \beta_0, \beta_1)$ in Example 3.2.1 by setting up the relevant matrices and quadratic forms.

13. Based on the information given in Example 4.3.1, is β_5 needed in a model that already contains the parameters β_0, β_1, β_2, β_3, and β_4? Explain based on the P value of the appropriate F statistic.

14. Use the information of Example 4.3.1 to test the null hypothesis that $\beta_3 = 0$. Can we conclude from this that the best model for explaining variability in squid weight should not contain β_3? Explain.

15. Use the information given in Example 4.3.1 to compute $R(\beta_1, \beta_2, \beta_3, \beta_4, \beta_5 | \beta_0)$.

a. Test $H_0: \gamma_1 = \mathbf{0}$ where

$$\gamma_1 = \begin{bmatrix} \beta_1 \\ \beta_2 \\ \beta_3 \\ \beta_4 \\ \beta_5 \end{bmatrix}$$

using the ANOVA in Table 4.5.

b. Test the same null hypothesis via the ANOVA in Table 4.6.

16. In mammals the toxicity of various drugs, pesticides, and chemical carcinogens can be altered by inducing liver enzyme activity. A study to investigate this process was reported in "Organophosphate Detoxification Related by Induced Hepatic Microsomal Enzymes in Chickens" by M. Ehrich, C. Larson, and J. Arnold, *American Journal of Veterinary Research* 45, 1983. A regression study was conducted to pinpoint the variables related to the detoxification of the insecticide malathion. The response y is the percentage of detoxification of malathion. Five enzyme activities are used as regressors. Denote these by x_1, x_2, x_3, x_4, and x_5.

a. Write the expression for the full model.

b. The ANOVA table for the data obtained is shown in Table 4.9. Portion A of the table is the ANOVA based on the "corrected" sum of squares. How many data points were used in the analysis? What null hypothesis is being tested by the F value 31.927? Can this null hypothesis be rejected? What is the P value of the test? Explain what this means in a physical sense. What is the numerical value of SS_{Res}? of $SS_{Res}/(n-p)$?

TABLE 4.9 **Partial and sequential tests: ANOVA for the detoxification study of Exercise 16**

Model: MODEL1
Dependent Variable: Y

Analysis of Variance

A

Source	DF	Sum of Squares	Mean Square	F Value	Prob>F
Model	5	8618.767	1723.753	31.927	0.0026
Error	4	215.9612	53.9903		
C Total	9	8834.728			

Root MSE	7.347809	R-square	0.9756
Dep Mean	153.5713	Adj R-sq	0.9450
C.V.	4.784624		

Parameter Estimates

C

Variable	DF	Parameter Estimate	Standard Error	T for H0: Parameter=0	Prob > |T|
INTERCEP	1	54.07869	29.67181	1.823	0.1425
X1	1	0.09681479	0.03418576	2.832*	0.0472
X2	1	0.03416486	0.01612155	2.119	0.1014
X3	1	0.5223269	0.08870192	5.889	0.0042
X4	1	-2.65508	0.3092585	-8.585	0.0010
X5	1	2.558968	0.3749683	6.824	0.0024

B

Variable	DF	Type I SS	Type II SS
INTERCEP	1	235841.4	179.3412
X1	1	1641.579	433.0209
X2	1	2116.087	242.4719
X3	1	423.3618	1872.123
X4	1	1923.211	3979.481
X5	1	2514.527	2514.527

*See Exercise 20.

c. The sequential sums of squares are listed in portion B of the table under the heading TYPE I SS. Give the numerical value of each of these:

$R(\beta_0)$
$R(\beta_1|\beta_0)$
$R(\beta_2|\beta_0, \beta_1)$
$R(\beta_3|\beta_0, \beta_1, \beta_2)$
$R(\beta_4|\beta_0, \beta_1, \beta_2, \beta_3)$
$R(\beta_5|\beta_0, \beta_1, \beta_2, \beta_3, \beta_4)$

Verify that $\Sigma_{i=1}^{5} R(\beta_i|\beta_0, \beta_1, \ldots, \beta_{i-1})$ is equal to the model sum of squares. What is the numerical value of $\mathbf{y}'\mathbf{y}$?

d. The partial sums of squares are listed in portion B under the heading TYPE II SS. Give the numerical value of each of these:

$R(\beta_0|\beta_1, \beta_2, \beta_3, \beta_4, \beta_5)$
$R(\beta_3|\beta_0, \beta_1, \beta_2, \beta_4, \beta_5)$

e. Suppose we want to test $H_0{:}\boldsymbol{\gamma}_1 = \mathbf{0}$ where

$$\boldsymbol{\gamma}_1 = \begin{bmatrix} \beta_3 \\ \beta_4 \\ \beta_5 \end{bmatrix}$$

Verify that the correct sum of squares for the numerator is

$$R(\beta_3|\beta_0, \beta_1, \beta_2) + R(\beta_4|\beta_0, \beta_1, \beta_2, \beta_3) + R(\beta_5|\beta_0, \beta_1, \beta_2, \beta_3, \beta_4)$$

Test this null hypothesis.

Section 4.4

***17.** Some of these algebraic results have been proved in Chapter 3. However, it will be helpful to review them at this point. Prove these theorems:

a. If M is positive definite, then M is nonsingular. (*Hint*: See Exercise 3.64.)
b. If M is symmetric, then M^{-1} is symmetric.
c. If M is symmetric and positive definite, then M^{-1} is positive definite.
d. If a symmetric matrix M is partitioned as

$$M = \left[\begin{array}{c|c} M_{11} & M_{12} \\ \hline M_{21} & M_{22} \end{array}\right]$$

where M_{11} and M_{22} are square, then $M'_{11} = M_{11}$, $M'_{22} = M_{22}$, $M'_{12} = M_{21}$, and $M'_{21} = M_{12}$.

***18.** Prove Theorem 4.4.1. (*Hint*: See Theorem 2.1.3 and Exercise 64 in Chapter 3.)

***19.** In this exercise you will prove the general result that the square of a t random variable with γ degrees of freedom follows an F distribution with 1 and γ degrees of freedom.

a. Show that if Z is a standard normal random variable, then Z^2 follows a chi-squared distribution with 1 degree of freedom. (*Hint*: Apply Definition 2.3.1.)

b. Let T denote a t random variable with γ degrees of freedom. Show that T^2 follows an F distribution with 1 and γ degrees of freedom. (See Exercise 59 in Chapter 2 and Definition 3.8.1.)

c. Suppose that the estimated variances and covariances among $\hat{\beta}_0$, $\hat{\beta}_1$, and $\hat{\beta}_2$ are

$$\widehat{\text{var}\,\hat{\beta}_0} = 28 \qquad \widehat{\text{cov}\,(\hat{\beta}_0, \hat{\beta}_1)} = -16$$
$$\widehat{\text{var}\,\hat{\beta}_1} = 24 \qquad \widehat{\text{cov}\,(\hat{\beta}_1, \hat{\beta}_2)} = 14$$
$$\widehat{\text{var}\,\hat{\beta}_2} = 18 \qquad \widehat{\text{cov}\,(\hat{\beta}_2, \hat{\beta}_3)} = -12$$

The estimated value of $\boldsymbol{\beta}'$ is $\mathbf{b}' = [3 \quad 5 \quad 2]$. Give the numerical value of the t statistic used to test $H_0{:}\beta_2 = 0$. (*Hint*: See Section 3.6.) Give the numerical value of the F statistic used to test this hypothesis.

20. Consider Exercise 16. Suppose we want to test

$$H_0{:}\beta_1 = 0 \quad \text{versus} \quad H_1{:}\beta_1 \neq 0$$

a. Use the SAS printout of Table 4.9 to find the F ratio used to test H_0.

b. The t value used to test H_0 is given in portion C of Table 4.9 and is indicated by an *. Show that $t^2 = F$ as expected.

c. Suppose we want to test

$$H_0{:}\beta_4 \geq 0 \quad \text{versus} \quad H_1{:}\beta_4 < 0$$

Which test is appropriate? What is the observed value of the test statistic? What is the P value of the test? Be careful! The P values shown are two-tailed.

Section 4.6

21. Show that $(\mathbf{b} - \boldsymbol{\beta}^*)'X'X(\mathbf{b} - \boldsymbol{\beta}^*) = \mathbf{q}'X(X'X)^{-1}X'\mathbf{q}$ where $\mathbf{q} = \mathbf{y} - X\boldsymbol{\beta}^*$. (*Hint*: Show that each of these expressions can be written as

$$\mathbf{y}'X(X'X)^{-1}X'\mathbf{y} - 2\boldsymbol{\beta}^{*\prime}X'\mathbf{y} + \boldsymbol{\beta}^{*\prime}X'X\boldsymbol{\beta}^*$$

Remember that $\mathbf{b} = (X'X)^{-1}X'\mathbf{y}$.)

22. Show that $SS_{\text{Res}} = \mathbf{q}'[I - X(X'X)^{-1}X']\mathbf{q}$ where $\mathbf{q} = \mathbf{y} - X\boldsymbol{\beta}^*$.

Section 4.7

23. Use Theorem 2.2.1 to show that

$$E[(C\mathbf{b} - \boldsymbol{\delta}^*)'[C(X'X)^{-1}C']^{-1}(C\mathbf{b} - \boldsymbol{\delta}^*)/r]$$
$$= \sigma^2 + (C\boldsymbol{\beta} - \boldsymbol{\delta}^*)'[C(X'X)^{-1}C']^{-1}(C\boldsymbol{\beta} - \boldsymbol{\delta}^*)/r$$

24. Find the matrix C and the vector $\boldsymbol{\delta}^*$ required to express each of the following in general linear form. In each case show that the F ratio reduces to that developed in previous sections.

a. $H_0{:}\boldsymbol{\gamma}_1 = \mathbf{0}$. (*Hint*: Write the original F ratio in the form given in Section 4.4.)

b. $H_0{:}\beta_j = 0$

c. $H_0{:}\boldsymbol{\beta} = \boldsymbol{\beta}^*$

d. $H_0{:}\boldsymbol{\gamma}_1 = \boldsymbol{\gamma}_1^*$

25. Consider the model

$$y_i = \beta_0 + \beta_1 x_{i1} + \beta_2 x_{i2} + \varepsilon_i$$

Find the matrix C and the vector $\boldsymbol{\delta}^*$ needed to test $H_0{:}\beta_0 = 0$ and $\beta_1 = \beta_2$.

Section 4.8

26. Show that in a mutually orthogonal model, $X'X$ is diagonal. Find $(X'X)^{-1}$.

27. Show that in a mutually orthogonal model

$$R(\boldsymbol{\beta}) = R(\beta_0) + R(\beta_1) + R(\beta_2) + \cdots + R(\beta_k)$$

28. Consider the simple linear regression model expressed in the form

$$y_i = \beta_0 + \beta_1(x_i - \bar{x}) + \varepsilon_i \qquad i = 1, 2, 3, \ldots n$$

where $\bar{x} = \sum_{i=1}^{n} x_i/n$. Show that this is a mutually orthogonal model. Find $R(\beta_0)$ and $R(\beta_1)$.

29. Consider the model

$$y_i = \beta_0 + \beta_1(x_{i1} - \bar{x}_1) + \beta_2(x_{i2} - \bar{x}_2) + \varepsilon_i$$

a. Show that the following experimental design yields a mutually orthogonal model.

x_1	x_2	y
10	50	25
100	50	29
10	100	30
100	100	40

b. Find $X'X$, $(X'X)^{-1}$, $R(\beta_0)$, $R(\beta_1)$, and $R(\beta_2)$ for these data. Find $R(\boldsymbol{\beta})$.

30. Find $R(\beta_2|\beta_0, \beta_1, \beta_3)$ for the data of Example 4.8.2.

31. a. Verify that the model whose X matrix is

$$X = \begin{bmatrix} 1 & -1 & -1 \\ 1 & -1 & 1 \\ 1 & 1 & -1 \\ 1 & 1 & 1 \\ 1 & 0 & 0 \\ 1 & \sqrt{2} & 0 \\ 1 & -\sqrt{2} & 0 \\ 1 & 0 & \sqrt{2} \\ 1 & 0 & -\sqrt{2} \end{bmatrix}$$

is mutually orthogonal.

b. Let

$$\mathbf{y} = \begin{bmatrix} 2 \\ 1 \\ 3 \\ 5 \\ 0 \\ -2 \\ 1 \\ 1 \\ 2 \end{bmatrix}$$

Find $R(\beta_1|\beta_0, \beta_2)$.

c. Find $X'X$, $(X'X)^{-1}$, and $X'\mathbf{y}$.

d. Test $H_0:\beta_1 = 0$.

Section 4.9

32. In Section 4.5 it was shown that the F ratio used to test

$$H_0:\boldsymbol{\beta} = \boldsymbol{\beta}^* \quad \text{versus} \quad H_1:\boldsymbol{\beta} \neq \boldsymbol{\beta}^*$$

is

$$F_{p,(n-p)} = \frac{(\mathbf{b} - \boldsymbol{\beta}^*)'X'X(\mathbf{b} - \boldsymbol{\beta}^*)/p}{SS_{\text{Res}}/(n-p)}$$

In this exercise you will show that the F test developed in Section 4.5 is a likelihood ratio test.

a. Show that

$$R = \left[\frac{(\mathbf{y} - X\boldsymbol{\beta}^*)'(\mathbf{y} - X\boldsymbol{\beta}^*)}{(\mathbf{y} - X\mathbf{b})'(\mathbf{y} - X\mathbf{b})}\right]^{n/2}$$

b. Verify that

$$(\mathbf{y} - X\boldsymbol{\beta}^*)'(\mathbf{y} - X\boldsymbol{\beta}^*) = Q_1^* + Q_0^*$$

where

$$Q_1^* = (\mathbf{y} - X\boldsymbol{\beta}^*)'X(X'X)^{-1}X'(\mathbf{y} - X\boldsymbol{\beta}^*)$$
$$Q_0^* = (\mathbf{y} - X\boldsymbol{\beta}^*)'[I - X(X'X)^{-1}X'](\mathbf{y} - X\boldsymbol{\beta}^*)$$

c. Verify that $Q_1^* = (\mathbf{b} - \boldsymbol{\beta}^*)'X'X(\mathbf{b} - \boldsymbol{\beta}^*)$. (*Hint*: Remember that $\mathbf{b} = (X'X)^{-1}X'\mathbf{y}$.)

d. Verify that $Q_0^* = \mathbf{y}'[I - X(X'X)^{-1}X']\mathbf{y} = SS_{\text{Res}}$.

e. Show that

$$R = \left[\frac{Q_1^* + Q_0^*}{Q_0^*}\right]^{n/2} = \left[1 + \frac{Q_1^*}{Q_0^*}\right]^{n/2}$$

f. Show that

$$F_{p,(n-p)} = \frac{Q_1^*/p}{Q_0^*/(n-p)}$$

g. Write R in terms of F.

h. Show that F is a monotonic function of R, thus showing that the given F test is a likelihood ratio test.

COMPUTING SUPPLEMENT E (MULTIPLE REGRESSION USING PROC REG)

Regression analysis can be conducted on SAS via PROC REG. The data of Example 4.3.1 are used to illustrate this procedure. An ANOVA to choose between a reduced model that contains only an intercept and the full model is run. Partial and sequential sums of square are generated. Finally, we test to see if notch to wing length (x_4) and width in inches (x_5) are useful in a model that already contains an intercept and information on rostral length (x_1), wing length (x_2), and rostral to notch length (x_3). The required program is shown below:

Statement	Purpose
`DATA SQUID;`	Names data set
`INPUT Y X1 X2 X3 X4 X5;`	Names variables
`CARDS;`	Signals that data follow
`1.95 1.31 1.07 0.44 0.75 0.35`	Data
`2.9 1.55 1.49 0.53 0.90 0.47`	
.	
.	
.	
`6.88 1.73 1.67 1.14 0.55`	
`;`	Signals end of data
`PROC REG;`	Calls for the regression procedure
`MODEL Y= X1 X2 X3 X4 X5/SS1 SS2;`	Identifies the model as a linear regression model with five variables, X1, X2, X3, X4, X5, and response Y; asks for sequential sums of squares (SS1) and partial sums of squares (SS2) to be printed
`TEST X4, X5;`	Tests for the importance of variables X4 and X5 in a model containing an intercept and variables X1, X2, and X3
`TITLE1 REGRESSION EXAMPLE;`	Titles output
`TITLE2 SQUID DATA;`	

The output of this program is shown in Figure 4.1.

Notice that the ANOVA table produced assumes the presence of an intercept. It is a table based on the "corrected" total sum of squares explained in Exercise 9 and follows the format shown in Table 4.6. The F value used to choose between the reduced and full models is shown in (1), with its P value given in (2). Since the P value is very small, we conclude that the regressors are useful in explaining variability in response in the presence of an intercept. Note that sequential sums of squares are listed as TYPE I SS on the printout and are shown by (3). The partial sums of squares, TYPE II SS, are given by (4). The F value used to test for the usefulness of x_4 and x_5 is given by (5); its P value is shown at (6). Note that these values agree with those given in Example 4.3.1 earlier apart from roundoff error.

FIGURE 4.1 **Regression analysis of squid data from Example 4.3.1**

DEP VARIABLE: Y

ANALYSIS OF VARIANCE

SOURCE	DF	SUM OF SQUARES	MEAN SQUARE	F VALUE	PROB>F
MODEL	5	208.00722737	41.60144547	84.070 (1)	0.0001 (2)
ERROR	16	7.91752263	0.49484516		
C TOTAL	21	215.92475000			

ROOT MSE	0.7034523	R-SQUARE	0.9633
DEP MEAN	4.195	ADJ R-SQ	0.9519
C.V.	16.76883		

PARAMETER ESTIMATES

VARIABLE	DF	PARAMETER ESTIMATE	STANDARD ERROR	T FOR H0: PARAMETER=0	PROB > |T|
INTERCEP	1	-6.51221	0.93356074	-6.976	0.0001
X1	1	1.99941331	2.57333840	0.777	0.4485
X2	1	-3.6751	2.77365962	-1.325	0.2038
X3	1	2.52448599	6.34749500	0.398	0.6961
X4	1	5.15808172	3.66028320	1.409	0.1779
X5	1	14.40116226	4.85599380	2.966	0.0091

VARIABLE	DF	TYPE I SS (3)	TYPE II SS (4)
INTERCEP	1	387.15655000	24.07917543
X1	1	199.14533555	0.29873127
X2	1	0.12666407	0.86876115
X3	1	4.11953880	0.07827274
X4	1	0.26349570	0.98268967
X5	1	4.35219326	4.35219326

TEST:	NUMERATOR:	2.30784	DF:	2	F VALUE:	4.6638 (5)
	DENOMINATOR:	0.494845	DF:	16	PROB >F :	0.0254 (6)

CHAPTER 5

ESTIMATION IN THE LESS THAN FULL RANK MODEL

In Chapters 3 and 4 the theory underlying the full rank model was developed. Recall that this model is of the form

$$\mathbf{y} = X\boldsymbol{\beta} + \boldsymbol{\varepsilon}$$

where X is $n \times p$ of rank p; that is, X is full rank. The primary implication of this assumption is that the matrix $X'X$ is $p \times p$ of rank p and is therefore nonsingular. This allows us to estimate each parameter in the model uniquely via **b**, the best linear unbiased estimator where

$$\mathbf{b} = (X'X)^{-1}X'\mathbf{y}$$

In this chapter we consider models that are not of full rank. These are models that can be written as

$$\mathbf{y} = X\boldsymbol{\beta} + \boldsymbol{\varepsilon}$$

where X is $n \times p$ of rank $r < p$. The implication of this is, of course, that $X'X$ is not of full rank and is therefore singular. Since this matrix has no inverse in the usual sense, the techniques developed for the full rank model no longer apply. To understand the problem more fully we begin by considering a less than full rank model that is probably familiar to you.

5.1

A LESS THAN FULL RANK MODEL

In courses in applied statistics, analysis of variance is often introduced by first considering the *one-way classification model with fixed effects.* This model arises in experiments in which interest centers on k populations. The populations might arise from natural groupings or they might be the result of applying k treatments to a group of similar objects at the outset of the experiment. Regardless of the manner in which the populations arise, they are of particular interest to the experimenter and are *preselected* for inclusion in the study rather than chosen randomly. For example, a medical researcher might be interested in comparing the three leading pain relievers for effectiveness in relieving arthritis pain; a biologist might study the effects of four experimental treatments used to enhance the growth of tomato plants; an engineer might want to investigate the sulfur content in the five major coal seams found in a particular geographic region. In each of these cases, one factor is being studied at k preselected levels. Each level is thought of as constituting a "treatment." The model in general is given by

$$y_{ij} = \mu + \tau_i + \varepsilon_{ij} \qquad \begin{array}{l} i = 1, 2, \ldots, k \\ j = 1, 2, \ldots, n_i \end{array}$$

where k denotes the number of treatments or levels involved in the experiment and n_i denotes the number of responses available at the ith level. In matrix notation, the model is expressed in the form

$$\mathbf{y} = X\boldsymbol{\beta} + \boldsymbol{\varepsilon}$$

where $\mathbf{y}$ is a vector of responses of dimension $\Sigma_{i=1}^{k} n_i$ by 1; $\boldsymbol{\beta}' = [\mu \;\; \tau_1 \;\; \tau_2 \;\; \ldots, \tau_k]$ is a vector of parameters; X is a matrix of dimension $\Sigma_{i=1}^{k} n_i$ by $(k + 1)$ called the *design matrix*; and $\boldsymbol{\varepsilon}$ is a vector of random errors.

There are some subtle and some not so subtle differences between such a model and the full rank models studied earlier:

1. In the full rank model it is assumed that the parameters specified in the model are unique. That is, it is assumed that there exists *exactly one* set of real numbers $\{\beta_0, \beta_1, \beta_2, \ldots, \beta_k\}$ that describes the system. In the general less than full rank model there are *infinitely many* sets of real numbers that describe the system. In our example there are infinitely many choices for $\mu, \tau_1, \tau_2, \ldots, \tau_k$ that all describe the same system. When this occurs we say that the model parameters are not "identifiable." (More will be said about this later. The idea is illustrated in Example 5.1.1.)
2. In the full rank model $X'X$ is nonsingular. The system of normal equations

$$(X'X)\mathbf{b} = X'\mathbf{y}$$

has only one solution, namely

$$\mathbf{b} = (X'X)^{-1}X'\mathbf{y}$$

In the general less than full rank model there are *infinitely many* solutions to the system of normal equations. This should not be surprising since in the less than full rank case there are infinitely many sets of real numbers that describe the system. Each solution to the normal equations estimates one of these possible parameter sets.

3. In the full rank model all linear functions of $\beta_0, \beta_1, \beta_2, \ldots, \beta_k$ can be estimated unbiasedly; in the less than full rank model this is not true. In the one-way classification model some linear functions of $\mu, \tau_1, \tau_2, \ldots, \tau_k$ can be estimated unbiasedly, whereas others cannot. Our task is to find those functions whose estimated values will be the same regardless of which estimators are used to estimate the model parameters and learn how to estimate them. Such functions are said to be *invariant* to the choice of estimators for $\mu, \tau_1, \tau_2, \ldots, \tau_k$. Estimators and tests must be constructed by means of these uniquely estimated functions.

Example 5.1.1 illustrates the differences listed above. The remainder of the chapter discusses how to circumvent the problems posed by the less than full rank model.

Example 5.1.1 Assume that $k = 3$ and that sample sizes are given by n_1, n_2, and n_3, respectively. In this case, the model is

$$\mathbf{y} = X\boldsymbol{\beta} + \boldsymbol{\varepsilon}$$

where

$$\mathbf{y} = \begin{bmatrix} y_{11} \\ y_{12} \\ \vdots \\ y_{1n_1} \\ \hline y_{21} \\ y_{22} \\ \vdots \\ y_{2n_2} \\ \hline y_{31} \\ y_{32} \\ \vdots \\ y_{3n_3} \end{bmatrix}, \quad X = \begin{bmatrix} 1 & 1 & 0 & 0 \\ 1 & 1 & 0 & 0 \\ \vdots & \vdots & \vdots & \vdots \\ 1 & 1 & 0 & 0 \\ \hline 1 & 0 & 1 & 0 \\ 1 & 0 & 1 & 0 \\ \vdots & \vdots & \vdots & \vdots \\ 1 & 0 & 1 & 0 \\ \hline 1 & 0 & 0 & 1 \\ 1 & 0 & 0 & 1 \\ \vdots & \vdots & \vdots & \vdots \\ 1 & 0 & 0 & 1 \end{bmatrix}, \quad \boldsymbol{\beta} = \begin{bmatrix} \mu \\ \tau_1 \\ \tau_2 \\ \tau_3 \end{bmatrix}, \quad \text{and} \quad \boldsymbol{\varepsilon} = \begin{bmatrix} \varepsilon_{11} \\ \varepsilon_{12} \\ \vdots \\ \varepsilon_{1n_1} \\ \hline \varepsilon_{21} \\ \varepsilon_{22} \\ \vdots \\ \varepsilon_{2n_2} \\ \hline \varepsilon_{31} \\ \varepsilon_{32} \\ \vdots \\ \varepsilon_{3n_3} \end{bmatrix}$$

The model expresses the idea that the variation in response is random about $\mu + \tau_i$. To see that the parameters are nonidentifiable, suppose that

$$\mu + \tau_1 = 10$$
$$\mu + \tau_2 = 12$$
$$\mu + \tau_3 = 8$$

If we let $\mu = 2$, then $\tau_1 = 8$, $\tau_2 = 10$, and $\tau_3 = 6$; if $\mu = 3$, then $\tau_1 = 7$, $\tau_2 = 9$, and $\tau_3 = 5$. It is evident that we can pick μ to be any real number and then determine τ_1, τ_2, and τ_3 to write a model that expresses the notion of random variation about $\mu + \tau_i$. There is no unique set of real numbers that describes the system; rather, there are infinitely many choices for μ, τ_1, τ_2, and τ_3 that all describe the same system.

Since the first column of X can be expressed as the sum of columns 2, 3, and 4, and the last three columns are linearly independent, the rank of X is 3 and $X'X$ is a 4×4 matrix of rank 3. That is, it is less than full rank. For this reason, it is singular and the normal equations

$$(X'X)\mathbf{b} = X'\mathbf{y}$$

have infinitely many solutions.

This example points out the problems posed by the less than full rank model. These problems will be overcome in a variety of ways in the sections to come.

5.2

A POSSIBLE SOLUTION — REPARAMETERIZATION

One often-used method of approaching the less than full rank model is *reparameterization.* In this method the model parameters are redefined by combining several parameters into one in such a way that the new design matrix is of full rank. In this way the normal equations can be solved uniquely using full rank techniques to estimate the new model parameters. This method works well in the one-way classification model introduced in Section 5.1 and illustrated in Example 5.2.1.

In Example 5.2.1 "dot" notation is used to simplify the expressions for sums and averages. The dot is used to indicate that summation has occurred relative to a given subscript. For example, $\Sigma_{i=1}^{n} y_{ij}$ is denoted by $y_{.j}$ and $\Sigma_{j=1}^{n} y_{ij}$ is given by $y_{i.}$. The corresponding averages for these data are denoted by $\bar{y}_{.j}$ and $\bar{y}_{i.}$, respectively.

Example 5.2.1 Consider the one-way classification model with $k = 3$,

$$y_{ij} = \mu + \tau_i + \varepsilon_{ij} \qquad i = 1, 2, 3$$
$$j = 1, 2, \ldots, n_i$$

The design matrix and vector of parameters for this model as stated are

$$X_{\text{old}} = \begin{bmatrix} 1 & 1 & 0 & 0 \\ 1 & 1 & 0 & 0 \\ \vdots & \vdots & \vdots & \vdots \\ 1 & 1 & 0 & 0 \\ \hline 1 & 0 & 1 & 0 \\ 1 & 0 & 1 & 0 \\ \vdots & \vdots & \vdots & \vdots \\ 1 & 0 & 1 & 0 \\ \hline 1 & 0 & 0 & 1 \\ 1 & 0 & 0 & 1 \\ \vdots & \vdots & \vdots & \vdots \\ 1 & 0 & 0 & 1 \end{bmatrix} \quad \text{and} \quad \boldsymbol{\beta}_{\text{old}} = \begin{bmatrix} \mu \\ \tau_1 \\ \tau_2 \\ \tau_3 \end{bmatrix}$$

X_{old} is less than full rank. The model is reparameterized by defining three new parameters μ_1, μ_2, μ_3 by

$$\mu_i = \mu + \tau_i \qquad i = 1, 2, 3$$

The new model is

$$y_{ij} = \mu_i + \varepsilon_{ij} \qquad \begin{array}{l} i = 1, 2, 3 \\ j = 1, 2, \ldots, n_i \end{array}$$

The design matrix and vector of parameters for the new model are

$$X_{\text{new}} = X = \begin{bmatrix} 1 & 0 & 0 \\ 1 & 0 & 0 \\ \vdots & \vdots & \vdots \\ 1 & 0 & 0 \\ \hline 0 & 1 & 0 \\ 0 & 1 & 0 \\ \vdots & \vdots & \vdots \\ 0 & 1 & 0 \\ \hline 0 & 0 & 1 \\ 0 & 0 & 1 \\ \vdots & \vdots & \vdots \\ 0 & 0 & 1 \end{bmatrix} \quad \text{and} \quad \boldsymbol{\beta}_{\text{new}} = \boldsymbol{\beta} = \begin{bmatrix} \mu_1 \\ \mu_2 \\ \mu_3 \end{bmatrix}$$

Note that X is $N \times 3$ of rank 3 where $N = \Sigma_{i=1}^{3} n_i$; it is now full rank. The theory developed in Chapter 3 can be applied to conclude that

$$\hat{\boldsymbol{\beta}} = \mathbf{b} = (X'X)^{-1}X'\mathbf{y}$$

Here

$$X'X = \begin{bmatrix} n_1 & 0 & 0 \\ 0 & n_2 & 0 \\ 0 & 0 & n_3 \end{bmatrix}, \quad (X'X)^{-1} = \begin{bmatrix} 1/n_1 & 0 & 0 \\ 0 & 1/n_2 & 0 \\ 0 & 0 & 1/n_3 \end{bmatrix},$$

$$X'\mathbf{y} = \begin{bmatrix} \sum_{j=1}^{n_1} y_{1j} \\ \sum_{j=1}^{n_2} y_{2j} \\ \sum_{j=1}^{n_3} y_{3j} \end{bmatrix}, \quad \text{and} \quad \mathbf{b} = (X'X)^{-1}X'\mathbf{y} = \begin{bmatrix} \sum_{j=1}^{n_1} y_{1j}/n_1 \\ \sum_{j=1}^{n_2} y_{2j}/n_2 \\ \sum_{j=1}^{n_3} y_{3j}/n_3 \end{bmatrix}$$

Thus

$$\hat{\mu}_1 = \sum_{j=1}^{n_1} y_{1j}/n_1 = \text{Mean of sample drawn from population } 1 = \bar{y}_{1.}$$

$$\hat{\mu}_2 = \sum_{j=1}^{n_2} y_{2j}/n_2 = \text{Mean of sample drawn from population } 2 = \bar{y}_{2.}$$

$$\hat{\mu}_3 = \sum_{j=1}^{n_3} y_{3j}/n_3 = \text{Mean of sample drawn from population } 3 = \bar{y}_{3.}$$

Linear functions of μ_1, μ_2, and μ_3, functions of the form $\mathbf{t}'\boldsymbol{\beta}$, are estimated via $\mathbf{t}'\mathbf{b}$, the corresponding linear function of the respective sample means. For example, the linear function $2\mu_1 - \mu_2 - \mu_3$ is estimated by $2\bar{y}_{1.} - \bar{y}_{2.} - \bar{y}_{3.}$. If the usual model assumptions are made here, then it is assumed that $\boldsymbol{\varepsilon}$ is normally distributed with mean $\mathbf{0}$ and variance $\sigma^2 I$. This is equivalent to assuming that sampling is from three normally distributed populations with means μ_1, μ_2, and μ_3, respectively, and *common* variance σ^2. This common variance can be estimated via s^2. Recall that

$$s^2 = [\mathbf{y}'\mathbf{y} - \mathbf{y}'X(X'X)^{-1}X'\mathbf{y}]/(N - p)$$

where p is the number of parameters in the model and $N = \Sigma_{i=1}^{3} n_i$ is the total number of responses. Here,

$$\mathbf{y}'X = \left[\sum_{j=1}^{n_1} y_{1j} \quad \sum_{j=1}^{n_2} y_{2j} \quad \sum_{j=1}^{n_3} y_{3j} \right]$$

$$\mathbf{y}'X(X'X)^{-1} = \left[\sum_{j=1}^{n_1} y_{1j}/n_1 \quad \sum_{j=1}^{n_2} y_{2j}/n_2 \quad \sum_{j=1}^{n_3} y_{3j}/n_3 \right]$$

$$\mathbf{y}'X(X'X)^{-1}X'\mathbf{y} = \left(\sum_{j=1}^{n_1} y_{1j} \right)^2 \bigg/ n_1 + \left(\sum_{j=1}^{n_2} y_{2j} \right)^2 \bigg/ n_2 + \left(\sum_{j=1}^{3} y_{3j} \right)^2 \bigg/ n_3$$

$$= \sum_{i=1}^{3} (y_{i.})^2/n_i$$

Hence

$$s^2 = [\mathbf{y}'\mathbf{y} - \mathbf{y}'X(X'X)^{-1}X'\mathbf{y}]/(N-3)$$
$$= \left[\sum_{i=1}^{3}\sum_{j=1}^{n_i} y_{ij}^2 - \sum_{i=1}^{3} y_{i.}^2/n_i\right] \Big/ (N-3)$$

You might not recognize this way of writing s^2 from your earlier courses in applied statistics. In such courses, s^2 is usually written as a "pooled" variance. That is, it is expressed as a weighted mean of the available sample variances. In this case, it can be shown that

$$s^2 = \frac{(n_1-1)s_1^2 + (n_2-1)s_2^2 + (n_3-1)s_3^2}{(n_1-1)+(n_2-1)+(n_3-1)}$$

where

$$s_i^2 = \sum_{j=1}^{n_i} (y_{ij} - \bar{y}_{i.})^2/(n_i - 1) \qquad i = 1, 2, 3$$

The proof of this fact is left as an exercise (see Exercise 3).

As you can see, it is possible to work with a less than full rank model by reparameterizing to a full rank model and then applying the theory already available. However, this approach is not necessary. A general method for handling the less than full rank model that can be applied to all such models can be developed. To do so we must review some results from the theory of equations and introduce the concept of a conditional inverse.

5.3

INTRODUCTION TO CONDITIONAL INVERSES

To find a general method for handling the less than full rank model, some theorems and definitions from the field of linear algebra must be reviewed. Recall that a system of the form

$$A\mathbf{x} = \mathbf{g}$$

where A is an $n \times p$ matrix of real numbers, $\mathbf{x}$ is a $p \times 1$ vector of unknowns, and $\mathbf{g}$ is a $p \times 1$ vector of real numbers represents a system of n linear equations in p unknowns. If the system has no solution it is said to be *inconsistent*; otherwise it is *consistent*. Given any such system, exactly one of three things must hold:

1. The system is inconsistent.
2. The system is consistent and has exactly one solution.
3. The system is consistent and has many solutions.

Theorem 5.3.1 is useful in checking a system for consistency.

Theorem 5.3.1 The system $A\mathbf{x} = \mathbf{g}$ is consistent if and only if the rank of $[A \mid \mathbf{g}]$ is equal to the rank of A.

Proof

Assume that the system is consistent. Then there exists at least one vector $\mathbf{x}_0$ such that $A\mathbf{x}_0 = \mathbf{g}$. It is evident that $r(A) \leqslant r([A \mid \mathbf{g}])$. Note that

$$\begin{aligned} r([A \mid \mathbf{g}]) &= r([A \mid A\mathbf{x}_0]) \\ &= r(A[I \mid \mathbf{x}_0]) \end{aligned}$$

However, by property 5 of rank given in Section 1.4,

$$r(A[I \mid \mathbf{x}_0]) \leqslant r(A)$$

Hence,

$$r(A) \leqslant r([A \mid \mathbf{g}]) \leqslant r(A)$$

and it can be concluded that $r([A \mid \mathbf{g}]) = r(A)$ as claimed. Now assume that $r([A \mid \mathbf{g}]) = r(A)$. This implies that $\mathbf{g}$ is a linear combination of the columns of A. That is, there exist constants $x_1, x_2, \ldots, x_p$ not all zero such that

$$x_1\mathbf{a}_1 + x_2\mathbf{a}_2 + \cdots x_p\mathbf{a}_p = \mathbf{g}$$

where $\mathbf{a}_i$ is the ith column of A. In matrix notation,

$$A\mathbf{x}_0 = \mathbf{g}$$

where

$$\mathbf{x}_0 = \begin{bmatrix} x_1 \\ x_2 \\ \vdots \\ x_p \end{bmatrix}$$

Hence $\mathbf{x}_0$ is a solution to the system showing that the system is consistent. ■

These ideas play an important role in the theory of linear models. In particular, the normal equations

$$(X'X)\mathbf{b} = X'\mathbf{y}$$

constitute a system of linear equations. It is easy to show that this system is consistent, thus showing that the normal equations associated with a linear model always have at least one solution.

Theorem 5.3.2 Let $X\boldsymbol{\beta} + \boldsymbol{\varepsilon}$ be a linear model. Then the system of normal equations

$$(X'X)\mathbf{b} = X'\mathbf{y}$$

is consistent.

Proof

To show that this system is consistent, it is sufficient to show that the conditions of Theorem 5.3.1 are satisfied. It is evident that $r(X'X) \leqslant r([X'X \mid X'\mathbf{y}])$. Now

$$\begin{aligned} r([X'X \mid X'\mathbf{y}]) &= r(X'[X \mid \mathbf{y}]) \\ &\leqslant r(X') \\ &= r(X'X) \end{aligned}$$

Thus it has been shown that

$$r(X'X) \leqslant r([X'X \mid X'\mathbf{y}]) \leqslant r(X'X)$$

This can only be true if

$$r([X'X \mid X'\mathbf{y}]) = r(X'X)$$

By Theorem 5.3.1 the system of normal equations is consistent. ■

As has been shown earlier, in the full rank model the system of normal equations has exactly one solution; in the less than full rank model, many solutions to the system exist. The immediate problem is to find a general method for solving the normal equations in the less than full rank model. To do so, the notion of a *conditional inverse* is needed.

Definition 5.3.1

Let A be an $n \times p$ matrix. A $p \times n$ matrix A^c such that

$$AA^cA = A$$

is called a conditional inverse for A.

Example 5.3.1 illustrates this definition and shows that conditional inverses are not necessarily unique.

Example 5.3.1 Consider the 3×3 matrix A of rank 2 given by

$$A = \begin{bmatrix} 2 & 4 & 2 \\ 1 & 0 & -1 \\ 3 & 1 & -2 \end{bmatrix}$$

Let

$$A_1 = \begin{bmatrix} 0 & +1 & 0 \\ \frac{1}{4} & -\frac{1}{2} & 0 \\ 0 & 0 & 0 \end{bmatrix}$$

Multiplication will show that $AA_1A = A$. Now consider the matrix A_2 where

$$A_2 = \begin{bmatrix} 0 & 1 & 0 \\ 0 & -3 & 1 \\ 0 & 0 & 0 \end{bmatrix}$$

It can also be verified by multiplication that $AA_2A = A$. Hence both A_1 and A_2 are conditional inverses for the matrix A. How such inverses are found will be demonstrated shortly.

Recall that for a matrix to have an ordinary inverse it must be $p \times p$ of rank p; it must be a full rank square matrix. This is not true of conditional inverses. Theorem 5.3.3 shows that every matrix has a conditional inverse.

Theorem 5.3.3 Let A be an $n \times p$ matrix. There exists a matrix A^c such that $AA^cA = A$. That is, A has a conditional inverse.

Proof

Let A be an $n \times p$ matrix of rank r. It is known that a series of elementary row and column operations can be performed on A to reduce it to the form

$$\begin{bmatrix} I_r & 0 \\ 0 & 0 \end{bmatrix}$$

where I_r is an $r \times r$ identity matrix [**11**]. This means that nonsingular matrices P and Q exist such that

$$PAQ = \begin{bmatrix} I_r & 0 \\ 0 & 0 \end{bmatrix} = B$$

Multiplication will show that the matrix

$$B^c = \begin{bmatrix} I_r & U \\ V & W \end{bmatrix}$$

where U, V, and W are arbitrary is a conditional inverse for B. Since $B = PAQ$, $A = P^{-1}BQ^{-1}$. Now consider the matrix $A^c = QB^cP$. By substitution,

$$\begin{aligned} AA^cA &= P^{-1}BQ^{-1}QB^cPP^{-1}BQ^{-1} \\ &= P^{-1}BB^cBQ^{-1} \\ &= P^{-1}BQ^{-1} \\ &= A \end{aligned}$$

By definition, A^c is a conditional inverse for A. ■

It is easy to show that if A is nonsingular, then $A^c = A^{-1}$. The verification of this result is left as an exercise (see Exercise 17).

How does one go about finding a conditional inverse for a matrix? Of course the method outlined in the existence proof could be used. However, this is not the easiest way to generate such an inverse. A computing algorithm was used to find the inverses of Example 5.3.1:

An Algorithm for Finding Conditional Inverses

Let A be $n \times p$ of rank r. To find a conditional inverse A^c,

1. Find *any* nonsingular $r \times r$ minor M.
2. Find M^{-1} and $(M^{-1})'$.
3. Replace M in A with $(M^{-1})'$.
4. Replace all other entries in A with zeros.
5. Transpose the resulting matrix.

You can verify that the matrix A of Example 5.3.1 was found by applying this algorithm with

$$M_1 = \begin{bmatrix} 2 & 4 \\ 1 & 0 \end{bmatrix}$$

A_2 was based on minor M_2 where

$$M_2 = \begin{bmatrix} 1 & 0 \\ 3 & 1 \end{bmatrix}$$

Conditional inverses are similar to traditional inverses in many ways. Some of their more important properties are summarized below for easy reference. Proofs are straightforward and are left as exercises (Exercises 8–16).

Properties of Conditional Inverses

Let A be an $n \times p$ matrix of rank r where $n \geq p \geq r$. Then

1. A^cA and AA^c are idempotent.
2. $r(AA^c) = r(A^cA) = r$.
3. If A^c is a conditional inverse of A, then $(A^c)'$ is a conditional inverse of A'. That is, $(A^c)' = (A')^c$.
4. $A = A(A'A)^c(A'A)$ and $A' = (A'A)(A'A)^cA'$.
5. $A(A'A)^cA'$ is unique, symmetric, and idempotent. To say that this matrix

is unique means that it is invariant to the choice of a conditional inverse. Furthermore, $r[A(A'A)^c A'] = r$.

6. $I - A(A'A)^c A'$ is unique, symmetric, and idempotent, and $r[I - A(A'A^c)A'] = n - r$.
7. If $n = r = p$, then $A^c = A^{-1}$. That is, in the full rank case the conditional inverse is the same as the traditional inverse.
8. $I - A^c A$ is idempotent.

CONDITIONAL INVERSES AND THE NORMAL EQUATIONS

Conditional inverses are particularly important because they can be used to solve consistent systems of the form

$$A\mathbf{x} = \mathbf{g}$$

Since the normal equations for any linear model are consistent and assume this form, conditional inverses provide a method for solving the normal equations in a less than full rank model. To see how a conditional inverse does this consider Theorem 5.3.4.

Theorem 5.3.4 Let $A\mathbf{x} = \mathbf{g}$ be consistent. Then $\mathbf{x} = A^c\mathbf{g}$ is a solution to the system where A^c is any conditional inverse for A.

Proof

Let $A\mathbf{x} = \mathbf{g}$ be consistent and let A^c denote any conditional inverse for A. By definition,

$$AA^cA\mathbf{x} = A\mathbf{x}$$

By assumption, $A\mathbf{x} = \mathbf{g}$. Substituting,

$$AA^c\mathbf{g} = \mathbf{g}$$

Let $\mathbf{x}_0 = A^c\mathbf{g}$. Then $A\mathbf{x}_0 = \mathbf{g}$, showing that $\mathbf{x}_0$ is a solution to the system. ■

Applying this theorem in the context of linear models, it is easy to see that

$$\mathbf{b} = (X'X)^c X'\mathbf{y}$$

is a solution to the system of normal equations $(X'X)\mathbf{b} = X'\mathbf{y}$. Any conditional inverse will generate a solution. However, in the less than full rank model the actual solution obtained depends on the choice of $(X'X)^c$; different conditional inverses result in different solutions.

Example 5.3.2 Suppose that for a particular linear model

$$X'X = \begin{bmatrix} 2 & 1 & 1 \\ 1 & 1 & 0 \\ 1 & 0 & 1 \end{bmatrix} \quad \text{and} \quad X'\mathbf{y} = \begin{bmatrix} 14 \\ 6 \\ 8 \end{bmatrix}$$

The normal equations are

$$\begin{bmatrix} 2 & 1 & 1 \\ 1 & 1 & 0 \\ 1 & 0 & 1 \end{bmatrix} \begin{bmatrix} b_0 \\ b_1 \\ b_2 \end{bmatrix} = \begin{bmatrix} 14 \\ 6 \\ 8 \end{bmatrix}$$

It is easy to see that $|X'X| = 0$, and hence the model is not of full rank. By inspection, it can be seen that column 3 is equal to column 1 minus column 2. Methods of Section 1.4 can be applied to show that the vectors of columns 1 and 2 are linearly independent. Thus $X'X$ is of rank 2. Consider the 2×2 nonsingular minor

$$M_1 = \begin{bmatrix} 2 & 1 \\ 1 & 1 \end{bmatrix}$$

Using the algorithm just developed, the conditional inverse based on this minor is

$$(X'X)_1^c = \begin{bmatrix} 1 & -1 & 0 \\ -1 & 2 & 0 \\ 0 & 0 & 0 \end{bmatrix}$$

Based on this inverse, a solution to the normal equations is

$$(X'X)_1^c X'\mathbf{y} = \begin{bmatrix} 1 & -1 & 0 \\ -1 & 2 & 0 \\ 0 & 0 & 0 \end{bmatrix} \begin{bmatrix} 14 \\ 6 \\ 8 \end{bmatrix} = \begin{bmatrix} 8 \\ -2 \\ 0 \end{bmatrix}$$

A quick calculation will verify that this is indeed a proper solution. To generate a different solution, we need only find a different conditional inverse. This time, let us begin with the minor

$$M_2 = \begin{bmatrix} 1 & 0 \\ 0 & 1 \end{bmatrix}$$

The conditional inverse based on this minor is

$$(X'X)_2^c = \begin{bmatrix} 0 & 0 & 0 \\ 0 & 1 & 0 \\ 0 & 0 & 1 \end{bmatrix}$$

Based on this inverse, another solution to the normal equations is

$$(X'X)_2^c X'\mathbf{y} = \begin{bmatrix} 0 & 0 & 0 \\ 0 & 1 & 0 \\ 0 & 0 & 1 \end{bmatrix} \begin{bmatrix} 14 \\ 6 \\ 8 \end{bmatrix} = \begin{bmatrix} 0 \\ 6 \\ 8 \end{bmatrix}$$

Multiplication will also verify this solution.

Theorem 5.3.5 allows us to conclude that any conditional inverse actually generates an uncountable number of solutions in the less than full rank model; otherwise it generates the usual least squares estimators.

Theorem 5.3.5 Let $A\mathbf{x} = \mathbf{g}$ be consistent and let A^c denote any conditional inverse for A. Then

$$\mathbf{x}_0 = A^c\mathbf{g} + (I - A^cA)\mathbf{z}$$

is a solution to the system where $\mathbf{z}$ is an arbitrary $p \times 1$ vector.

Proof

Assume that $A\mathbf{x} = \mathbf{g}$ is consistent and let A^c denote any conditional inverse for A. Let

$$\mathbf{x}_0 = A^c\mathbf{g} + (I - A^cA)\mathbf{z}$$

where $\mathbf{z}$ is arbitrarily chosen. Then

$$\begin{aligned} A\mathbf{x}_0 &= A[A^c\mathbf{g} + (I - A^cA)\mathbf{z}] \\ &= AA^c\mathbf{g} + A(I - A^cA)\mathbf{z} \end{aligned}$$

It is known that $A^c\mathbf{g}$ is a solution to the system; hence $AA^c\mathbf{g} = \mathbf{g}$. Substituting, we obtain

$$A\mathbf{x}_0 = \mathbf{g} + (A - AA^cA)\mathbf{z}$$

By definition, $AA^cA = A$. Hence

$$A\mathbf{x}_0 = \mathbf{g} + (A - A)\mathbf{z} = \mathbf{g}$$

From this, it can be concluded that $\mathbf{x}_0$ is a solution to the system. ■

In the linear models context, Theorem 5.3.5 implies that every vector of the form

$$\mathbf{b}_0 = (X'X)^c X'\mathbf{y} + [I - (X'X)^c(X'X)]\mathbf{z}$$

where $(X'X)^c$ is a conditional inverse for $X'X$ and $\mathbf{z}$ is arbitrary is a solution to the normal equations. If $\mathbf{z} = \mathbf{0}$, then we get the solution obtained via Theorem 5.3.4; if $X'X$ is of full rank, then we obtain the usual least squares estimators developed in Chapter 3.

Example 5.3.3 Reconsider the linear system of Example 5.3.2. One solution to the normal equations

$$\begin{bmatrix} 2 & 1 & 1 \\ 1 & 1 & 0 \\ 1 & 0 & 1 \end{bmatrix} \begin{bmatrix} b_0 \\ b_1 \\ b_2 \end{bmatrix} = \begin{bmatrix} 14 \\ 6 \\ 8 \end{bmatrix}$$

was found to be

$$\begin{bmatrix} 8 \\ -2 \\ 0 \end{bmatrix}$$

Here

$$(X'X)^c = \begin{bmatrix} 1 & -1 & 0 \\ -1 & 2 & 0 \\ 0 & 0 & 0 \end{bmatrix}$$

and

$$(X'X)^c(X'X) = \begin{bmatrix} 1 & -1 & 0 \\ -1 & 2 & 0 \\ 0 & 0 & 0 \end{bmatrix} \begin{bmatrix} 2 & 1 & 1 \\ 1 & 1 & 0 \\ 1 & 0 & 1 \end{bmatrix} = \begin{bmatrix} 1 & 0 & 1 \\ 0 & 1 & -1 \\ 0 & 0 & 0 \end{bmatrix}$$

Since $\mathbf{z}$ is arbitrary, let

$$\mathbf{z} = \begin{bmatrix} 1 \\ 1 \\ 1 \end{bmatrix}$$

Substituting, we see that another solution to the normal equations is

$$\begin{aligned} \mathbf{b}_0 &= \begin{bmatrix} 8 \\ -2 \\ 0 \end{bmatrix} + \left(\begin{bmatrix} 1 & 0 & 0 \\ 0 & 1 & 0 \\ 0 & 0 & 1 \end{bmatrix} - \begin{bmatrix} 1 & 0 & 1 \\ 0 & 1 & -1 \\ 0 & 0 & 0 \end{bmatrix} \right) \begin{bmatrix} 1 \\ 1 \\ 1 \end{bmatrix} \\ &= \begin{bmatrix} 8 \\ -2 \\ 0 \end{bmatrix} + \begin{bmatrix} 0 & 0 & -1 \\ 0 & 0 & 1 \\ 0 & 0 & 1 \end{bmatrix} \begin{bmatrix} 1 \\ 1 \\ 1 \end{bmatrix} \\ &= \begin{bmatrix} 8 \\ -2 \\ 0 \end{bmatrix} + \begin{bmatrix} -1 \\ 1 \\ 1 \end{bmatrix} = \begin{bmatrix} 7 \\ -1 \\ 1 \end{bmatrix} \end{aligned}$$

Theorem 5.3.6 allows us to conclude that every solution to a consistent system can be expressed in terms of any conditional inverse. Hence, in practice, only one conditional inverse need be found.

Theorem 5.3.6 Let $A\mathbf{x} = \mathbf{g}$ be consistent and let A^c be any conditional inverse for A. Let $\mathbf{x}_0$ be any solution to the system. Then

$$\mathbf{x}_0 = A^c\mathbf{g} + (I - A^cA)\mathbf{z}$$

where $\mathbf{z} = (I - A^cA)\mathbf{x}_0$.

Proof

Let $A\mathbf{x} = \mathbf{g}$ be consistent and let $\mathbf{x}_0$ be a solution to the system. Let

$$\mathbf{z} = (I - A^cA)\mathbf{x}_0$$

Consider the vector $A^c\mathbf{g} + (I - A^cA)\mathbf{z}$. Substituting,

$$\begin{aligned} A^c\mathbf{g} + (I - A^cA)\mathbf{z} &= A^c\mathbf{g} + (I - A^cA)(I - A^cA)\mathbf{x}_0 \\ &= A^c\mathbf{g} + (I - A^cA)^2\mathbf{x}_0 \end{aligned}$$

By property 8 of conditional inverses, $I - A^cA$ is idempotent. Hence $(I - A^cA)^2 = I - A^cA$. Thus

$$\begin{aligned} A^c\mathbf{g} + (I - A^cA)\mathbf{z} &= A^c\mathbf{g} + (I - A^cA)\mathbf{x}_0 \\ &= A^c\mathbf{g} + \mathbf{x}_0 - A^cA\mathbf{x}_0 \end{aligned}$$

Since $\mathbf{x}_0$ is assumed to be a solution to the system, $A\mathbf{x}_0 = \mathbf{g}$. Substituting,

$$A^c\mathbf{g} + (I - A^cA)\mathbf{z} = A^c\mathbf{g} + \mathbf{x}_0 - A^c\mathbf{g} = \mathbf{x}_0$$

The solution $\mathbf{x}_0$ has been expressed in the desired form. ■

The implication to linear models is that every solution to the normal equations can be expressed in terms of any conditional inverse $(X'X)^c$. This idea is illustrated numerically in Example 5.3.4.

Example 5.3.4 Consider the system of normal equations

$$(X'X)\mathbf{b} = X'\mathbf{y}$$

where

$$(X'X) = \begin{bmatrix} 2 & 1 & 1 \\ 1 & 1 & 0 \\ 1 & 0 & 1 \end{bmatrix} \quad \text{and} \quad X'\mathbf{y} = \begin{bmatrix} 14 \\ 6 \\ 8 \end{bmatrix}$$

of Example 5.3.2. Two conditional inverses for $X'X$ are

$$(X'X)^c_1 = \begin{bmatrix} 1 & -1 & 0 \\ -1 & 2 & 0 \\ 0 & 0 & 0 \end{bmatrix} \quad \text{and} \quad (X'X)^c_2 = \begin{bmatrix} 0 & 0 & 0 \\ 0 & 1 & 0 \\ 0 & 0 & 1 \end{bmatrix}$$

In Example 5.3.3 it was found that

$$\mathbf{b}_0 = \begin{bmatrix} 8 \\ -2 \\ 0 \end{bmatrix}$$

is a solution to the system generated by $(X'X)_1^c$. This solution can also be written in terms of $(X'X)_2^c$. Note that

$$\begin{aligned}(X'X)_2^c(X'X) &= \begin{bmatrix} 0 & 0 & 0 \\ 0 & 1 & 0 \\ 0 & 0 & 1 \end{bmatrix}\begin{bmatrix} 2 & 1 & 1 \\ 1 & 1 & 0 \\ 1 & 0 & 1 \end{bmatrix} \\ &= \begin{bmatrix} 0 & 0 & 0 \\ 1 & 1 & 0 \\ 1 & 0 & 1 \end{bmatrix}\end{aligned}$$

and that

$$I - (X'X)_2^c(X'X) = \begin{bmatrix} 1 & 0 & 0 \\ -1 & 0 & 0 \\ -1 & 0 & 0 \end{bmatrix}$$

Let

$$\begin{aligned}\mathbf{z} &= [I - (X'X)_2^c(X'X)]\mathbf{b}_0 \\ &= \begin{bmatrix} 1 & 0 & 0 \\ -1 & 0 & 0 \\ -1 & 0 & 0 \end{bmatrix}\begin{bmatrix} 8 \\ -2 \\ 0 \end{bmatrix} = \begin{bmatrix} 8 \\ -8 \\ -8 \end{bmatrix}\end{aligned}$$

Now

$$(X'X)_2^c X'\mathbf{y} = \begin{bmatrix} 0 & 0 & 0 \\ 0 & 1 & 0 \\ 0 & 0 & 1 \end{bmatrix}\begin{bmatrix} 14 \\ 6 \\ 8 \end{bmatrix} = \begin{bmatrix} 0 \\ 6 \\ 8 \end{bmatrix}$$

Substituting into the expression of Theorem 5.3.6,

$$\begin{aligned}(X'X)_2^c X'\mathbf{y} + [I - (X'X)_2^c(X'X)]\mathbf{z} &= \begin{bmatrix} 0 \\ 6 \\ 8 \end{bmatrix} + \begin{bmatrix} 1 & 0 & 0 \\ -1 & 0 & 0 \\ -1 & 0 & 0 \end{bmatrix}\begin{bmatrix} 8 \\ -8 \\ -8 \end{bmatrix} \\ &= \begin{bmatrix} 0 \\ 6 \\ 8 \end{bmatrix} + \begin{bmatrix} 8 \\ -8 \\ -8 \end{bmatrix} = \begin{bmatrix} 8 \\ -2 \\ 0 \end{bmatrix} = \mathbf{b}_0\end{aligned}$$

The results of this section underlie much of the theoretical development of the estimation and hypothesis-testing techniques for the less than full rank linear statistical model. At this point the major statistical result to keep in mind is that in the less than full rank model the normal equations have many solutions and that these solutions can be found via conditional inverses.

5.4

INTRODUCTION TO ESTIMABILITY

In the less than full rank model the vector $\boldsymbol{\beta}$ cannot be estimated uniquely. There are many choices for $(X'X)^c$, each of which generates a number of solutions to the normal equations. For this reason, in the less than full rank statistical model interest centers not on $\boldsymbol{\beta}$ but on linear functions of $\boldsymbol{\beta}$; that is, functions of the form $\mathbf{t}'\boldsymbol{\beta}$ where $\mathbf{t}'$ is a vector of real numbers. The natural estimator for $\mathbf{t}'\boldsymbol{\beta}$ is $\mathbf{t}'\mathbf{b}$ where $\mathbf{b}$ is any solution to the normal equations. This presents a problem. It is conceivable that different choices for $\mathbf{b}$ will result in different estimates for $\mathbf{t}'\boldsymbol{\beta}$. We shall be concerned only with linear functions that are invariant to the choice of $\mathbf{b}$; that is, their value remains the same regardless of which solution to the normal equations is used. As you shall see, this property is tied to the notion of *estimability*.

Definition 5.4.1

Let $\mathbf{y} = X\boldsymbol{\beta} + \boldsymbol{\varepsilon}$ where X is $n \times p$ of rank $r \leqslant p$, $E[\boldsymbol{\varepsilon}] = \mathbf{0}$, and var $\boldsymbol{\varepsilon} = \sigma^2 I$. A function $\mathbf{t}'\boldsymbol{\beta}$ is said to be estimable if there exists a vector $\mathbf{c}$ such that $E[\mathbf{c}'\mathbf{y}] = \mathbf{t}'\boldsymbol{\beta}$.

To be estimable, there must be some linear combination of the responses $y_1, y_2, \ldots y_n$ whose expected value is $\mathbf{t}'\boldsymbol{\beta}$; there must be a *linear unbiased estimator* for $\mathbf{t}'\boldsymbol{\beta}$. To say that $\boldsymbol{\beta}$ itself is estimable means that each element of this vector of parameters is estimable. Theorem 5.4.1 provides a method for testing a linear function for estimability.

Theorem 5.4.1 Let $\mathbf{y} = X\boldsymbol{\beta} + \boldsymbol{\varepsilon}$ where X is $n \times p$ of rank $r \leqslant p$, $E[\boldsymbol{\varepsilon}] = \mathbf{0}$, and var $\boldsymbol{\varepsilon} = \sigma^2 I$. A necessary and sufficient condition for $\mathbf{t}'\boldsymbol{\beta}$ to be estimable is that there exists a solution to the system $(X'X)\mathbf{z} = \mathbf{t}$.

Proof

Assume that there exists a solution $\mathbf{z}_0$ to the system $(X'X)\mathbf{z} = \mathbf{t}$ so that $(X'X)\mathbf{z}_0 = \mathbf{t}$ and $\mathbf{z}_0'X'X = \mathbf{t}'$. Let $\mathbf{c}' = \mathbf{z}_0'X'$. Then

$$\begin{aligned} E[\mathbf{c}'\mathbf{y}] &= E[\mathbf{z}_0'X'\mathbf{y}] \\ &= \mathbf{z}_0'X'E[\mathbf{y}] \end{aligned}$$

Via the model assumptions, $E[\mathbf{y}] = X\boldsymbol{\beta}$. Substituting, we see that

$$E[\mathbf{c}'\mathbf{y}] = \mathbf{z}_0' X'X\boldsymbol{\beta} = \mathbf{t}'\boldsymbol{\beta}$$

By definition, $\mathbf{t}'\boldsymbol{\beta}$ is estimable. To prove the converse, assume that $\mathbf{t}'\boldsymbol{\beta}$ is estimable. By definition, there exists a vector $\mathbf{c}_0$ such that $E[\mathbf{c}_0'\mathbf{y}] = \mathbf{t}'\boldsymbol{\beta}$ regardless of the choice of $\boldsymbol{\beta}$. This implies that $\mathbf{c}_0' E[\mathbf{y}] = \mathbf{c}_0' X\boldsymbol{\beta} = \mathbf{t}'\boldsymbol{\beta}$ for every choice of $\boldsymbol{\beta}$. This can occur only if $\mathbf{c}_0' X = \mathbf{t}'$ or $X'\mathbf{c}_0 = \mathbf{t}$. Thus $\mathbf{c}_0$ is a solution to the system $X'\mathbf{c} = \mathbf{t}$. By Theorem 5.3.1,

$$r(X') = r([X' \mid \mathbf{t}])$$

Note that

$$[X'X \mid \mathbf{t}] = [X' \mid \mathbf{t}] \left[\begin{array}{c|c} X_{n\times p} & \mathbf{0}_{n\times 1} \\ \hline \mathbf{0}_{1\times p}' & 1 \end{array}\right]$$

Since the rank of a product is less than or equal to the rank of each of its factors, it can be concluded that

$$\begin{aligned} r([X'X \mid \mathbf{t}]) &\leqslant r([X' \mid \mathbf{t}]) \\ &= r(X') \\ &= r(X'X) \end{aligned}$$

Since adding a column to a matrix cannot reduce its rank, $r([X'X \mid \mathbf{t}]) \geqslant r(X'X)$. This can occur only if

$$r([X'X \mid \mathbf{t}]) = r(X'X)$$

By Theorem 5.3.1, the system $(X'X)\mathbf{z} = \mathbf{t}$ has a solution. ■

Although this theorem can be used to test for estimability, an easier method exists. This method is based on the fact that the test just derived can be shown to be equivalent to the statement that $\mathbf{t}'(X'X)^c(X'X) = \mathbf{t}'$. This result is shown in Theorem 5.4.2.

Theorem 5.4.2 Let $\mathbf{y} = X\boldsymbol{\beta} + \boldsymbol{\varepsilon}$ where X is $n \times p$ of rank $r \leqslant p$, $E[\boldsymbol{\varepsilon}] = \mathbf{0}$, and $\text{var}\,\boldsymbol{\varepsilon} = \sigma^2 I$. The function $\mathbf{t}'\boldsymbol{\beta}$ is estimable if and only if $\mathbf{t}'(X'X)^c(X'X) = \mathbf{t}'$ where $(X'X)^c$ is any conditional inverse for $X'X$.

Proof

Assume that $\mathbf{t}'(X'X)^c(X'X) = \mathbf{t}'$ and hence that $(X'X)[(X'X)^c]'\mathbf{t} = \mathbf{t}$. By property 3 of conditional inverses,

$$(X'X)(X'X)^c\mathbf{t} = \mathbf{t}$$

This implies that $(X'X)^c\mathbf{t}$ is a solution to the system $(X'X)\mathbf{z} = \mathbf{t}$. By Theorem 5.4.1 $\mathbf{t}'\boldsymbol{\beta}$ is estimable. To prove the converse, assume that $\mathbf{t}'\boldsymbol{\beta}$ is estimable. Then by Theorem 5.4.1 there exists a solution to the system $(X'X)\mathbf{z} = \mathbf{t}$. By Theorem 5.3.4, $(X'X)^c\mathbf{t}$ is a solution to the system. Substituting, we obtain

$$(X'X)(X'X)^c\mathbf{t} = \mathbf{t}$$

Taking transposes and noting that $[(X'X)^c]' = (X'X)^c$, it can be concluded that

$$\mathbf{t}'(X'X)^c(X'X) = \mathbf{t}'$$

as desired. ■

Example 5.4.1 illustrates the use of this theorem.

Example 5.4.1 Consider the linear system of Example 5.3.4. Here

$$(X'X) = \begin{bmatrix} 2 & 1 & 1 \\ 1 & 1 & 0 \\ 1 & 0 & 1 \end{bmatrix} \quad \text{and} \quad (X'X)^c = \begin{bmatrix} 1 & -1 & 0 \\ -1 & 2 & 0 \\ 0 & 0 & 0 \end{bmatrix}$$

Multiplying, we obtain

$$(X'X)^c(X'X) = \begin{bmatrix} 1 & 0 & 1 \\ 0 & 1 & -1 \\ 0 & 0 & 0 \end{bmatrix}$$

To see if β_0 is estimable, we write this parameter as $\beta_0 = \mathbf{t}'\boldsymbol{\beta}$ where $\mathbf{t}' = [1 \quad 0 \quad 0]$. Note that

$$\mathbf{t}'(X'X)^c(X'X) = [1 \quad 0 \quad 0]\begin{bmatrix} 1 & 0 & 1 \\ 0 & 1 & -1 \\ 0 & 0 & 0 \end{bmatrix} = [1 \quad 0 \quad 1] \neq \mathbf{t}'$$

From this it can be concluded that β_0 is not estimable and hence that $\boldsymbol{\beta}$ is not estimable. To see if β_1 is estimable, write β_1 as $\beta_1 = \mathbf{t}'\boldsymbol{\beta}$ where $\mathbf{t}' = [0 \quad 1 \quad 0]$.

$$\mathbf{t}'(X'X)^c(X'X) = [0 \quad 1 \quad 0]\begin{bmatrix} 1 & 0 & 1 \\ 0 & 1 & -1 \\ 0 & 0 & 0 \end{bmatrix} = [0 \quad 1 \quad -1] \neq \mathbf{t}'$$

This parameter is not estimable. A quick calculation will show that β_2 is also nonestimable. Now consider the function $\beta_1 - \beta_2$. This function is written as

$$\beta_1 - \beta_2 = \mathbf{t}'\boldsymbol{\beta}$$

where $\mathbf{t}' = [0 \quad 1 \quad -1]$. Multiplying, we see that

$$\mathbf{t}'(X'X)^c(X'X) = [0 \quad 1 \quad -1]\begin{bmatrix} 1 & 0 & 1 \\ 0 & 1 & -1 \\ 0 & 0 & 0 \end{bmatrix} = [0 \quad 1 \quad -1] = \mathbf{t}'$$

Theorem 5.4.2 implies that $\beta_1 - \beta_2$ is estimable.

As indicated earlier, in this section linear functions of real interest to statisticians are those that are invariant to the choice of a solution to the normal equations. We said that this property is tied to the notion of estimability. Lemma 5.4.2 allows us to establish this result.

Lemma 5.4.1 Let $\mathbf{y} = X\boldsymbol{\beta} + \boldsymbol{\varepsilon}$ where X is $n \times p$ of rank $r \leqslant p$, $E[\boldsymbol{\varepsilon}] = \mathbf{0}$, and var $\boldsymbol{\varepsilon} = \sigma^2 I$. The best linear unbiased estimator for any estimable function $\mathbf{t}'\boldsymbol{\beta}$ is $\mathbf{z}'X'\mathbf{y}$ where $\mathbf{z}$ is a solution to the system $(X'X)\mathbf{z} = \mathbf{t}$.

Proof

Assume that the best linear unbiased estimator for $\mathbf{t}'\boldsymbol{\beta}$ is $\mathbf{k}'\mathbf{y}$. Let $\mathbf{z}_0$ denote a solution to the system $(X'X)\mathbf{z} = \mathbf{t}$. Write $\mathbf{k}'$ in the form

$$\mathbf{k}' = \mathbf{z}_0'X' + \mathbf{c}'$$

It must be shown that $\mathbf{c} = \mathbf{0}$. Since $\mathbf{k}'\mathbf{y}$ is an unbiased estimator for $\mathbf{t}'\boldsymbol{\beta}$,

$$E[\mathbf{k}'\mathbf{y}] = \mathbf{k}'E[\mathbf{y}] = \mathbf{k}'X\boldsymbol{\beta} = \mathbf{t}'\boldsymbol{\beta}$$

Substituting for $\mathbf{k}'$, we obtain

$$(\mathbf{z}_0'X' + \mathbf{c}')X\boldsymbol{\beta} = (\mathbf{z}_0'X'X + \mathbf{c}'X)\boldsymbol{\beta} = \mathbf{t}'\boldsymbol{\beta}$$

Since $\mathbf{z}_0$ is a solution to the system $(X'X)\mathbf{z} = \mathbf{t}$, $(X'X)\mathbf{z}_0 = \mathbf{t}$ and $\mathbf{z}_0'X'X = \mathbf{t}'$. Hence

$$(\mathbf{t}' + \mathbf{c}'X)\boldsymbol{\beta} = \mathbf{t}'\boldsymbol{\beta}$$

From this it can be concluded that $\mathbf{c}'X = \mathbf{0}'$. For $\mathbf{k}'\mathbf{y}$ to be the best linear unbiased estimator for $\mathbf{t}'\boldsymbol{\beta}$, its variance is no larger than that of any other linear unbiased estimator for $\mathbf{t}'\boldsymbol{\beta}$. Now

$$\begin{aligned} \text{var}\,(\mathbf{k}'\mathbf{y}) &= E[(\mathbf{k}'\mathbf{y} - \mathbf{t}'\boldsymbol{\beta})^2] \\ &= E[(\mathbf{k}'\mathbf{y} - \mathbf{t}'\boldsymbol{\beta})(\mathbf{k}'\mathbf{y} - \mathbf{t}'\boldsymbol{\beta})'] \end{aligned}$$

Since $\mathbf{y} = X\boldsymbol{\beta} + \boldsymbol{\varepsilon}$,

$$\text{var}\,(\mathbf{k}'\mathbf{y}) = E\{[\mathbf{k}'(X\boldsymbol{\beta} + \boldsymbol{\varepsilon}) - \mathbf{t}'\boldsymbol{\beta}][\mathbf{k}'(X\boldsymbol{\beta} + \boldsymbol{\varepsilon}) - \mathbf{t}'\boldsymbol{\beta}]'\}$$

Letting $\mathbf{k}' = \mathbf{z}_0' X' + \mathbf{c}'$ where $\mathbf{c}'X = \mathbf{0}'$, this variance becomes

$$\begin{aligned}\text{var}(\mathbf{k}'\mathbf{y}) &= E\{[\mathbf{z}_0'X'X\boldsymbol{\beta} + \mathbf{c}'X\boldsymbol{\beta} + \mathbf{z}_0'X'\boldsymbol{\varepsilon} + \mathbf{c}'\boldsymbol{\varepsilon} - \mathbf{t}'\boldsymbol{\beta}]\\ &\quad \cdot [\mathbf{z}_0'X'X\boldsymbol{\beta} + \mathbf{c}'X\boldsymbol{\beta} + \mathbf{z}_0'X'\boldsymbol{\varepsilon} + \mathbf{c}'\boldsymbol{\varepsilon} - \mathbf{t}'\boldsymbol{\beta}]'\}\\ &= E\{[\mathbf{z}_0'X'X\boldsymbol{\beta} + \mathbf{z}_0'X'\boldsymbol{\varepsilon} + \mathbf{c}'\boldsymbol{\varepsilon} - \mathbf{t}'\boldsymbol{\beta}]\\ &\quad \cdot [\mathbf{z}_0'X'X\boldsymbol{\beta} + \mathbf{z}_0'X'\boldsymbol{\varepsilon} + \mathbf{c}'\boldsymbol{\varepsilon} - \mathbf{t}'\boldsymbol{\beta}]'\}\\ &= E\{[\mathbf{z}_0'X'X\boldsymbol{\beta} + \mathbf{k}'\boldsymbol{\varepsilon} - \mathbf{t}'\boldsymbol{\beta}][\mathbf{z}_0'X'X\boldsymbol{\beta} + \mathbf{k}'\boldsymbol{\varepsilon} - \mathbf{t}'\boldsymbol{\beta}]'\}\end{aligned}$$

Using the fact that $\mathbf{z}_0'X'X = \mathbf{t}'$, we obtain

$$\begin{aligned}\text{var}(\mathbf{k}'\mathbf{y}) &= E\{[\mathbf{t}'\boldsymbol{\beta} + \mathbf{k}'\boldsymbol{\varepsilon} - \mathbf{t}'\boldsymbol{\beta}][\mathbf{t}'\boldsymbol{\beta} + \mathbf{k}'\boldsymbol{\varepsilon} - \mathbf{t}'\boldsymbol{\beta}]'\}\\ &= E[(\mathbf{k}'\boldsymbol{\varepsilon})(\mathbf{k}'\boldsymbol{\varepsilon})']\\ &= E[\mathbf{k}'\boldsymbol{\varepsilon}\boldsymbol{\varepsilon}'\mathbf{k}]\\ &= \mathbf{k}'E[\boldsymbol{\varepsilon}\boldsymbol{\varepsilon}']\mathbf{k}\end{aligned}$$

By definition, $E[\boldsymbol{\varepsilon}\boldsymbol{\varepsilon}'] = \text{var}\,\boldsymbol{\varepsilon} = \sigma^2 I$. Hence

$$\text{var}(\mathbf{k}'\mathbf{y}) = \sigma^2\mathbf{k}'\mathbf{k}$$

where $\mathbf{k}' = \mathbf{z}_0'X' + \mathbf{c}'$. Substituting, we conclude that

$$\begin{aligned}\text{var}(\mathbf{k}'\mathbf{y}) &= \sigma^2(\mathbf{z}_0'X' + \mathbf{c}')(X\mathbf{z}_0 + \mathbf{c})\\ &= \sigma^2(\mathbf{z}_0'X'X\mathbf{z}_0 + \mathbf{z}_0'X'\mathbf{c} + \mathbf{c}'X\mathbf{z}_0 + \mathbf{c}'\mathbf{c})\end{aligned}$$

Since $\mathbf{c}'X = \mathbf{0}'$,

$$\text{var}(\mathbf{k}'\mathbf{y}) = \sigma^2(\mathbf{z}_0'X'X\mathbf{z}_0 + \mathbf{c}'\mathbf{c})$$

Since $\mathbf{z}_0'X'X\mathbf{z}_0$ is fixed in value, var $(\mathbf{k}'\mathbf{y})$ is at a minimum when $\mathbf{c}'\mathbf{c}$ is at a minimum. Since $\mathbf{c}'\mathbf{c} = \Sigma_{i=1}^n c_i^2$, this minimum occurs when $c_i = 0$, $i = 1, 2, \ldots, n$. That is, var $(\mathbf{k}'\mathbf{y})$ is at a minimum when $\mathbf{c} = \mathbf{0}$ as claimed. ∎

GAUSS–MARKOFF THEOREM

Lemma 5.4.1 is used primarily to establish Theorem 5.4.3. This theorem is a Gauss–Markoff theorem applicable to the less than full rank model, with the full rank model being a special case (see Theorem 3.2.3). It allows us to conclude two things. First, it is shown that if $\mathbf{t}'\boldsymbol{\beta}$ is estimable then the best linear unbiased estimate $\widehat{\mathbf{t}'\boldsymbol{\beta}} = \mathbf{z}'X'\mathbf{y}$ is invariant to the choice of $\mathbf{z}$ where $\mathbf{z}$ is a solution to the system $(X'X)\mathbf{z} = \mathbf{t}$. That is, the best linear unbiased estimate for $\mathbf{t}'\boldsymbol{\beta}$ is *unique*. Second, in the course of the proof it is shown that $\widehat{\mathbf{t}'\boldsymbol{\beta}} = \mathbf{t}'\mathbf{b}$ where $\mathbf{b}$ is any solution to the normal equations. Thus, the best linear unbiased estimate for $\mathbf{t}'\boldsymbol{\beta}$ is indeed $\mathbf{t}'\mathbf{b}$ and is invariant to the choice of $\mathbf{b}$ as claimed earlier.

Theorem 5.4.3 (A Gauss–Markoff theorem) Let $\mathbf{y} = X\boldsymbol{\beta} + \boldsymbol{\varepsilon}$ where X is $n \times p$ of rank $r \leqslant p$, $E[\boldsymbol{\varepsilon}] = \mathbf{0}$, and $\text{var}\,\boldsymbol{\varepsilon} = \sigma^2 I$. Let $\mathbf{t}'\boldsymbol{\beta}$ be estimable. Then any solution to the system $(X'X)\mathbf{z} = \mathbf{t}$ yields the same estimate for $\mathbf{t}'\boldsymbol{\beta}$. Furthermore, this best linear unbiased estimate is $\mathbf{t}'\mathbf{b}$ where $\mathbf{b}$ is any solution to the normal equations.

Proof

Assume that $\mathbf{z}_0$ and $\mathbf{z}_1$ are each solutions to the system $(X'X)\mathbf{z} = \mathbf{t}$ so that $(X'X)\mathbf{z}_0 = \mathbf{t}$, $(X'X)\mathbf{z}_1 = \mathbf{t}$, and $\mathbf{z}_0'X'X = \mathbf{z}_1'X'X = \mathbf{t}'$. Let $\mathbf{b}$ denote *any* solution to the normal equations and note that $(X'X)\mathbf{b} = X'\mathbf{y}$. Consider the estimates $\mathbf{z}_0'X'\mathbf{y}$ and $\mathbf{z}_1'X'\mathbf{y}$. Note that

$$\mathbf{z}_0'X'\mathbf{y} = \mathbf{z}_0'(X'X)\mathbf{b} = \mathbf{t}'\mathbf{b}$$

and

$$\mathbf{z}_1'X'\mathbf{y} = \mathbf{z}_1'(X'X)\mathbf{b} = \mathbf{t}'\mathbf{b}$$

From this it can be concluded that

$$\mathbf{z}_0'X'\mathbf{y} = \mathbf{z}_1'X'\mathbf{y}$$

The best linear unbiased estimate for $\mathbf{t}'\boldsymbol{\beta}$ is unique. Furthermore, the unique estimate for $\mathbf{t}'\boldsymbol{\beta}$ is $\mathbf{t}'\mathbf{b}$ where $\mathbf{b}$ is any solution to the normal equations. ■

Example 5.4.2 Consider the system of Example 5.4.1. It was shown that the linear function

$$\mathbf{t}'\boldsymbol{\beta} = [0 \quad 1 \quad -1]\boldsymbol{\beta} = \beta_1 - \beta_2$$

is estimable. In Example 5.3.3 it was shown that

$$\mathbf{b} = \begin{bmatrix} 8 \\ -2 \\ 0 \end{bmatrix} \quad \text{and} \quad \mathbf{b}_0 = \begin{bmatrix} 7 \\ -1 \\ 1 \end{bmatrix}$$

are each solutions to the normal equations. Theorem 5.4.3 guarantees that the best linear unbiased estimate for $\mathbf{t}'\boldsymbol{\beta}$ is unique. Furthermore, $\mathbf{b}$ or $\mathbf{b}_0$ or any other solution to the normal equations can be used to generate this estimate. Here, note that

$$\mathbf{t}'\mathbf{b} = [0 \quad 1 \quad -1]\begin{bmatrix} 8 \\ -2 \\ 0 \end{bmatrix} = -2 \quad \text{and} \quad \mathbf{t}'\mathbf{b}_0 = [0 \quad 1 \quad -1]\begin{bmatrix} 7 \\ -1 \\ 1 \end{bmatrix} = -2$$

as expected.

In this section the notion of estimability has been developed and a test for estimability derived. The most important points to remember in future applications follow:

1. In the less than full rank model interest centers on estimable functions of the form $\mathbf{t}'\boldsymbol{\beta}$.
2. Estimable functions can be estimated uniquely.
3. The unique estimate for such functions is $\mathbf{t}'\mathbf{b}$ where $\mathbf{b}$ is *any* solution to the normal equations.
4. $\mathbf{t}'\mathbf{b}$ is the best linear unbiased estimator for $\mathbf{t}'\boldsymbol{\beta}$.

In the next section some general theorems on estimability are presented.

5.5

SOME ESTIMABILITY THEOREMS

Although it is not hard to test a linear function for estimability via the criterion just developed, in practice it is convenient to have certain general information concerning the less than full rank model at your fingertips. In this section we present a few such results and illustrate their use in the context of the one-way classification model with fixed effects.

Recall that in the full rank model much interest centers on estimating the expected response, $E[y]$, for a given *arbitrary* set of x values. In the less than full rank model, primary interest centers on estimating the expected response at a set of x values that constitute a *row* of the X matrix itself. That is, we want to estimate $E[y_i]$ for $i = 1, 2, \ldots, n$. Since $\mathbf{y} = X\boldsymbol{\beta} + \boldsymbol{\varepsilon}$ where $E[\boldsymbol{\varepsilon}] = \mathbf{0}$, $E[\mathbf{y}] = X\boldsymbol{\beta}$. Hence we want to estimate each of the elements of $X\boldsymbol{\beta}$. Theorem 5.5.2 shows that these elements are in fact estimable.

Theorem 5.5.1 Let $\mathbf{y} = X\boldsymbol{\beta} + \boldsymbol{\varepsilon}$ where X is $n \times p$ of rank $r \leqslant p$, $E[\boldsymbol{\varepsilon}] = \mathbf{0}$, and var $\boldsymbol{\varepsilon} = \sigma^2 I$. Elements of $X\boldsymbol{\beta}$ are estimable.

Proof

To show that each element of $X\boldsymbol{\beta}$ is estimable, it must be shown that

$$\mathbf{x}_i'(X'X)^c(X'X) = \mathbf{x}_i'$$

where $\mathbf{x}_i' = [x_{i1} \quad x_{i2} \quad \ldots \quad x_{ip}]$ for $i = 1, 2, \ldots, n$. This is true if and only if

$$X(X'X)^c(X'X) = X$$

Property 4 of conditional inverses yields this result. ■

Example 5.5.1 Consider the one-way classification model with fixed effects where $k = 3$. For this model

$$X = \begin{bmatrix} 1 & 1 & 0 & 0 \\ 1 & 1 & 0 & 0 \\ \vdots & \vdots & \vdots & \vdots \\ 1 & 1 & 0 & 0 \\ \hline 1 & 0 & 1 & 0 \\ 1 & 0 & 1 & 0 \\ \vdots & \vdots & \vdots & \vdots \\ 1 & 0 & 1 & 0 \\ \hline 1 & 0 & 0 & 1 \\ 1 & 0 & 0 & 1 \\ \vdots & \vdots & \vdots & \vdots \\ 1 & 0 & 0 & 1 \end{bmatrix}, \quad \boldsymbol{\beta} = \begin{bmatrix} \mu \\ \tau_1 \\ \tau_2 \\ \tau_3 \end{bmatrix}, \quad \text{and} \quad \mathbf{y} = \begin{bmatrix} y_{11} \\ y_{12} \\ \vdots \\ y_{1n_1} \\ \hline y_{21} \\ y_{22} \\ \vdots \\ y_{2n_2} \\ \hline y_{31} \\ y_{32} \\ \vdots \\ y_{3n_3} \end{bmatrix}$$

It was shown earlier that $\boldsymbol{\beta}$ is not estimable. Recall that in this model the real parameters of interest are $\mu_1 = \mu + \tau_1$, $\mu_2 = \mu + \tau_2$, and $\mu_3 = \mu + \tau_3$, the mean responses for the three populations from which the samples are drawn, respectively. Are these means estimable? It is easy to see that they are by applying Theorem 5.5.1. We need only note that each of these means can be expressed as elements of $X\boldsymbol{\beta}$. In particular,

$$\mu_1 = [1 \quad 1 \quad 0 \quad 0]\boldsymbol{\beta}$$
$$\mu_2 = [1 \quad 0 \quad 1 \quad 0]\boldsymbol{\beta}$$
$$\mu_3 = [1 \quad 0 \quad 0 \quad 1]\boldsymbol{\beta}$$

Estimators for these parameters are

$$\hat{\mu}_1 = [1 \quad 1 \quad 0 \quad 0]\mathbf{b}$$
$$\hat{\mu}_2 = [1 \quad 0 \quad 1 \quad 0]\mathbf{b}$$
$$\hat{\mu}_3 = [1 \quad 0 \quad 0 \quad 1]\mathbf{b}$$

where $\mathbf{b}$ is any solution to the system of normal equations.

In the previous example, $k = 3$. This choice was arbitrary. It is easy to show that in the one-way classification model with fixed effects, functions of the form $\mu + \tau_i$, $i = 1, 2, \ldots, k$, are estimable regardless of the value chosen for k.

Intuitively we would expect a linear combination of estimable functions itself to be estimable. Theorem 5.5.2 shows that this is indeed the case.

Theorem 5.5.2 Let $\mathbf{t}_1'\boldsymbol{\beta}, \mathbf{t}_2'\boldsymbol{\beta}, \ldots, \mathbf{t}_k'\boldsymbol{\beta}$ be a collection of estimable functions. Let $z = a_1\mathbf{t}_1'\boldsymbol{\beta} + a_2\mathbf{t}_2'\boldsymbol{\beta} + \cdots + a_k\mathbf{t}_k'\boldsymbol{\beta}$ be a linear combination of these functions. Then z is estimable. Furthermore, the best linear unbiased estimator for z is $z = a_1\mathbf{t}_1'\mathbf{b} + a_2\mathbf{t}_2'\mathbf{b} + \cdots + a_k\mathbf{t}_k'\mathbf{b}$ where $\mathbf{b}$ is any solution to the system of normal equations.

Proof

Let

$$\begin{aligned} z &= a_1\mathbf{t}_1'\boldsymbol{\beta} + a_2\mathbf{t}_2'\boldsymbol{\beta} + \cdots + a_k\mathbf{t}_k'\boldsymbol{\beta} \\ &= (a_1\mathbf{t}_1' + a_2\mathbf{t}_2' + \cdots + a_k\mathbf{t}_k')\boldsymbol{\beta} \end{aligned}$$

Consider the vector

$$(a_1\mathbf{t}_1' + a_2\mathbf{t}_2' + \cdots + a_k\mathbf{t}_k')(X'X)^c(X'X) = \\ a_1\mathbf{t}_1'(X'X)^c X'X + a_2\mathbf{t}_2'(X'X)^c(X'X) + \cdots + a_k\mathbf{t}_k'(X'X)^c(X'X)$$

Since $\mathbf{t}_1'\boldsymbol{\beta}, \mathbf{t}_2'\boldsymbol{\beta}, \ldots, \mathbf{t}_k'\boldsymbol{\beta}$ are each estimable, Theorem 5.4.2 can be applied to conclude that

$$(a_1\mathbf{t}_1' + a_2\mathbf{t}_2' + \cdots + a_k\mathbf{t}_k')(X'X)^c(X'X) = a_1\mathbf{t}_1' + a_2\mathbf{t}_2' + \cdots + a_k\mathbf{t}_k'$$

and hence that z is estimable. From Theorem 5.4.3, the best linear unbiased estimator for z is

$$\begin{aligned} \hat{z} &= (a_1\mathbf{t}_1' + a_2\mathbf{t}_2' + \cdots + a_k\mathbf{t}_k')\mathbf{b} \\ &= a_1\mathbf{t}_1'\mathbf{b} + a_2\mathbf{t}_2'\mathbf{b} + \cdots + a_k\mathbf{t}_k'\mathbf{b} \end{aligned}$$

where $\mathbf{b}$ is any solution to the normal equations. ■

Of course, some linear combinations of estimable functions are of more interest to statisticians than others because they are of scientific significance. They convey practical information about the study being conducted. In the one-way classification model with fixed effects, important comparisons can be made by employing *treatment contrasts*. These are linear combinations of the parameters $\tau_1, \tau_2, \ldots, \tau_k$ of the form

$$a_1\tau_1 + a_2\tau_2 + \cdots + a_k\tau_k$$

where $\Sigma_{i=1}^k a_i = 0$. Example 5.5.2 shows that any such contrast is estimable.

Example 5.5.2 Consider the one-way classification model with fixed effects. Let

$$z = a_1\tau_1 + a_2\tau_2 + \cdots + a_k\tau_k$$

denote a treatment contrast. Note that

$$c = a_1(\mu + \tau_1) + a_2(\mu + \tau_2) + \cdots + a_k(\mu + \tau_k)$$

is estimable since it is a linear combination of the estimable functions $\mu + \tau_i$, $i = 1, 2, \ldots, k$. This function can be rewritten as

$$c = a_1\mu + a_1\tau_1 + a_2\mu + a_2\tau_2 + \cdots + a_k\mu + a_k\tau_k = \left(\sum_{i=1}^{k} a_i\right)\mu + a_1\tau_1 + a_2\tau_2 + \cdots + a_k\tau_k$$

Since z is a contrast, $\Sigma_{i=1}^{k} a_i = 0$, and hence the estimable function c is equal to z, implying that z is estimable. Contrasts of particular interest in this model are those of the form

$$\tau_i - \tau_j$$

for $i \neq j$. To see why this is true, note that

$$\begin{aligned} \tau_i - \tau_j &= (\mu + \tau_i) - (\mu + \tau_j) \\ &= \mu_i - \mu_j \end{aligned}$$

This contrast gives us a way to compare the mean of population i to that of population j.

You are probably aware of the fact that in the one-way classification model the difference in population means, $\mu_i - \mu_j$, is estimated via the difference in corresponding sample means, $\bar{y}_{i.} - \bar{y}_{j.}$. This intuitive procedure now can be fully justified based on the theory of linear models just developed. To see how this is done, consider a typical example.

Example 5.5.3 Consider the one-way classification model with fixed effects where $k = 3$ and $N = n_1 + n_2 + n_3$. Let us estimate $\mu_1 - \mu_2$. For this model

$$X = \begin{bmatrix} 1 & 1 & 0 & 0 \\ 1 & 1 & 0 & 0 \\ \vdots & \vdots & \vdots & \vdots \\ 1 & 1 & 0 & 0 \\ \hline 1 & 0 & 1 & 0 \\ 1 & 0 & 1 & 0 \\ \vdots & \vdots & \vdots & \vdots \\ 1 & 0 & 1 & 0 \\ \hline 1 & 0 & 0 & 1 \\ 1 & 0 & 0 & 1 \\ \vdots & \vdots & \vdots & \vdots \\ 1 & 0 & 0 & 1 \end{bmatrix}, \quad \mathbf{y} = \begin{bmatrix} y_{11} \\ y_{12} \\ \vdots \\ y_{1n_1} \\ \hline y_{21} \\ y_{22} \\ \vdots \\ y_{2n_2} \\ \hline y_{31} \\ y_{32} \\ \vdots \\ y_{3n_3} \end{bmatrix}, \quad \boldsymbol{\beta} = \begin{bmatrix} \mu \\ \tau_1 \\ \tau_2 \\ \tau_3 \end{bmatrix},$$

and

$$X'\mathbf{y} = \begin{bmatrix} \sum_{i=1}^{3} \sum_{j=1}^{n_i} y_{ij} \\ \sum_{j=1}^{n_1} y_{1j} \\ \sum_{j=1}^{n_2} y_{2j} \\ \sum_{j=1}^{n_3} y_{3j} \end{bmatrix}$$

The matrix $X'X$ is given by

$$X'X = \begin{bmatrix} N & n_1 & n_2 & n_3 \\ n_1 & n_1 & 0 & 0 \\ n_2 & 0 & n_2 & 0 \\ n_3 & 0 & 0 & n_3 \end{bmatrix}$$

and a conditional inverse for this matrix is

$$(X'X)^c = \begin{bmatrix} 0 & 0 & 0 & 0 \\ 0 & \dfrac{1}{n_1} & 0 & 0 \\ 0 & 0 & \dfrac{1}{n_2} & 0 \\ 0 & 0 & 0 & \dfrac{1}{n_3} \end{bmatrix}$$

A solution to the normal equations based on this conditional inverse is

$$\mathbf{b} = (X'X)^c X'\mathbf{y} = \begin{bmatrix} 0 \\ \bar{y}_{1.} \\ \bar{y}_{2.} \\ \bar{y}_{3.} \end{bmatrix}$$

(See Exercise 24.)

The difference in means, $\mu_1 - \mu_2$, can be written as

$$\mu_1 - \mu_2 = \tau_1 - \tau_2 = [1 \quad 1 \quad 0 \quad 0]\boldsymbol{\beta} - [1 \quad 0 \quad 1 \quad 0]\boldsymbol{\beta}$$

Applying Theorem 5.5.2, the best linear unbiased estimator for $\mu_1 - \mu_2$ is

$$\begin{aligned} \widehat{\mu_1 - \mu_2} &= [1 \quad 1 \quad 0 \quad 0]\mathbf{b} - [1 \quad 0 \quad 1 \quad 0]\mathbf{b} \\ &= \bar{y}_{1\cdot} - \bar{y}_{2\cdot} \end{aligned}$$

as expected.

A few other general results concerning estimable functions are outlined in Exercises 27 and 28.

5.6

ESTIMATING σ^2 IN THE LESS THAN FULL RANK MODEL

Recall that in the full rank model the variance, σ^2, is estimated unbiasedly by

$$s^2 = \frac{SS_{\text{Res}}}{(n-p)}$$

where n is the sample size, p is the *rank* of X as well as the number of parameters in the model, and SS_{Res} is given by

$$\begin{aligned} SS_{\text{Res}} &= (\mathbf{y} - X\mathbf{b})'(\mathbf{y} - X\mathbf{b}) \\ &= \mathbf{y}'[I - X(X'X)^{-1}X']\mathbf{y} \end{aligned}$$

Here $\mathbf{b}$ is the unique solution to the system of normal equations. In the less than full rank model, s^2 is defined similarly. In particular, the *residual sum of squares* is defined by

$$SS_{\text{Res}} = (\mathbf{y} - X\mathbf{b})'(\mathbf{y} - X\mathbf{b})$$

where $\mathbf{b}$ is any solution to the normal equations. Since elements of $X\boldsymbol{\beta}$ are estimable, $X\mathbf{b}$ is invariant to the choice of $\mathbf{b}$ and SS_{Res} is unique. This sum of squares reflects the variability in response about the average or expected response. To investigate the distributional properties of functions of SS_{Res}, it is useful to rewrite this sum of squares in an alternative form that parallels that used in the full rank model. This form is given in Theorem 5.6.1.

Theorem 5.6.1 The residual sum of squares can be written as

$$SS_{\text{Res}} = \mathbf{y}'[I - X(X'X)^c X']\mathbf{y}$$

where $(X'X)^c$ is any conditional inverse for $X'X$.

Proof

By definition,

$$SS_{\text{Res}} = (\mathbf{y} - X\mathbf{b})'(\mathbf{y} - X\mathbf{b})$$

Expanding the right-hand side, we obtain

$$\begin{aligned} SS_{\text{Res}} &= (\mathbf{y}' - \mathbf{b}'X')(\mathbf{y} - X\mathbf{b}) \\ &= \mathbf{y}'\mathbf{y} - \mathbf{b}'X'\mathbf{y} - \mathbf{y}'X\mathbf{b} + \mathbf{b}'X'X\mathbf{b} \\ &= \mathbf{y}'\mathbf{y} - 2\mathbf{y}'X\mathbf{b} + \mathbf{b}'X'X\mathbf{b} \end{aligned}$$

Since SS_{Res} is invariant to the choice of a solution to the normal equations, let $\mathbf{b} = (X'X)^c X'\mathbf{y}$ and $\mathbf{b}' = \mathbf{y}'X(X'X)^c$ where $(X'X)^c$ is any conditional inverse for $X'X$. Substituting,

$$SS_{\text{Res}} = \mathbf{y}'\mathbf{y} - 2\mathbf{y}'X(X'X)^c X'\mathbf{y} + \mathbf{y}'X(X'X)^c X'X(X'X)^c X'\mathbf{y}$$

By property 4 of conditional inverses,

$$X(X'X)^c X'X = X$$

Hence,

$$\begin{aligned} SS_{\text{Res}} &= \mathbf{y}'\mathbf{y} - 2\mathbf{y}'X(X'X)^c X'\mathbf{y} + \mathbf{y}'X(X'X)^c X'\mathbf{y} \\ &= \mathbf{y}'\mathbf{y} - \mathbf{y}'X(X'X)^c X'\mathbf{y} \\ &= \mathbf{y}'[I - X(X'X)^c X']\mathbf{y} \end{aligned}$$

as claimed. ■

In the less than full rank model, s^2 is given by

$$s^2 = \frac{SS_{\text{Res}}}{n - r}$$

where n denotes the overall sample size and r denotes the *rank* of X but does not indicate the number of parameters in the model. It is easy to show that s^2 is an unbiased estimator for σ^2.

Theorem 5.6.2 Let $\mathbf{y} = X\boldsymbol{\beta} + \boldsymbol{\varepsilon}$ where X is $n \times p$ of rank $r \leqslant p$, $E[\boldsymbol{\varepsilon}] = \mathbf{0}$, and var $\boldsymbol{\varepsilon} = \sigma^2 I$. Then s^2 is an unbiased estimator for σ^2.

Proof

From Theorem 5.6.1,

$$\begin{aligned} E[s^2] = E\left[\frac{SS_{\text{Res}}}{n-r}\right] &= \frac{1}{n-r} E[SS_{\text{Res}}] \\ &= \frac{1}{n-r} E[\mathbf{y}'(I - X(X'X)^c X')\mathbf{y}] \end{aligned}$$

By Theorem 2.2.1,

$$E[s^2] = \frac{1}{n-r}\{\text{tr}\,[I - X(X'X)^c X']\sigma^2 + (X\boldsymbol{\beta})'[I - X(X'X)^c X']X\boldsymbol{\beta}\}$$

Since $I - X(X'X)^c X'$ is symmetric and idempotent, Theorem 1.5.2 can be applied to conclude that

$$\text{tr}\,[I - X(X'X)^c X'] = r[I - X(X'X)^c X']$$

By property 6 of conditional inverses,

$$r[I - X(X'X)^c X'] = n - r$$

Substituting and expanding, we see that

$$E[s^2] = \frac{1}{n-r}[(n-r)\sigma^2 + \boldsymbol{\beta}'X'X\boldsymbol{\beta} - \boldsymbol{\beta}'X'X(X'X)^c X'X\boldsymbol{\beta}]$$

By definition of conditional inverses,

$$X'X(X'X)^c X'X = X'X$$

It can be concluded that

$$E[s^2] = \frac{1}{n-r}[(n-r)\sigma^2] = \sigma^2$$

as claimed. ■

In the one-way classification model with fixed effects, the estimator for σ^2 usually given is a pooled estimator. It is a weighted average of the k available sample variances. This was the estimator obtained in Section 5.2 by means of reparameterization. It is easy to show that this is also the estimator that results when the theory developed here is applied. Exercises 31 and 32 outline the idea.

5.7

INTERVAL ESTIMATION IN THE LESS THAN FULL RANK MODEL

Once point estimators have been found for estimable functions of interest, the natural question to ask is, Can confidence intervals be found for these functions? As with full rank models, this can be done only if we are willing to make a further assumption about the vector of random errors. Thus far we have assumed only that $\boldsymbol{\varepsilon}$ is a random vector with mean $\mathbf{0}$ and variance $\sigma^2 I$. We now make the additional assumption that this vector is normally distributed. Under these assumptions the distributions of certain important quadratic forms can be determined using the theorems developed in Chapter 2.

We begin by considering the distribution of SS_{Res}/σ^2 Since the derivation is similar to that given in the full rank case, the details of the proof are left as an exercise (see Exercise 34).

Theorem 5.7.1 Let $\mathbf{y} = X\boldsymbol{\beta} + \boldsymbol{\varepsilon}$ where X is $n \times p$ of rank $r \leq p$, $\boldsymbol{\beta}$ is a $p \times 1$ vector of parameters, and $\boldsymbol{\varepsilon}$ is an $n \times 1$ normally distributed random vector with mean $\mathbf{0}$ and variance $\sigma^2 I$.

Then

$$\frac{(n-r)s^2}{\sigma^2} = \frac{SS_{\text{Res}}}{\sigma^2}$$

follows a chi-squared distribution with $n-r$ degrees of freedom. ■

If previous trends persist, the t distribution can be used to generate confidence bounds on estimable functions $\mathbf{t}'\boldsymbol{\beta}$. Furthermore, these bounds should parallel those found in the full rank case with $(X'X)^{-1}$ replaced by the conditional inverse $(X'X)^c$. Thus, the bounds on $\mathbf{t}'\boldsymbol{\beta}$ are given by

Confidence bounds on $\mathbf{t}'\boldsymbol{\beta}$ $(n-r \text{ df})$: $\mathbf{t}'\mathbf{b} \pm t_{\alpha/2} s \sqrt{\mathbf{t}'(X'X)^c\mathbf{t}}$

To derive this result, two things must be done. First, the distribution of the estimator $\mathbf{t}'\mathbf{b}$ must be determined. Second, it must be shown that this estimator is independent of SS_{Res} so that a proper t random variable can be formed. The derivation then follows the usual procedure for creating confidence bounds. It is instructive to derive these bounds for yourself. Exercises 35–38 lead you through the derivation step by step. The theorem is illustrated by deriving a result that is probably familiar to you concerning the one-way classification model.

Example 5.7.1 Consider the one-way classification model with fixed effects where $k=3$. Let us find a general expression for a $100(1-\alpha)\%$ confidence interval on $\mu_1 - \mu_2$. Since $\mu_1 - \mu_2 = \tau_1 - \tau_2$, we want to find a confidence interval on the estimable function $\mathbf{t}'\boldsymbol{\beta}$ where $\mathbf{t}' = [0 \quad 1 \quad -1 \quad 0]$. From Example 5.5.3 it is known that

$$\widehat{\mu_1 - \mu_2} = \bar{y}_{1.} - \bar{y}_{2.}$$

and that

$$(X'X)^c = \begin{bmatrix} 0 & 0 & 0 & 0 \\ 0 & \dfrac{1}{n_1} & 0 & 0 \\ 0 & 0 & \dfrac{1}{n_2} & 0 \\ 0 & 0 & 0 & \dfrac{1}{n_3} \end{bmatrix}$$

Hence

$$\mathbf{t}'(X'X)^c\mathbf{t} = [0 \quad 1 \quad -1 \quad 0] \begin{bmatrix} 0 & 0 & 0 & 0 \\ 0 & \frac{1}{n_1} & 0 & 0 \\ 0 & 0 & \frac{1}{n_2} & 0 \\ 0 & 0 & 0 & \frac{1}{n_3} \end{bmatrix} \begin{bmatrix} 0 \\ 1 \\ -1 \\ 0 \end{bmatrix}$$

$$= \frac{1}{n_1} + \frac{1}{n_2}$$

Substituting into the expression given for confidence bounds on $\mathbf{t}'\boldsymbol{\beta}$, the desired confidence bounds are

$$(\bar{y}_{1.} - \bar{y}_{2.}) \pm t_{\alpha/2} s \sqrt{\frac{1}{n_1} + \frac{1}{n_2}}$$

where the t distribution involved has $n - 3$ degrees of freedom.

Computing Supplement F illustrates some of the ideas presented in this chapter.

EXERCISES

Section 5.1

1. Consider the one-way classification model

$$y_{ij} = \mu + \tau_i + \varepsilon_{ij} \qquad i = 1, 2 \quad j = 1, 2, 3$$

a. Find the design matrix for this model. What are its dimensions? What is its rank?
b. Suppose that $\mu + \tau_1 = 8$ and $\mu + \tau_2 = -4$. Find the numerical values of τ_1 and τ_2 if $\mu = 3$; $\mu = 5$; $\mu = 0$.

2. Consider the model

$$y_{ij} = \mu + \tau_i + \beta_j + \varepsilon_{ij} \qquad i = 1, 2 \quad j = 1, 2$$

Let

$$\boldsymbol{\beta} = \begin{bmatrix} \mu \\ \tau_1 \\ \tau_2 \\ \beta_1 \\ \beta_2 \end{bmatrix} \quad \text{and} \quad \mathbf{y} = \begin{bmatrix} y_{11} \\ y_{12} \\ y_{21} \\ y_{22} \end{bmatrix}$$

a. Find the design matrix for this model. What are its dimensions? What is its rank?
b. Suppose that

$$\mu + \tau_1 + \beta_1 = 3$$
$$\mu + \tau_1 + \beta_2 = 5$$
$$\mu + \tau_2 + \beta_1 = 8$$
$$\mu + \tau_2 + \beta_2 = 10$$

Find two different sets of real numbers $\{\mu, \tau_1, \tau_2, \beta_1, \beta_2\}$ that each describe the system.

Section 5.2

3. Prove that in the one-way classification model with $k = 3$,

$$s^2 = \frac{(n_1 - 1)s_1^2 + (n_2 - 1)s_2^2 + (n_3 - 1)s_3^2}{(n_1 - 1) + (n_2 - 1) + (n_3 - 1)}$$

4. A study of the five major coal seams in a particular region is conducted. The model assumed is the one-way classification model with $k = 5$.

a. Write the model in its less than full rank form and find the design matrix, X_{old}, for the model. What are the dimensions of X_{old}? What is its rank?
b. Write the model in reparameterized form and find the design matrix, X_{new}, for the new model. What are the dimensions of X_{new}? What is its rank?
c. The following data are obtained on the sulfur content of samples drawn from each of the five seams. The response is the percentage sulfur per core sample.

		Seam		
1	2	3	4	5
1.51	1.69	1.56	1.30	0.73
1.92	0.64	1.22	0.75	0.80
1.08	0.90	1.32	1.26	0.90
2.04	1.41	1.39	0.69	1.24
2.14	1.01	1.33	0.62	0.82
1.76	0.84	1.54	0.90	0.72
1.17	1.28	1.04	1.20	0.57
	1.59	2.25	0.32	1.18
		1.49		0.54
				1.30

Estimate μ_1, μ_2, μ_3, μ_4, and μ_5. Use the method derived in Section 5.2 to estimate σ^2. Verify that the value obtained is equal to the pooled variance usually seen in applied statistics courses.

d. Estimate the linear function $\mathbf{t}'\boldsymbol{\beta}$ where $\mathbf{t}' = [1 \quad 1 \quad 1 \quad 1 \quad -4]$ and

$$\boldsymbol{\beta} = \begin{bmatrix} \mu_1 \\ \mu_2 \\ \mu_3 \\ \mu_4 \\ \mu_5 \end{bmatrix}$$

5. Consider the less than full rank model

$$y_{ij} = \mu + \tau_i + \beta_j + \varepsilon_{ij} \qquad i = 1, 2 \quad j = 1, 2$$

This model can be reparameterized by defining

$$\mu_{ij} = \mu + \tau_i + \beta_j$$

a. Let

$$\boldsymbol{\beta}_{\text{new}} = \begin{bmatrix} \mu_{11} \\ \mu_{12} \\ \mu_{21} \\ \mu_{22} \end{bmatrix}$$

Find the design matrix X_{new} for the reparameterized model. What are the dimensions of X_{new}? What is its rank?

b. Estimate μ_{11}, μ_{12}, μ_{21}, and μ_{22}.

c. Can σ^2 be estimated in this model?

Section 5.3

6. Let

$$A = \begin{bmatrix} 1 & 2 & 5 & 2 \\ 3 & 7 & 12 & 4 \\ 0 & 1 & -3 & -2 \end{bmatrix}$$

a. Show that $r(A) = 2$.

b. Let

$$M = \begin{bmatrix} 1 & 2 \\ 3 & 7 \end{bmatrix}$$

Show that the conditional inverse based on this minor is

$$A^c = \begin{bmatrix} 7 & -2 & 0 \\ -3 & 1 & 0 \\ 0 & 0 & 0 \\ 0 & 0 & 0 \end{bmatrix}$$

c. Find another conditional inverse for A.

7. Let

$$A = \begin{bmatrix} 4 & 1 & 2 & 0 \\ 1 & 1 & 5 & 15 \\ 3 & 1 & 3 & 5 \end{bmatrix}$$

a. Show that $r(A) = 2$.

b. Let

$$M = \begin{bmatrix} 1 & 5 \\ 1 & 3 \end{bmatrix}$$

Show that the conditional inverse based on this minor is

$$A^c = \begin{bmatrix} 0 & 0 & 0 \\ 0 & -\frac{3}{2} & \frac{5}{2} \\ 0 & \frac{1}{2} & -\frac{1}{2} \\ 0 & 0 & 0 \end{bmatrix}$$

c. Find another conditional inverse for A.

*8. Prove that A^cA and AA^c are idempotent.

*9. Let A be an $n \times p$ matrix of rank r. Prove that $r(A^cA) = r$. (*Hint*: Consider the matrix AA^cA. Use property 5 of rank [Section 1.4] and the definition of conditional inverse to argue that $r \leqslant r(A^cA) \leqslant r$.)

*10. Prove that if A^c is a conditional inverse for A, then $(A^c)'$ is a conditional inverse for A'. That is, show that $(A^c)' = (A')^c$.

*11. A lemma from linear algebra states that if $M'M = 0$ then $M = 0$ [11] [8].

a. Prove this lemma. (*Hint*: Show that the jth diagonal element of $M'M = \Sigma_{i=1}^{n} m_{ij}^2$. Hence each element of M appears as a squared term on a main diagonal. Show that if $M \neq 0$, then $M'M \neq 0$, thus proving the contrapositive of the lemma.)

b. Use this lemma to show that $A = A(A'A)^cA'A$. (*Hint*: Let $M = A - A(A'A)^cA'A$.)

*12. Prove that $A' = A'A(A'A)^cA'$.

*13. Prove that $A(A'A)^cA'$ is symmetric and idempotent.

*14. Prove that $A(A'A)^cA'$ is unique. That is, prove that this product is invariant to the choice of a conditional inverse. (*Hint*: Let $(A'A)_1^c$ and $(A'A)_2^c$ each denote conditional inverses for $A'A$. Let $M = A(A'A)_1^cA' - A(A'A)_2A'$. Show that $M'M = 0$ and apply Exercises 11 and 12.)

*15. Let A be an $n \times p$ matrix of rank r. Prove that $r[A(A'A)^cA'] = r$. (*Hint*: Use the properties of rank to argue that $r[A(A'A)^cA'] \leqslant r$. Use property 4 of conditional inverses to show that $r = r[A(A'A)^cA'A] \leqslant r[A(A'A)^cA']$.)

*16. Prove that $I - A(A'A)^cA'$ is unique, symmetric, and idempotent and that $r[I - A(A'A)^cA'] = n - r$. (*Hint*: See Theorem 1.5.5.)

*17. Prove that in the full rank case, $A^c = A^{-1}$.

*18. Prove that $I - A^cA$ is idempotent.

19. Suppose that for a particular linear model,

$$X = \begin{bmatrix} 1 & 1 & 0 \\ 1 & 1 & 0 \\ 1 & 0 & 1 \\ 1 & 0 & 1 \end{bmatrix} \quad \text{and} \quad X'\mathbf{y} = \begin{bmatrix} 5 \\ 3 \\ 2 \end{bmatrix}$$

a. Show that $|X'X| = 0$ and hence that $(X'X)$ is singular.

b. Show that $r(X'X) = 2$.

c. Show that the system of normal equations is consistent.

d. Find a conditional inverse for $(X'X)$ based on the minor

$$M = \begin{bmatrix} 2 & 0 \\ 0 & 2 \end{bmatrix}$$

e. Use Theorem 5.3.4 to find a solution to the normal equations.
f. Use Theorem 5.3.5 to find two more solutions to the normal equations.
g. Find a conditional inverse for $X'X$ based on the minor

$$M = \begin{bmatrix} 4 & 2 \\ 2 & 2 \end{bmatrix}$$

h. Find a solution to the normal equations based on the conditional inverse of part g.
i. Use Theorem 5.3.6 to express the solution found in part e in terms of the conditional inverse found in part g.
j. Let $(X'X)_1^c$ and $(X'X)_2^c$ denote the conditional inverses of parts d and g, respectively. Show that $X(X'X)_1^c X' = X(X'X)_2^c X'$, thus illustrating property 5 of conditional inverses.

20. Suppose that for a particular linear model,

$$X = \begin{bmatrix} 1 & 1 \\ 1 & 1 \\ 1 & 1 \end{bmatrix} \quad \text{and} \quad X'\mathbf{y} = \begin{bmatrix} 6 \\ 6 \end{bmatrix}$$

a. Show that $|X'X| = 0$ and hence that $X'X$ is singular.
b. Show that $r(X'X) = 1$.
c. Show that the system of normal equations is consistent.
d. Find a conditional inverse for $X'X$ based on the minor

$$M_1 = m_{11} = 3$$

e. Use Theorem 5.3.4 to find a solution to the normal equations based on the conditional inverse of part d.
f. Let

$$\mathbf{z} = \begin{bmatrix} 0 \\ 1 \end{bmatrix}$$

Use Theorem 5.3.5 to find an additional solution to the normal equations.
g. Find a conditional inverse for $X'X$ based on the minor

$$M_2 = m_{22} = 3$$

h. Find a solution to the normal equation based on the conditional inverse of part g and Theorem 5.3.4.
i. Use Theorem 5.3.6 to express the solution of part h in terms of the conditional inverse found in part d.
j. Can the minor

$$M_3 = m_{12} = 3$$

be used to generate a conditional inverse for $X'X$? Explain.
k. Let $(X'X)_1^c$ and $(X'X)_2^c$ denote the conditional inverses found in parts d and g, respectively. Show that $X(X'X)_1^c X' = X(X'X)_2^c X'$, thus illustrating property 5 of conditional inverses.

Section 5.4

21. Let $\mathbf{y} = X\boldsymbol{\beta} + \boldsymbol{\varepsilon}$ where

$$X = \begin{bmatrix} 1 & 1 & 0 & 0 \\ 1 & 1 & 0 & 0 \\ 1 & 0 & 1 & 0 \\ 1 & 0 & 1 & 0 \\ 1 & 0 & 0 & 1 \\ 1 & 0 & 0 & 1 \end{bmatrix} \quad \text{and} \quad \mathbf{y} = \begin{bmatrix} 3 \\ 1 \\ 2 \\ 2 \\ 0 \\ 4 \end{bmatrix}$$

a. Find $X'X$ and show that $r(X'X) = 3$.
b. Show that the system of normal equations is consistent.
c. Find the conditional inverse for $X'X$ based on the minor M where

$$M = \begin{bmatrix} 2 & 0 & 0 \\ 0 & 2 & 0 \\ 0 & 0 & 2 \end{bmatrix}$$

d. Show that

$$(X'X)^c(X'X) = \begin{bmatrix} 0 & 0 & 0 & 0 \\ 1 & 1 & 0 & 0 \\ 1 & 0 & 1 & 0 \\ 1 & 0 & 0 & 1 \end{bmatrix}$$

e. Note that $\beta_0 = \mathbf{t}'\boldsymbol{\beta}$ where $\mathbf{t}' = [1 \quad 0 \quad 0 \quad 0]$. Show that β_0 is not estimable, thus showing that $\boldsymbol{\beta}$ is not estimable.
f. Write β_1 as a linear function of $\boldsymbol{\beta}$ and show that β_1 is *not* estimable.
g. Write β_3 as a linear function of $\boldsymbol{\beta}$ and show that β_3 is *not* estimable.
h. Consider the linear function $\beta_3 - \beta_1$. Is this function estimable?
i. Find two different solutions to the normal equations. Use each of them to estimate $\beta_3 - \beta_1$. Are these estimates identical as indicated in Theorem 5.4.3?

22. Consider the one-way classification model with fixed effects presented in Section 5.1. Assume that $k = 3$ and $n_1 = n_2 = n_3 = n$.
a. Show that

$$X'X = \begin{bmatrix} 3n & n & n & n \\ n & n & 0 & 0 \\ n & 0 & n & 0 \\ n & 0 & 0 & n \end{bmatrix}$$

b. Show that the conditional inverse for $X'X$ based on the minor M where

$$M = \begin{bmatrix} n & 0 & 0 \\ 0 & n & 0 \\ 0 & 0 & n \end{bmatrix}$$

is

$$(X'X)^c = \begin{bmatrix} 0 & 0 & 0 & 0 \\ 0 & \frac{1}{n} & 0 & 0 \\ 0 & 0 & \frac{1}{n} & 0 \\ 0 & 0 & 0 & \frac{1}{n} \end{bmatrix}$$

c. Show that

$$(X'X)^c(X'X) = \begin{bmatrix} 0 & 0 & 0 & 0 \\ 1 & 1 & 0 & 0 \\ 1 & 0 & 1 & 0 \\ 1 & 0 & 0 & 1 \end{bmatrix}$$

d. Show that $\boldsymbol{\beta}$ is not estimable.

e. Consider the linear function $\tau_1 - \tau_2$. Is this function estimable?

Section 5.5

23. Consider the one-way classification model with fixed effects, $k = 3$, and $n_1 = n_2 = n_3 = n$.

a. Find $X'\mathbf{y}$.

b. Use the conditional inverse of Exercise 22, part b, to find a solution to the normal equations.

c. Use the solution of part b to find estimators for μ_1, μ_2, and μ_3, respectively. Are these the same as those found via reparameterization in Section 5.2?

24. Verify that $X'\mathbf{y}$ and $X'X$ are as given in Example 5.5.3. Also verify by direct multiplication that the conditional inverse given is in fact a conditional inverse for $X'X$. Show that $\mathbf{b}$ is as claimed in the example.

25. Consider the model of Example 5.5.3.

a. Show that $\tau_1 + \tau_2 - 2\tau_3$ is a contrast.

b. Write this contrast in terms of μ_1, μ_2, and μ_3.

c. Use Theorem 5.5.2 to verify that

$$\widehat{\mu_1 + \mu_2 - 2\mu_3} = \bar{y}_{1.} + \bar{y}_{2.} - 2\bar{y}_{3.}$$

26. An experiment is conducted to compare two drugs for effectiveness in reducing high blood pressure. The drug can be bought in either pill or liquid form. Thus four possible treatments are available to a physician:

Treatment 1: Drug A in pill form
Treatment 2: Drug B in pill form
Treatment 3: Drug A in liquid form
Treatment 4: Drug B in liquid form

The measured response is the percentage decrease in blood pressure 15 minutes after administration of the drug. Assume the one-way classification model with fixed effects with $k = 4$ and $n_1 = n_2 = n_3 = n_4 = 5$.

a. Find the X matrix for the model.
b. Find $X'X$.
c. Find the conditional inverse for $X'X$ based on the minor M where

$$M = \begin{bmatrix} n_1 & 0 & 0 & 0 \\ 0 & n_2 & 0 & 0 \\ 0 & 0 & n_3 & 0 \\ 0 & 0 & 0 & n_4 \end{bmatrix}$$

d. Assume that when the experiment is conducted,

$$\sum_{j=1}^{5} y_{1j} = 35, \quad \sum_{j=1}^{5} y_{2j} = 30, \quad \sum_{j=1}^{5} y_{3j} = 51, \quad \text{and} \quad \sum_{j=1}^{5} y_{4j} = 32$$

Find $X'\mathbf{y}$.
e. Find the solution to the normal equations based on the conditional inverse of part c.
f. Estimate the difference in average response using treatments 1 and 2.
g. The contrast $(\tau_1 + \tau_2) - (\tau_3 + \tau_4)$ can be used to compare pills to liquids. For this contrast $\mathbf{t}' = [0 \quad 1 \quad 1 \quad -1 \quad -1]$. Find $(X'X)^c(X'X)$ and verify that this contrast is estimable using Theorem 5.4.2. Estimate this contrast.
h. Find a contrast that can be used to compare drug A to drug B. Estimate this contrast.

***27.** Show that all elements of $(X'X)\boldsymbol{\beta}$ are estimable. (*Hint*: Write $(X'X)\boldsymbol{\beta}$ as $X'(X\boldsymbol{\beta})$ and apply Theorems 5.5.1 and 5.5.2.)

***28.** Show that all estimable functions can be written as linear combinations of elements of $X\boldsymbol{\beta}$. (*Hint*: Let $\mathbf{t}'\boldsymbol{\beta}$ be estimable. Apply Theorem 5.4.2 to see how to write $\mathbf{t}'$ in the form $\mathbf{c}'X\boldsymbol{\beta}$ as required.)

Section 5.6

29. a. Find the numerical value of $X\mathbf{b}$ for each of the solutions to the normal equations found in Exercise 21. Did you get the same answer each time?
b. Use the data of Exercise 21 to estimate σ^2.

***30.** Show that $SS_{\text{Res}} = \mathbf{y}'\mathbf{y} - \mathbf{b}'X'\mathbf{y}$ where $\mathbf{b}$ is any solution to the normal equations.

31. Consider the data of Exercise 26. Assume that $\Sigma_{i=1}^{4} \Sigma_{j=1}^{5} y_{ij}^2 = 1210.28$. Use the results of Exercise 30 to find s^2.

32. Consider the one-way classification model with fixed effects and $k = 3$ of Example 5.5.3.
a. Show that $r(X) = 3$.
b. It is known that the respective sample variances are given by

$$s_i^2 = \left[\sum_{j=1}^{n_i} y_{ij}^2 - \frac{\left(\sum_{j=1}^{n_i} y_{ij}\right)^2}{n_i}\right] \Big/ (n_i - 1) \qquad i = 1, 2, 3$$

Show that

$$s^2 = \frac{(n_1 - 1)s_1^2 + (n_2 - 1)s_2^2 + (n_3 - 1)s_3^2}{n_1 + n_2 + n_3 - 3}$$

Compare this to the estimator found via reparameterization in Section 5.2. (*Hint*: Apply Exercise 30.)

***33.** Show that in the one-way classification model with fixed effects $r(X) = k$. Show that in this case,

$$s^2 = \sum_{i=1}^{k} \frac{(n_i - 1)s_i^2}{N - k} = \frac{SS_{\text{Res}}}{N - k}$$

where $N = n_1 + n_2 + \cdots + n_k$.

Section 5.7

34. Prove Theorem 5.7.1. (*Hint*: See Corollary 2.3.2 and property 6 of conditional inverses.)

***35.** Show that in the less than full rank model any solution to the system of normal equations can be expressed in the form

$$\mathbf{b} = (X'X)^c X'\mathbf{y} + [I - (X'X)^c X'X]\mathbf{z}$$

where $(X'X)^c$ is an arbitrary conditional inverse for $X'X$ and $\mathbf{z} = [I - (X'X)^c X'X]\mathbf{b}$.

***36.** In this exercise you will derive the distribution of $\mathbf{t}'\mathbf{b}$, the best linear unbiased estimator of $\mathbf{t}'\boldsymbol{\beta}$ under the assumption that $\boldsymbol{\varepsilon}$ is normally distributed with mean $\mathbf{0}$ and variance $\sigma^2 I$. Remember throughout that $\mathbf{t}'\boldsymbol{\beta}$ is assumed to be estimable so that $\mathbf{t}'(X'X)^c X'X = \mathbf{t}'$ and that $\mathbf{b}$ denotes any solution to the normal equations.

a. Argue that $\mathbf{t}'\mathbf{b}$ is normally distributed by applying Exercise 35 to express $\mathbf{t}'\mathbf{b}$ in terms of $\mathbf{y}$.
b. Argue that $E[\mathbf{t}'\mathbf{b}] = \mathbf{t}'\boldsymbol{\beta}$.
c. Show that var $\mathbf{t}'\mathbf{b} = \mathbf{t}'(X'X)^c\mathbf{t}\sigma^2$.
d. Show that the random variable

$$\frac{\mathbf{t}'\mathbf{b} - \mathbf{t}'\boldsymbol{\beta}}{\sigma\sqrt{\mathbf{t}'(X'X)^c\mathbf{t}}}$$

follows a standard normal distribution.

***37.** Assume that $\mathbf{t}'\boldsymbol{\beta}$ is estimable. Show that $\mathbf{t}'\mathbf{b}$ is independent of SS_{Res}. (*Hint*: Let $\mathbf{t}'\mathbf{b} = \mathbf{t}'(X'X)^c X'\mathbf{y}$ and write the residual sum of squares as

$$SS_{\text{Res}} = \mathbf{y}'[I - X(X'X)^c X']\mathbf{y}$$

Show that $\mathbf{t}'(X'X)^c X'\sigma^2 I[I - X(X'X)^c X'] = 0$ and apply Theorem 2.4.2.)

***38.** Show that the random variable

$$\frac{\mathbf{t}'\mathbf{b} - \mathbf{t}'\boldsymbol{\beta}}{s\sqrt{\mathbf{t}'(X'X)^c\mathbf{t}}}$$

follows a t distribution with $n - r$ degrees of freedom.

***39.** Derive the $100(1-\alpha)\%$ confidence bounds on the estimable function $\mathbf{t}'\boldsymbol{\beta}$.

40. Consider the one-way classification model with fixed effects and $k = 3$. Suppose that a $100(1-\alpha)\%$ confidence interval on $\mu_2 - \mu_3$ is desired.

a. Write $\mu_2 - \mu_3$ in the form $\mathbf{t}'\boldsymbol{\beta}$ where $\boldsymbol{\beta}' = [\mu \quad \tau_1 \quad \tau_2 \quad \tau_3]$.
b. Find the confidence bounds for $\mu_2 - \mu_3$.

41. Use the results of Exercises 21 and 29 to find a 95% confidence interval on $\beta_3 - \beta_1$.

42. Use the results of Exercises 26 and 31 to find 95% confidence intervals on $\mu_1 - \mu_2$ and on $(\mu_1 + \mu_2) - (\mu_3 + \mu_4)$.

43. Consider the data of Exercise 4. Find a 90% confidence interval on the difference in average sulfur content between seams 1 and 3. Based on the interval obtained, would you be surprised to hear someone claim that there is no difference in these two averages? Explain.

44. It is known that a toxic material was dumped in a river that flows into a large salt-water commercial fishing area. Civil engineers are interested in the amount of toxic material in parts per million found in oysters harvested at three different locations ranging from the estuary to the bay itself. These data result:

Site 1 (river)		Site 2 (bay)		Site 3 (estuary)	
15	29	19	20	22	15
26	28	15	13	26	17
20	21	10	15	24	24
20	26	26	18	26	
		11			

a. Assume that the model for the experiment is

$$y_{ij} = \mu + \tau_i + \varepsilon_{ij}$$

Find the expressions for X, $\mathbf{y}$, and $\boldsymbol{\beta}$.

b. Find $X'X$ and $X'\mathbf{y}$.

c. Via SAS, a conditional inverse for $X'X$ is

$$(X'X)^c = \begin{bmatrix} 0.14285714 & -0.14285714 & -0.14285714 & 0 \\ -0.14285714 & 0.26785714 & 0.14285714 & 0 \\ -0.14285714 & 0.14285714 & 0.25396825 & 0 \\ 0 & 0 & 0 & 0 \end{bmatrix}$$

Use this to find a solution to the normal equations.

d. The engineers want to compare the amount of toxic material found in the river to that found in the bay. That is, they want to compare μ_1 with μ_2. Find an estimable function, $\mathbf{t}'\boldsymbol{\beta}$, that will enable this comparison to be made. Estimate $\mu_1 - \mu_2$ via $\mathbf{t}'\mathbf{b}$ where $\mathbf{b}$ is the solution to the normal equations found in part c.

e. Using SAS, s^2 is found to be 22.80357143. Use this information to find a 95% confidence interval on $\mu_1 - \mu_2$. Based on this interval, can it be concluded that there is a difference in the average amounts of toxic material found in these two locations? Explain.

f. Use the results of Exercise 3 to verify the estimated variance found by SAS.

COMPUTING SUPPLEMENT F (Finding Conditional Inverses)

The SAS procedure PROC GLM can be used to find conditional inverses and to estimate σ^2 and $\boldsymbol{\beta}$ in the less than full rank model. The information gained can then be used easily to generate point or interval estimates for estimable functions $\mathbf{t}'\boldsymbol{\beta}$. The data of Exercise 44 are used to illustrate the idea.

Statement	Purpose
`DATA TOXIC;`	Names data set
`INPUT Y X1 X2 X3;`	Names variables
`CARDS;`	Signals that the data follow
`15 1 0 0`	Data lines
`26 1 0 0`	
`20 1 0 0`	
`20 1 0 0`	
`29 1 0 0`	
`28 1 0 0`	
`21 1 0 0`	
`26 1 0 0`	
`19 0 1 0`	
`15 0 1 0`	
`10 0 1 0`	
`26 0 1 0`	
`11 0 1 0`	
`20 0 1 0`	
`13 0 1 0`	
`15 0 1 0`	
`18 0 1 0`	
`22 0 0 1`	
`26 0 0 1`	
`24 0 0 1`	
`26 0 0 1`	
`15 0 0 1`	
`17 0 0 1`	
`24 0 0 1`	
`;`	Signals end of data
`PROC GLM;`	Calls for general linear models procedure
`MODEL Y=X1 X2 X3 / XPX I;`	Identifies the independent variables as X1, X2, and X3; names Y as the response; asks for the matrices $X'X$ and $(X'X)^c$ to be found and printed
`TITLE1 FINDING A CONDITIONAL;`	Titles the output
`TITLE2 INVERSE AND ESTIMATING;`	
`TITLE3 THE VARIANCE IN THE LESS;`	
`TITLE4 THAN FULL RANK MODEL;`	
`ESTIMATE 'T1-T2' X11 X2 -1;`	'$T_1 - T_2$' is a label for the contrast $\tau_1 - \tau_2$; X1 1 X2 −1 forms the vector $\mathbf{t}' = [0 \quad 1 \quad -1 \quad 0]$ that forms the contrast $\tau_1 - \tau_2$
`ESTIMATE 'T1' X1 1;`	'T1' is a label that says we are trying to estimate τ_1; X1 1 forms the vector $\mathbf{t}' = [0 \quad 1 \quad 0 \quad 0]$ used to express τ_1 in the form $\mathbf{t}'\boldsymbol{\beta}$

The output of this program is shown in Figure 5.1.

The matrix $X'X$ is shown by ①. A conditional inverse is found at ②. SAS refers to this inverse as being a generalized inverse. SAS also notes that $X'X$ is singular in its user NOTE: shown by ③. The estimated variance is given by ④. The estimated difference between μ_1 and μ_2 is given by ⑤ and its standard error, $s\sqrt{\mathbf{t}'(X'X)^c\mathbf{t}}$, is given by ⑥. When SAS is asked to estimate a nonestimable function, the message NON-EST is printed as in ⑦. Estimates for μ, τ_1, τ_2, and τ_3 are given in ⑧. As indicated in the user note ③, these estimates are not unique. They are based on the conditional inverse found here.

FIGURE 5.1

```
                         FINDING A CONDITIONAL                              1
                          INVERSE AND ESTIMATING
                         THE VARIANCE IN THE LESS
                           THAN FULL RANK MODEL

                      GENERAL LINEAR MODELS PROCEDURE

                             THE X'X MATRIX

DEPENDENT VARIABLE : Y

                 INTERCEPT          X1              X2              X3

   INTERCEPT    24.00000000     8.00000000      9.00000000      7.00000000
   X1            8.00000000     8.00000000      0.00000000      0.00000000   ①
   X2            9.00000000     0.00000000      9.00000000      0.00000000
   X3            7.00000000     0.00000000      0.00000000      7.00000000

                         FINDING A CONDITIONAL                              2
                          INVERSE AND ESTIMATING
                         THE VARIANCE IN THE LESS
                           THAN FULL RANK MODEL

                      GENERAL LINEAR MODELS PROCEDURE

                     X'X GENERALIZED INVERSE (G2)

DEPENDENT VARIABLE : Y

                 INTERCEPT          X1              X2              X3

   INTERCEPT     0.14285714    -0.14285714     -0.14285714      0.00000000
   X1           -0.14285714     0.26785714      0.14285714      0.00000000
   X2           -0.14285714     0.14285714      0.25396825      0.00000000   ②
   X3            0.00000000     0.00000000      0.00000000      0.00000000
```

FIGURE 5.1 Continued

FIGURE 5.1 Continued

```
                    FINDING A CONDITIONAL                              3
                    INVERSE AND ESTIMATING
                    THE VARIANCE IN THE LESS
                    THAN FULL RANK MODEL

               GENERAL LINEAR MODELS PROCEDURE

DEPENDENT VARIABLE: Y

SOURCE                 DF      SUM OF SQUARES        MEAN SQUARE        F VALUE

MODEL                   2        225.62500000       112.81250000           4.95

ERROR                  21        478.87500000        22.80357143 (4)     PR > F

CORRECTED TOTAL        23        704.50000000                            0.0174


R-SQUARE             C.V.            ROOT MSE             Y MEAN

0.320263          23.5818          4.77530852        20.25000000


SOURCE                 DF         TYPE I SS    F VALUE     PR > F

X1                      1       99.18750000       4.35     0.0494
X2                      1      126.43750000       5.54     0.0283
X3                      0        0.00000000          .          .


SOURCE                 DF       TYPE III SS    F VALUE     PR > F

X1                      0        0.00000000          .          .
X2                      0        0.00000000          .          .
X3                      0        0.00000000          .          .


                                        T FOR H0:      PR > |T|      STD ERROR OF
PARAMETER              ESTIMATE       PARAMETER=0                     ESTIMATE

INTERCEPT          22.00000000 B            12.19        0.0001        1.80489697
X1                  1.12500000 B             0.46        0.6536        2.47145696
X2             (8) -5.66666667 B            -2.35        0.0283        2.40652929
X3                  0.00000000 B              .            .              .


NOTE: THE X'X MATRIX HAS BEEN DEEMED SINGULAR AND A GENERALIZED
      INVERSE HAS BEEN EMPLOYED TO SOLVE THE NORMAL EQUATIONS.
      THE ABOVE ESTIMATES REPRESENT ONLY ONE OF MANY POSSIBLE
      SOLUTIONS TO THE NORMAL EQUATIONS. ESTIMATES FOLLOWED BY
      THE LETTER B ARE BIASED AND DO NOT ESTIMATE THE PARAMETER
 (3)  BUT ARE BLUE FOR SOME LINEAR COMBINATION OF PARAMETERS
      (OR ARE ZERO). THE EXPECTED VALUE OF THE BIASED ESTIMATORS
      MAY BE OBTAINED FROM THE GENERAL FORM OF ESTIMABLE
      FUNCTIONS. FOR THE BIASED ESTIMATORS, THE STD ERR IS THAT
      OF THE BIASED ESTIMATOR AND THE T VALUE TESTS
      H0: E(BIASED ESTIMATOR) = 0. ESTIMATES NOT FOLLOWED BY THE
      LETTER B ARE BLUE FOR THE PARAMETER.
```

FIGURE 5.1 Continued

```
                 FINDING A CONDITIONAL                          4
                 INVERSE AND ESTIMATING
                THE VARIANCE IN THE LESS
                 THAN FULL RANK MODEL

            GENERAL LINEAR MODELS PROCEDURE

DEPENDENT VARIABLE: Y

                                T FOR H0:     PR > |T|       STD ERROR OF
PARAMETER         ESTIMATE    PARAMETER=0                      ESTIMATE

t1-t2          6.79166667 (5)        2.93       0.0081     (6) 2.32038285
t1                NON-EST (7)
```

CHAPTER 6

HYPOTHESIS TESTING IN THE LESS THAN FULL RANK MODEL

Hypothesis testing in the less than full rank model is approached in two different ways. First, the theory developed in Section 4.6 can be used to test hypotheses on estimable functions of $\boldsymbol{\beta}$. Second, specific models can be reparameterized to form full rank models. Hypotheses concerning the reparameterized models can then be tested via the methods developed in Chapter 4. We begin by considering the first approach.

6.1 HYPOTHESIS TESTING IN A GENERAL SETTING

Consider the less than full rank model

$$\mathbf{y} = X\boldsymbol{\beta} + \boldsymbol{\varepsilon}$$

where X is $n \times p$ of rank $r < p$. Assume that $\boldsymbol{\varepsilon}$ is a normally distributed random vector with mean $\mathbf{0}$ and variance $\sigma^2 I$. Various hypotheses concerning the model parameters can be posed. However, as you should suspect, since not all linear functions of $\boldsymbol{\beta}$ are estimable, it is not always possible to test a given hypothesis. When a test is possible, the hypothesis is said to be *testable*. More formally, a testable hypothesis is one that satisfies the following definition.

Definition 6.1.1

A hypothesis H_0 is said to be testable whenever there exists a set of estimable functions $\mathbf{c}_1'\boldsymbol{\beta}, \mathbf{c}_2'\boldsymbol{\beta}, \ldots, \mathbf{c}_m'\boldsymbol{\beta}$ such that H_0 is true if and only if

$$\mathbf{c}_1'\boldsymbol{\beta} = \mathbf{c}_2'\boldsymbol{\beta} = \cdots = \mathbf{c}_m'\boldsymbol{\beta} = \mathbf{0}$$

where $\mathbf{c}_1', \mathbf{c}_2', \ldots, \mathbf{c}_m'$ are linearly independent.

It can be shown that in the less than full rank model the number of linearly independent estimable functions is r, the rank of X. Hence it is easy to see that a testable hypothesis assumes the general form

$$H_0: C\boldsymbol{\beta} = \mathbf{0}$$

where C is an $n \times p$ matrix of rank $m \leqslant r$. Recall that an individual linear function $\mathbf{c}'\boldsymbol{\beta}$ is estimable if and only if $\mathbf{c}'(X'X)^c X'X = \mathbf{c}'$. If this fact is applied to each row of the matrix C associated with a testable hypothesis, then it can be concluded that the hypothesis is testable if and only if $C(X'X)^c X'X = C$. Example 6.1.1 illustrates the notion of testability in a familiar setting.

Example 6.1.1 Consider the one-way classification model with fixed effects where $k = 3$ and $N = n_1 + n_2 + n_3$. The desired model is

$$y_{ij} = \mu + \tau_i + \varepsilon_{ij} \qquad \begin{aligned} i &= 1, 2, 3 \\ j &= 1, 2, \ldots, n_i \end{aligned}$$

The most useful hypothesis to test concerning this model is probably the null hypothesis that $\tau_1 = \tau_2 = \tau_3$. This, in effect, tests the null hypothesis of equality of means among the three normal populations from which the samples are drawn. It is easy to see that $\tau_1 = \tau_2 = \tau_3$ if and only if

$$\tau_1 - \tau_2 = 0$$

and

$$\tau_2 - \tau_3 = 0$$

In matrix form, H_0 is expressed as

$$H_0: C\boldsymbol{\beta} = \mathbf{0}$$

where

$$C = \begin{bmatrix} 0 & 1 & -1 & 0 \\ 0 & 1 & 0 & -1 \end{bmatrix} \quad \text{and} \quad \boldsymbol{\beta} = \begin{bmatrix} \mu \\ \tau_1 \\ \tau_2 \\ \tau_3 \end{bmatrix}$$

Since

$$\mathbf{c}_1'\boldsymbol{\beta} = [0 \quad 1 \quad -1 \quad 0]\boldsymbol{\beta} = \tau_1 - \tau_2$$

and

$$\mathbf{c}_2'\boldsymbol{\beta} = [0 \quad 1 \quad 0 \quad -1]\boldsymbol{\beta} = \tau_1 - \tau_3$$

are each contrasts, they are known to be estimable (see Example 5.5.2). To verify that these contrasts are independent, we must show that the rows of C are independent. To do so, consider the equation

$$a_1[0 \quad 1 \quad -1 \quad 0] + a_2[0 \quad 1 \quad 0 \quad -1] = \mathbf{0}'$$

This equation implies that

$$\begin{aligned} a_1 + a_2 &= 0 \\ -a_1 &= 0 \\ -a_2 &= 0 \end{aligned}$$

For this to be true, we must have $a_1 = a_2 = 0$. By Definition 1.4.2, the rows of C are independent and $H_0{:}\tau_1 = \tau_2 = \tau_3$ is testable.

Once a hypothesis has been found to be testable, the question remains: What is the appropriate statistic for testing H_0? It is not hard to guess the form of the test statistic. A testable hypothesis assumes the same form as the general linear hypothesis discussed in Section 4.6 with $\delta^* = \mathbf{0}$. The test statistic developed there to test the hypothesis

$$H_0{:}C\boldsymbol{\beta} = \mathbf{0}$$

in the full rank case is

$$F_{r,n-p} = \frac{(C\mathbf{b})'[C(X'X)^{-1}C']^{-1}(C\mathbf{b})/r}{SS_{\text{Res}}/(n-p)}$$

where r is the rank of C and $SS_{\text{Res}}/(n-p) = s^2$. It is reasonable to suspect that the test statistic needed here assumes the same general form with $(X'X)^{-1}$ replaced by a conditional inverse, $(X'X)^c$. Thus, the proposed statistic to use in testing a testable hypothesis is

$$\frac{(C\mathbf{b})'[C(X'X)^cC']^{-1}(C\mathbf{b})/m}{s^2}$$

In the less than full rank model,

$$s^2 = SS_{\text{Res}}/(n-r)$$

To show that this statistic follows an F distribution whenever H_0 is true, it must be shown that it can be expressed as the ratio of two independent chi-squared random variables, each divided by their respective degrees of freedom. To begin, we consider the distribution of the random variable

$$(C\mathbf{b})'[C(X'X)^c C']^{-1}(C\mathbf{b})/\sigma^2$$

Theorem 6.1.1 Let $\mathbf{y} = X\boldsymbol{\beta} + \boldsymbol{\varepsilon}$ where X is $n \times p$ of rank $r < p$ and $\boldsymbol{\varepsilon}$ is a normally distributed random vector with mean $\mathbf{0}$ and variance $\sigma^2 I$. Let $C\boldsymbol{\beta} = \mathbf{0}$ be testable where C is $m \times p$ of rank $m \leqslant r$. Then

$$(C\mathbf{b})'[C(X'X)^c C']^{-1}(C\mathbf{b})/\sigma^2$$

follows a noncentral chi-squared distribution with m degrees of freedom and noncentrality parameter λ where

$$\lambda = (C\boldsymbol{\beta})'[C(X'X)^c C']^{-1}(C\boldsymbol{\beta})/2\sigma^2$$

Proof

Recall that in the less than full rank model, $\mathbf{b}$ can be expressed as

$$\mathbf{b} = (X'X)^c X'\mathbf{y} + [I - (X'X)^c X'X]\mathbf{z}$$

where $(X'X)^c$ is any conditional inverse for $X'X$ and $\mathbf{z} = [I - (X'X)^c X'X]\mathbf{b}$ (see Exercise 35 in Chapter 5). By substituting and using the fact that $I - (X'X)^c X'X$ is idempotent, $C\mathbf{b}$ can be written as

$$\begin{aligned} C\mathbf{b} &= C\{(X'X)^c X'\mathbf{y} + [I - (X'X)^c X'X]\mathbf{b}\} \\ &= C(X'X)^c X'\mathbf{y} + C\mathbf{b} - C(X'X)^c X'X\mathbf{b} \end{aligned}$$

Since $C\boldsymbol{\beta} = \mathbf{0}$ is testable, $C(X'X)^c X'X = C$ and therefore

$$C\mathbf{b} = C(X'X)^c X'\mathbf{y}$$

From this it can be seen that $C\mathbf{b}$ is a normally distributed random vector with mean $C\boldsymbol{\beta}$ and variance

$$C(X'X)^c X'\sigma^2 I X(X'X)^c C' = C(X'X)^c C'\sigma^2$$

By Corollary 2.3.4,

$$\frac{(C\mathbf{b})'[C(X'X)^c C']^{-1}(C\mathbf{b})}{\sigma^2}$$

follows a noncentral chi-squared distribution with m degrees of freedom and noncentrality parameter

$$\lambda = (C\boldsymbol{\beta})'[C(X'X)^c C']^{-1}(C\boldsymbol{\beta})/2\sigma^2$$

as claimed. ■

When $H_0\!:\!C\boldsymbol{\beta} = \mathbf{0}$ is true, the noncentrality parameter associated with the random variable

$$\frac{(C\mathbf{b})'[C(X'X)^c C']^{-1}(C\mathbf{b})}{\sigma^2}$$

assumes the value 0. Hence, under H_0, this random variable follows a chi-squared distribution with m degrees of freedom. It has already been shown that the random variable SS_{Res}/σ^2 follows a chi-squared distribution with $n-r$ degrees of freedom (see Theorem 5.7.1). Consider the ratio

$$\frac{(C\mathbf{b})'[C(X'X)^c C']^{-1}(C\mathbf{b})/\sigma^2 m}{SS_{\text{Res}}/(n-r)\sigma^2} = \frac{(C\mathbf{b})'[C(X'X)^c C']^{-1}(C\mathbf{b})/m}{s^2}$$

If H_0 is true, this is the ratio of two chi-squared random variables, each divided by their respective degrees of freedom. It is also our proposed test statistic. To complete the argument that this statistic follows an F distribution under H_0, we need show only that the chi-squared random variables involved are independent. Since the numerator is a function only of $C\mathbf{b}$, we need argue only that $C\mathbf{b}$ is independent of s^2 in the less than full rank case.

Theorem 6.1.2 Let $\mathbf{y} = X\boldsymbol{\beta} + \boldsymbol{\varepsilon}$ where X is $n \times p$ of rank $r < p$ and $\boldsymbol{\varepsilon}$ is a normally distributed random vector with mean $\mathbf{0}$ and variance $\sigma^2 I$. Let $C\boldsymbol{\beta} = \mathbf{0}$ be testable where C is $m \times p$ of rank $m \leqslant r$. Then $C\mathbf{b}$ is independent of s^2.

Proof

Let $\mathbf{b}$ be a solution to the normal equations. It has been shown that

$$C\mathbf{b} = C(X'X)^c X'\mathbf{y}$$

where $(X'X)^c$ is any conditional inverse for $X'X$. Furthermore, it is known that

$$SS_{\text{Res}} = \mathbf{y}'[I - X(X'X)^c X']\mathbf{y}$$

Hence s^2 can be written as

$$s^2 = \frac{\mathbf{y}'[I - X(X'X)^c X']\mathbf{y}}{n-r}$$

By Theorem 2.4.2, to show that $C\mathbf{b}$ is independent of s^2 it is sufficient to show that

$$u = C(X'X)^c X'\sigma^2 I[I - X(X'X)^c X']/(n-r) = 0$$

Multiplying, we obtain

$$\begin{aligned} u &= [C(X'X)^c X' - C(X'X)^c X'X(X'X)^c X']\sigma^2/(n-r) \\ &= [C(X'X)^c X' - C(X'X)^c X']\sigma^2/(n-r) \\ &= 0 \end{aligned}$$

as desired. ■

By definition, it can now be concluded that the proposed test statistic does in fact follow an F distribution. That is, to test a testable hypothesis the statistic

$$F_{m,n-r} = \frac{(C\mathbf{b})'[C(X'X)^c C']^{-1}(C\mathbf{b})/m}{s^2}$$

is used.

A few subtle points about this F ratio must be considered. First, is the matrix $C(X'X)^c C'$ unique? Second, under the given conditions, is $C(X'X)^c C'$ nonsingular? Finally, does this F ratio behave as others have? Do values near 1 indicate that H_0 is probably true and values substantially larger than 1 indicate that H_0 is probably false? As you should suspect, the answer to each of these questions is "yes". You are asked to verify this in Exercises 3–5.

A numerical example will help clarify these ideas.

Example 6.1.2 Three different treatment methods for removing organic carbon from tar sand wastewater are to be compared. The methods are airflotation (AF), foam separation (FS), and ferric-chloride coagulation (FCC). These data, based on a study reported in "Statistical Planning and Analysis for Treatments of Tar Sand Wastewater," by W.R. Pirie, are obtained:

AF (I)	FS (II)	FCC (III)
34.6	38.8	26.7
35.1	39.0	26.7
35.3	40.1	27.0
35.8	40.9	27.1
36.1	41.0	27.5
36.5	43.2	28.1
36.8	44.9	28.1
37.2	46.9	28.7
37.4	51.6	30.7
37.7	53.6	31.2

Assume the one-way classification model with $n_1 = n_2 = n_3 = 10$ and $N = 30$. We want to test

$$H_0 : \tau_1 = \tau_2 = \tau_3$$

In matrix form, we are testing

$$H_0 : C\boldsymbol{\beta} = \mathbf{0}$$

where

$$C = \begin{bmatrix} 0 & 1 & -1 & 0 \\ 0 & 1 & 0 & -1 \end{bmatrix} \quad \text{and} \quad \boldsymbol{\beta} = \begin{bmatrix} \mu \\ \tau_1 \\ \tau_2 \\ \tau_3 \end{bmatrix}$$

The F statistic used to test H_0 is

$$F_{2,27} = \frac{(C\mathbf{b})'[C(X'X)^c C']^{-1}(C\mathbf{b})/2}{SS_{\text{Res}}/27}$$

For this model,

$$X = \begin{bmatrix} 1 & 1 & 0 & 0 \\ 1 & 1 & 0 & 0 \\ \vdots & \vdots & \vdots & \vdots \\ 1 & 1 & 0 & 0 \\ \hline 1 & 0 & 1 & 0 \\ 1 & 0 & 1 & 0 \\ \vdots & \vdots & \vdots & \vdots \\ 1 & 0 & 1 & 0 \\ \hline 1 & 0 & 0 & 1 \\ 1 & 0 & 0 & 1 \\ \vdots & \vdots & \vdots & \vdots \\ 1 & 0 & 0 & 1 \end{bmatrix} \quad \text{and} \quad X'X = \begin{bmatrix} 30 & 10 & 10 & 10 \\ 10 & 10 & 0 & 0 \\ 10 & 0 & 10 & 0 \\ 10 & 0 & 0 & 10 \end{bmatrix}$$

A conditional inverse for $X'X$ is

$$(X'X)^c = \begin{bmatrix} 0 & 0 & 0 & 0 \\ 0 & \frac{1}{10} & 0 & 0 \\ 0 & 0 & \frac{1}{10} & 0 \\ 0 & 0 & 0 & \frac{1}{10} \end{bmatrix}$$

Via this conditional inverse,

$$\mathbf{b} = (X'X)^c X'\mathbf{y} = \begin{bmatrix} 0 \\ 36.25 \\ 44.0 \\ 28.18 \end{bmatrix}$$

$$C(X'X)^c C' = \begin{bmatrix} 0.2 & 0.1 \\ 0.1 & 0.2 \end{bmatrix}$$

$$[C(X'X)^c C']^{-1} = \frac{1}{0.03}\begin{bmatrix} 0.2 & -0.1 \\ -0.1 & 0.2 \end{bmatrix}$$

$$C\mathbf{b} = \begin{bmatrix} -7.75 \\ 8.07 \end{bmatrix}$$

The numerator of the F ratio used to test H_0 is

$$\frac{(C\mathbf{b})'[C(X'X)^c C']^{-1}(C\mathbf{b})}{2} = \frac{1251.533}{2} = 625.766$$

The residual sum of squares for these data can be shown to be 278.661. Thus, the F ratio for testing $H_0: \tau_1 = \tau_2 = \tau_3$ is

$$F_{2,27} = \frac{625.766}{278.661/27} = 60.63$$

The P value associated with this F ratio is less than 0.01 and the null hypothesis can be rejected.

As you can see, the data of Example 6.1.2 can be analyzed by hand with relative ease because the matrices involved are rather simple in form. In particular, because the matrix $C(X'X)^c C'$ is 2×2, its inverse can be found easily. This is not always the case. When models become more complex, a computer is used to test hypotheses of the form $C\boldsymbol{\beta} = \mathbf{0}$. Computing Supplement G explains how to use SAS to test such hypotheses.

6.2

REPARAMETERIZATION: ONE-WAY CLASSIFICATION MODEL

As mentioned at the beginning of the chapter, there are two ways to approach the problem of hypothesis testing in the less than full rank model. The first, which treats H_0 as a special case of the general linear hypothesis, was explained in Section 6.1. The second, reparameterization to a full rank model, was discussed in

Chapter 5. Reparameterization will be explored in more detail in this section and throughout the chapter.

Basically, the objective of reparameterization is to transform a less than full rank model to one of full rank by redefining the parameters in a meaningful way. More specifically, consider the less than full rank model

$$\mathbf{y} = X\boldsymbol{\beta} + \boldsymbol{\varepsilon}$$

where X is $n \times p$ of rank $r < p$. When this model is written in the form

$$\mathbf{y} = Z\boldsymbol{\alpha} + \boldsymbol{\varepsilon}^*$$

where Z is $n \times r$ of rank r and $\boldsymbol{\alpha}$ is an $r \times 1$ vector of linearly independent estimable functions of $\boldsymbol{\beta}$, then the latter is called a reparameterization of the original model. Methods of Chapter 4 can be used to test hypotheses concerning $\boldsymbol{\alpha}$. Hypotheses of interest are those that are equivalent to testable hypotheses on $\boldsymbol{\beta}$. This idea is illustrated in Example 6.2.1 by reconsidering the reparameterization of the one-way classification model presented in Section 5.2.

Example 6.2.1 Consider the one-way classification model with fixed effects,

$$y_{ij} = \mu + \tau_i + \varepsilon_{ij} \qquad \begin{aligned} i &= 1, 2, 3, \ldots, k \\ j &= 1, 2, 3, \ldots, n_i \end{aligned}$$

Let $N = n_1 + n_2 = n_3 + \cdots + n_k$. The design matrix and vector of parameters for this model are

$$X = \begin{bmatrix} 1 & 1 & 0 & \cdots & 0 \\ 1 & 1 & 0 & & 0 \\ \vdots & \vdots & \vdots & & \vdots \\ 1 & 1 & 0 & & 0 \\ \hline 1 & 0 & 1 & \cdots & 0 \\ 1 & 0 & 1 & & 0 \\ \vdots & \vdots & \vdots & & \vdots \\ 1 & 0 & 1 & & 0 \\ \hline \vdots & & & & \\ \hline 1 & 0 & 0 & & 1 \\ 1 & 0 & 0 & & 1 \\ \vdots & \vdots & \vdots & & \vdots \\ 1 & 0 & 0 & & 1 \end{bmatrix} \quad \text{and} \quad \boldsymbol{\beta} = \begin{bmatrix} \mu \\ \tau_1 \\ \tau_2 \\ \vdots \\ \tau_k \end{bmatrix}$$

The matrix X is $N \times (k+1)$. Since columns 2 through $k+1$ are linearly independent and sum to column 1, the rank of the matrix is k. Hence the model

$$\mathbf{y} = X\boldsymbol{\beta} + \boldsymbol{\varepsilon}$$

is less than full rank. To reparameterize in a meaningful way, define k new parameters by letting $\mu_i = \mu + \tau_i$ for $i = 1, 2, \ldots, k$. The new model is

$$y_{ij} = \mu_i + \varepsilon_{ij} \qquad \begin{array}{l} i = 1, 2, \ldots, k \\ j = 1, 2, \ldots, n_i \end{array}$$

or

$$\mathbf{y} = Z\boldsymbol{\alpha} + \boldsymbol{\varepsilon}$$

where

$$Z = \begin{bmatrix} 1 & 0 & \cdots & 0 \\ 1 & 0 & & 0 \\ \vdots & \vdots & & \vdots \\ 1 & 0 & & 0 \\ \hline 0 & 1 & \cdots & 0 \\ 0 & 1 & & 0 \\ \vdots & \vdots & & \vdots \\ 0 & 1 & & 0 \\ \hline \vdots & & & \\ \hline 0 & 0 & \cdots & 1 \\ 0 & 0 & & 1 \\ \vdots & \vdots & & \vdots \\ 0 & 0 & & 1 \end{bmatrix} \quad \text{and} \quad \boldsymbol{\alpha} = \begin{bmatrix} \mu_1 \\ \mu_2 \\ \vdots \\ \mu_k \end{bmatrix}$$

Note that Z is $N \times k$ of rank k; it is of full rank. To see that $\mathbf{y} = Z\boldsymbol{\alpha} + \boldsymbol{\varepsilon}$ is truly a reparameterization of the original model, it must be shown that elements of $\boldsymbol{\alpha}$ are linearly independent estimable functions of $\boldsymbol{\beta}$. To do so, note that for each i, μ_i can be expressed as a row of X multiplied by $\boldsymbol{\beta}$. That is, each element of $\boldsymbol{\alpha}$ is an element of $X\boldsymbol{\beta}$. By Theorem 5.5.1, each element of $\boldsymbol{\alpha}$ is estimable. The ith element of $\boldsymbol{\alpha}$ can be written in vector form as $\mathbf{c}_i'\boldsymbol{\beta}$ where $\mathbf{c}_i'$ is a row vector of zeros and ones with the ones appearing in columns 1 and $i+1$. When each element of $\boldsymbol{\alpha}$ is written in vector form, it is easy to apply the definition of independence to argue that these elements are indeed independent estimable functions as required. For the reparameterization to serve its purpose, the new model parameters should be interpretable. That is the case here since μ_i represents the average response for the population from which the ith sample is drawn. In practice, most inferences made in the one-way classification model concern these mean values and the relationships that exist among them. The reparameterized model can be used to make these inferences using the full rank techniques developed in Chapters 3 and 4.

Most courses in applied statistics include a detailed discussion of the one-way classification model in the reparameterized form just developed. Usually a major portion of the discussion is devoted to developing an analysis of variance table appropriate for testing the null hypothesis of equality of means. This hypothesis is expressed as

$$H_0:\mu_1 = \mu_2 = \cdots = \mu_k$$

It is not hard to see that this hypothesis is equivalent to the testable hypothesis

$$H_0:\tau_1 = \tau_2 = \cdots = \tau_k$$

A typical layout for testing H_0 is shown in Table 6.1 [**18**] [**13**]. In this table, $y_{i.}$ denotes the sum of the responses for sample i, and $y_{..}$ denotes the sum of all responses. The table can now be justified via the theory of linear models.

TABLE 6.1 **ANOVA table for the one-way classification design with fixed effects**

Source of Variation	Degrees of Freedom	Sum of Squares	Mean Square	F
Treatment	$k-1$	$\sum_{i=1}^{k} \frac{y_{i.}^2}{n_i} - \frac{y_{..}^2}{N}$	$SS_{Tr}/(k-1)$	MS_{Tr}/MSE
Error	$N-k$	Subtraction	$SS_E/(N-k)$	
Total (corrected)	$N-1$	$\sum_{i=1}^{k} \sum_{j=1}^{n_i} y_{ij}^2 - \frac{y_{..}^2}{N}$		

Although the setting is not exactly the same as the one encountered when testing a null hypothesis on a subset of parameters in the full rank case, the ideas developed to perform such tests can be used here. In particular, we work with two models. The first is a full model that contains all the parameters $\mu_1, \mu_2, \ldots, \mu_k$. The second is a reduced model. It is the model that results when $H_0:\mu_1 = \mu_2 = \cdots = \mu_k$ is assumed to be true. We want to choose between these two models with the philosophy that the reduced model will be retained unless it is shown to be inadequate. That is, we will assume that the population means are identical unless we find evidence to the contrary. The method used to choose between the two parallels that employed in Section 4.2. We form the regression sum of squares for each model and then find the difference between them. A small difference tends to support the null hypothesis; a large difference indicates that H_0 is probably not true and that, in fact, there are differences among the k population means.

The full model assumes the form shown in Example 6.1.1; namely,

$$y_{ij} = \mu_i + \varepsilon_{ij} \qquad \begin{array}{l} i = 1, 2, \ldots, k \\ j = 1, 2, \ldots, n_i \end{array}$$

or

$$\mathbf{y} = Z\boldsymbol{\alpha} + \boldsymbol{\varepsilon}$$

The regression sum of squares for this model is denoted by $SS_{\text{Reg(Full)}}$ and is given by

$$SS_{\text{Reg(Full)}} = \mathbf{y}'[Z(Z'Z)^{-1}Z']\mathbf{y}$$

Note that

$$Z'Z = \begin{bmatrix} n_1 & 0 & \cdots & 0 \\ 0 & n_2 & & 0 \\ \vdots & \vdots & & \vdots \\ 0 & 0 & & n_k \end{bmatrix} \quad \text{and} \quad (Z'Z)^{-1} = \begin{bmatrix} \frac{1}{n_1} & 0 & \cdots & 0 \\ 0 & \frac{1}{n_2} & & 0 \\ \vdots & \vdots & & \vdots \\ 0 & 0 & & \frac{1}{n_k} \end{bmatrix}$$

Letting $y_{i.}$ denote the sum of the responses of the ith sample, it is easy to verify that

$$SS_{\text{Reg(Full)}} = \sum_{i=1}^{k} y_{i.}^2/n_i$$

The reduced model is written assuming that the null hypothesis is true. Letting $\mu_1 = \mu_2 = \cdots = \mu_k = \mu$, this model assumes the form

$$y_{ij} = \mu + \varepsilon_{ij} \qquad \begin{array}{l} i = 1, 2, \ldots, k \\ j = 1, 2, \ldots, n_i \end{array}$$

or

$$\mathbf{y} = Z_2\boldsymbol{\alpha}_2 + \boldsymbol{\varepsilon}$$

where Z_2 is an $N \times 1$ vector of ones and $\boldsymbol{\alpha}_2$ is the real number μ. That is,

$$Z_2 = \begin{bmatrix} 1 \\ 1 \\ \vdots \\ 1 \end{bmatrix} \quad \text{and} \quad \boldsymbol{\alpha}_2 = \mu$$

The regression sum of squares for this model is denoted by $SS_{\text{Reg(Reduced)}}$ and is given by

$$SS_{\text{Reg(Reduced)}} = \mathbf{y}'[Z_2(Z_2'Z_2)^{-1}Z_2']\mathbf{y}$$

It can be verified that

$$SS_{\text{Reg(Reduced)}} = y_{..}^2/N$$

where $y_{..}$ denotes the sum of all responses.

The difference between these regression sums of squares is denoted by $SS_{\text{Reg(Hypothesis)}}$ and is given by

$$\begin{aligned} SS_{\text{Reg(Hypothesis)}} &= SS_{\text{Reg(Full)}} - SS_{\text{Reg(Reduced)}} \\ &= \mathbf{y}'[Z(Z'Z)^{-1}Z' - Z_2(Z_2'Z_2)^{-1}Z_2']\mathbf{y} \\ &= \sum_{i=1}^{k} y_{i.}^2/n_i - y_{..}^2/N \end{aligned}$$

This is the amount of variation in response that is not random but that cannot be accounted for by the reduced model. Logic suggests that this quadratic form should play an important role in testing H_0. In particular, to parallel the techniques used in Section 4.2, the test statistic should be an F ratio with $SS_{\text{Reg(Full)}} - SS_{\text{Reg(Reduced)}}$ being the sum of squares of its numerator. The test statistic that is proposed is

$$\frac{(SS_{\text{Reg(Full)}} - SS_{\text{Reg(Reduced)}})/(k-1)}{SS_{\text{Res(Full)}}/(N-k)} = \frac{\left(\sum_{i=1}^{k} y_{i.}^2/n_i - y_{..}^2/N\right)\Big/(k-1)}{SS_E/(N-k)}$$

This is the test statistic given in Table 6.1.

To justify this statistic fully, it must be argued that it can be expressed as the ratio of two independent chi-squared random variables, each divided by their respective degrees of freedom. To do so, we parallel the argument given in Section 4.2 and begin by considering the identity

$$\frac{\mathbf{y}'\mathbf{y}}{\sigma^2} = \frac{\mathbf{y}'Z_2(Z_2'Z_2)^{-1}Z_2'\mathbf{y}}{\sigma^2} + \frac{\mathbf{y}'[Z(Z'Z)^{-1}Z' - Z_2(Z_2'Z_2)^{-1}Z_2']\mathbf{y}}{\sigma^2} + \frac{\mathbf{y}'[I - Z(Z'Z)^{-1}Z']\mathbf{y}}{\sigma^2}$$

Since $\mathbf{y}'Z_2(Z_2'Z_2)^{-1}Z_2'\mathbf{y}$ is the regression sum of squares for the reduced model, which has only one parameter, the first term on the right is known to follow a noncentral chi-squared distribution with 1 degree of freedom. It can be concluded from this that

$$r[Z_2(Z_2'Z_2)^{-1}Z_2'] = 1$$

The quadratic form $\mathbf{y}'[I - Z(Z'Z)^{-1}Z']\mathbf{y} = SS_E$ is the residual sum of squares for the full model. It is known to follow a chi-squared distribution with $N-k$ degrees of freedom. Hence

$$r([I - Z(Z'Z)^{-1}Z']) = N - k$$

Although it is not obvious mathematically, it can be shown that

$$r([Z(Z'Z)^{-1}Z' - Z_2(Z_2'Z_2)^{-1}Z_2']) = k - 1$$

Exercises 10 and 11 lead you through the proof of this statement step by step. The Cochran–Fisher theorem of Section 4.2 guarantees that the three quadratic forms on the right of the above identity are independently distributed noncentral chi-squared random variables. In particular, the quadratic forms

$$\frac{SS_{\text{Reg(Full)}} - SS_{\text{Reg(Reduced)}}}{\sigma^2} \quad \text{and} \quad \frac{SS_E}{\sigma^2}$$

are independent. To show that the proposed statistic follows an F distribution, it is necessary only to show that the noncentrality parameter λ associated with the random variable

$$\frac{\mathbf{y}'[Z(Z'Z)^{-1}Z' - Z_2(Z_2'Z_2)^{-1}Z_2']\mathbf{y}}{\sigma^2}$$

has value 0 whenever $\mu_1 = \mu_2 = \cdots = \mu_k$. By the Cochran–Fisher theorem,

$$\lambda = \frac{1}{2\sigma^2}(Z\boldsymbol{\alpha})'[Z(Z'Z)^{-1}Z' - Z_2(Z_2'Z_2)^{-1}Z_2'](Z\boldsymbol{\alpha})$$

When H_0 is true, $\mu_1 = \mu_2 = \cdots = \mu_k = \mu$ and $\boldsymbol{\alpha}' = [\mu \quad \mu \quad \cdots \quad \mu]$. It is not obvious that $\lambda = 0$ in this case. However, it is not hard to verify that this is true. Exercises 12–14 outline the derivation. It can be concluded that if H_0 is true, the statistic

$$\frac{\left[\sum_{i=1}^{k} y_{i.}^2/n_i - y_{..}^2/N\right]\Big/(k-1)}{SS_E/(N-k)}$$

follows an F distribution with $k-1$ and $N-k$ degrees of freedom.

It is known that the expected value of the denominator is σ^2. The expected value of the numerator can be shown to be equal to σ^2 when H_0 is true and to exceed σ^2 otherwise (see Exercise 15). Thus, as in the past, a right-tailed F test is appropriate.

The ANOVA shown in Table 6.2 summarizes the results developed here. This parallels the ANOVA given in Table 4.3 for testing hypotheses on a subset of $\boldsymbol{\beta}$ in the full rank model. Note that $SS_{\text{Reg(Reduced)}} + SS_{\text{Reg(Hypothesis)}} + SS_{\text{Res}} = SS_{\text{Total}}$ as expected and that their corresponding degrees of freedom are additive. As you can see, this table differs slightly from Table 6.1. Table 6.1 is based on a corrected total sum of squares, whereas Table 6.2 uses an uncorrected total.

Example 6.2.2 illustrates this method of comparing means in the one-way classification model by reconsidering the data analyzed in Section 6.1.

TABLE 6.2
ANOVA table for the one-way classification design with fixed effects

Source of Variation	Degrees of Freedom	Sum of Squares	Mean Square	F
Regression (full)	k	$\sum_{i=1}^{k} y_{i.}^2/n_i$		
Reduced model	1	$y_{..}^2/N$		
Hypothesis	$k-1$	$\sum_{i=1}^{k} y_{i.}^2/n_i - y_{..}^2/N$	$\frac{SS_{\text{Reg (Hypothesis)}}}{k-1}$	$\frac{MS_{\text{Hypothesis}}}{MSE}$
Residual	$N-k$	$\sum_{i=1}^{k}\sum_{j=1}^{n_i} y_{ij}^2 - \sum_{i=1}^{k} y_{i.}^2/n_i$	$\frac{SS_{\text{Res}}}{N-k}$	
Total (uncorrected)	N	$\sum_{i=1}^{k}\sum_{j=1}^{n_i} y_{ij}^2$		

Example 6.2.2 In Example 6.1.2 three different methods for removing organic carbon from tar sand wastewater were compared using the general linear models procedure for testing a hypothesis of the form $C\boldsymbol{\beta} = \mathbf{0}$. To test this hypothesis via the reparameterized model, the hypothesis

$$H_0: \mu_1 = \mu_2 = \mu_3$$

is tested. For the given data

$$y_{1.} = 362.5 \qquad y_{..} = 1084.3 \qquad \sum_{i=1}^{3} y_{i.}^2/n_i = 40441.749$$

$$y_{2.} = 440 \qquad n_1 = n_2 = n_3 = 10 \qquad y_{..}^2/N = 39190.216$$

$$y_{3.} = 281.8 \qquad N = 30 \qquad \sum_{i=1}^{3}\sum_{j=1}^{10} y_{ij}^2 = 40720.41$$

The ANOVA is given in Table 6.3.

As you can see,

$$SS_{\text{Reg(Hypothesis)}} = 1251.533 = (C\mathbf{b})'[C(X'X)^c C']^{-1}(C\mathbf{b})$$

as expected. As a result, the F ratio obtained via the full rank reparameterization agrees with that found by working with the less than full rank model in the general linear form. Computing Supplement H demonstrates the use of SAS in conjunction with the reparameterized model.

TABLE 6.3 **ANOVA for tar sand data**

Source of Variation	Degrees of Freedom	Sum of Squares	Mean Square	F
Regression (full)	3	40441.749		
Reduced model	1	39190.216		
Hypothesis	2	1251.533	625.7665	60.63
Residual	27	278.661	10.32	
Total (uncorrected)	30	40720.41		

6.3

TESTING A HYPOTHESIS ON A TREATMENT CONTRAST

In the one-way classification model, researchers often want to make specific comparisons among the parameters $\tau_1, \tau_2, \ldots, \tau_k$. Comparisons assume the form of treatment contrasts. Recall that a treatment contrast is a linear function $\Sigma_{i=1}^k a_i\tau_i$ such that $\Sigma_{i=1}^k a_i = 0$.

In Chapter 5 it was shown that these contrasts are estimable, and hence hypotheses of the form

$$H_0\colon \sum_{i=1}^{k} a_i\tau_i = 0$$

are testable. They can be tested as stated or can be rewritten in terms of the parameters $\mu_1, \mu_2, \ldots, \mu_k$ where $\mu_i = \mu + \tau_i$. In this reparameterized form, the null hypothesis to be tested is

$$H_0\colon \sum_{i=1}^{k} a_i\mu_i = 0$$

or

$$H_0\colon \mathbf{a}'\boldsymbol{\alpha} = 0$$

where $\mathbf{a}' = [a_1 \quad a_2 \quad \ldots \quad a_k]$ and $\boldsymbol{\alpha}' = [\mu_1 \quad \mu_2 \quad \ldots \quad \mu_k]$. The theory needed to develop an appropriate test statistic has already been developed. Recall that in the reparameterized model

$$(Z'Z)^{-1} = \begin{bmatrix} \frac{1}{n_1} & 0 & \cdots & 0 \\ 0 & \frac{1}{n_2} & & 0 \\ \vdots & \vdots & & \vdots \\ 0 & 0 & & \frac{1}{n_k} \end{bmatrix}$$

Since the reparameterized model is of full rank,

$$\hat{\boldsymbol{\alpha}} = (Z'Z)^{-1}Z'\mathbf{y} = \begin{bmatrix} \bar{y}_{1.} \\ \bar{y}_{2.} \\ \vdots \\ \bar{y}_{k.} \end{bmatrix}$$

Applying Exercise 38 in Chapter 5, it can be concluded that the statistic

$$\frac{\mathbf{a}'\hat{\boldsymbol{\alpha}}}{s\sqrt{\mathbf{a}'(Z'Z)^{-1}\mathbf{a}}}$$

follows a t distribution with $n-k$ degrees of freedom. This statistic is used to conduct either a right-, left-, or two-tailed test of H_0. It is easy to verify that the above statistic can be written as

$$\frac{\sum_{i=1}^{k} a_i \bar{y}_{i.}}{s\sqrt{\sum_{i=1}^{k} a_i^2/n_i}}$$

This is the form usually seen in beginning courses in applied statistics.

Before illustrating the use of this statistic numerically, one other point should be made. Since the square of a t random variable with n degrees of freedom follows an F distribution with 1 and n degrees of freedom (see Exercise 19 in Chapter 4), it can be concluded that the statistic

$$\frac{\left[\dfrac{\left(\sum_{i=1}^{k} a_i \bar{y}_{i.}\right)^2}{\sum_{i=1}^{k} a_i^2/n_i}\right]}{s^2}$$

follows an F distribution with 1 and $n-k$ degrees of freedom. The numerator of this F ratio is called a *contrast sum of squares.* It is easy to anticipate the form of a contrast sum of squares. The population means $\mu_1, \mu_2, \ldots, \mu_k$ of the contrast are replaced by their respective estimates $\bar{y}_{i.}, \bar{y}_{2.}, \ldots, \bar{y}_{k.}$; the resulting sum is squared and divided by the sum of squares of the coefficients of the contrast, each divided by their respective sample sizes. The F statistic is used to perform a *single degree of freedom test* on a contrast. It is appropriate only when testing against a two-tailed alternative.

Example 6.3.1 illustrates the use of the F ratio in testing a contrast.

Example 6.3.1 A tennis ball manufacturer is studying the life span of a newly developed tennis ball. The purpose of the study is to compare the behavior of the ball on five different court surfaces: clay, grass, composition, wood, and asphalt. An overall test of equality is to be conducted. In addition, the manufacturer is particularly interested in comparing the behavior of the ball on soft surface (clay, grass, composition) to that on hard surfaces (wood, asphalt). The following data are obtained based on samples of sizes 20, 22, 24, 21, and 25, respectively. The response is the time in hours that the ball is used before it is judged to be completely dead.

Clay	Grass	Composition	Wood	Asphalt
$\bar{y}_{1.} = 6.2$	$\bar{y}_{2.} = 6.8$	$\bar{y}_{3.} = 6.4$	$\bar{y}_{4.} = 5$	$\bar{y}_{5.} = 4.4$

For these data, $s^2 = 8.87$. The ANOVA is shown in Table 6.4.

Since the P value associated with an F ratio of 2.7 is small $(0.05 < P < 0.10)$, the hypothesis

$$H_0: \mu_1 = \mu_2 = \mu_3 = \mu_4 = \mu_5$$

can be rejected. The contrast used to compare the ball behavior on soft surfaces to that on hard surfaces is

$$\sum_{i=1}^{5} a_i \mu_i = (\tfrac{1}{3}\mu_1 + \tfrac{1}{3}\mu_2 + \tfrac{1}{3}\mu_3) - (\tfrac{1}{2}\mu_4 + \tfrac{1}{2}\mu_5)$$

Since no directional preference is indicated, an F test is used to test

$$H_0: \sum_{i=1}^{5} a_i \mu_i = 0 \quad \text{versus} \quad H_1: \sum_{i=1}^{5} a_i \mu_i \neq 0$$

The contrast sum of squares is given by

$$\frac{\left(\sum_{i=1}^{5} a_i \bar{y}_{i.}\right)^2}{\sum_{i=1}^{5} a_i^2/n_i} = \frac{[(\frac{1}{3})(6.2 + 6.8 + 6.4) - (\frac{1}{2})(5 + 4.4)]^2}{(\frac{1}{3})^2/20 + (\frac{1}{3})^2/22 + (\frac{1}{3})^2/24 + (-\frac{1}{2})^2/21 + (-\frac{1}{2})^2/25}$$

$$= 84.035$$

The F ratio used to test the two-tailed hypothesis is

$$F = \frac{84.035}{8.87} = 9.47$$

Based on this value, H_0 can be rejected with $P < 0.01$.

TABLE 6.4 **ANOVA for tennis ball data**

Source of Variation	Degrees of Freedom	Sum of Squares	Mean Square	F
Regression (full)	5	3778.12		
Reduced model	1	3682.33		
Hypothesis	4	95.79	23.95	2.7
Residual	107	993.44	8.87	
Total (uncorrected)	112	4771.56		

An F or a t test can be performed on any contrast. However, it is helpful if the contrasts are orthogonal. The term orthogonal is defined below.

Definition 6.3.1

Let $\sum_{i=1}^{k} a_i\mu_i$ and $\sum_{i=1}^{k} b_i\mu_i$ denote treatment contrasts. These contrasts are *orthogonal* if and only if

$$\sum_{i=1}^{k} a_i b_i / n_i = 0$$

When sample sizes are equal, the above condition reduces to the requirement that $\sum_{i=1}^{k} a_i b_i = 0$. It is relatively easy to design orthogonal contrasts in this case. When sample sizes are unequal the task becomes more difficult. An example will clarify the idea.

Example 6.3.2 Consider a one-way classification model with fixed effects in which $k = 5$ and sample sizes are equal. The contrasts $\omega_1 = \mu_1 + \mu_2 - \mu_3 - \mu_5$ and $\omega_2 = \mu_1 + \mu_2 + \mu_3 - 4\mu_4 + \mu_5$ are orthogonal. Neither is orthogonal to the contrast $\mu_1 - \mu_4$. However, in the context of Example 6.3.1, in which $n_1 = 20$, $n_2 = 22$, $n_3 = 24$, $n_4 = 21$, and $n_5 = 25$, the contrasts ω_1 and ω_2 are not orthogonal since

$$\sum_{i=1}^{5} \frac{a_i b_i}{n_i} \neq 0$$

Orthogonal contrasts are important in that it can be shown that their associated contrast sums of squares are statistically independent. Furthermore, if a set of $k-1$ orthogonal contrasts is chosen, their contrast sums of squares $SS_{\omega_1}, SS_{\omega_2}, \ldots, SS_{\omega_k}$ sum to the regression sum of squares associated with the hypothesis. That is, given a set of $k-1$ orthogonal contrasts,

$$\sum_{i=1}^{k} SS_{\omega_i} = SS_{\text{Reg(Hypotheses)}}$$

See [**17**]. In this case, $k-1$ single degree of freedom F tests can be conducted as part of the ANOVA.

Another advantage of orthogonal contrasts is that because of their statistical independence no duplication of effort is involved. Each contrast deals with a distinctly different aspect of the experiment and provides a different insight into the relationship among treatment means. Example 6.3.3 illustrates the idea in an applied setting.

Example 6.3.3 At a certain manufacturing plant, filters are used to remove pollutants. These filters must be replaced as soon as they fail due to cracking or forming holes. An experiment is conducted to compare five different types of filters. Six filters of each type are used under identical conditions and the time in hours to failure is noted for each. These data are obtained:

		Filter Type		
I	II	III	IV	V
261	221	201	600	160
186	188	146	301	135
239	167	96	608	455
243	224	173	283	402
296	178	280	193	457
270	147	100	159	559

For these data,

$y_{1.} = 1495 \quad \bar{y}_{1.} = 249.1667 \qquad y_{..} = 7928 \quad y_{..}^2/30 = 2{,}095{,}106.1$

$y_{2.} = 1125 \quad \bar{y}_{2.} = 187.50$

$y_{3.} = 996 \quad \bar{y}_{3.} = 166.00 \qquad \sum_{i=1}^{5}\sum_{j=1}^{6} y_{ij}^2 = 2{,}680{,}876$

$y_{4.} = 2144 \quad \bar{y}_{4.} = 357.3333$

$y_{5.} = 2168 \quad \bar{y}_{5.} = 361.3333 \qquad \sum_{i=1}^{5} y_{i.}^2/6 = 2{,}298{,}271$

The usual ANOVA for these data is shown in Table 6.5.

TABLE 6.5 **ANOVA for filter data**

Source of Variation	Degrees of Freedom	Sum of Squares	Mean Square	F
Regression	5	2,298,271		
Reduced model	1	2,095,106.1		
Hypothesis	4	203,164.9	50,791.225	3.3187
Residual	25	382,605	1,530.42	
Total (uncorrected)	30	2,680,876		

Based on the $F_{4,25}$ distribution, the null hypothesis of equality of treatment means can be rejected with $0.025 < P < 0.05$. Now consider the orthogonal contrasts

$$\omega_1 = \mu_1 + \mu_2 - \mu_3 - \mu_5$$
$$\omega_2 = \tfrac{1}{4}(\mu_1 + \mu_2 + \mu_3 + \mu_5) - \mu_4$$
$$\omega_3 = \mu_1 - \mu_2$$
$$\omega_4 = \mu_3 - \mu_5$$

The contrast sum of squares for ω_1 is

$$SS_{\omega_1} = \frac{(249.1667 + 187.50 - 166.00 - 361.3338)^2}{\dfrac{(1)^2 + (1)^2 + (-1)^2 + (0)^2 + (-1)^2}{6}} = 12{,}330.6485$$

The other contrast sums of squares are

$$SS_{\omega_2} = 64{,}960.4961$$
$$SS_{\omega_3} = 11{,}408.3457$$
$$SS_{\omega_4} = 114{,}465.2943$$

Note that

$$\sum_{i=1}^{4} SS_{\omega_i} = 203{,}164.7846$$

which agrees with $SS_{\text{Reg(Hypothesis)}}$, apart from roundoff error, as expected. These sums of squares are used to form the four F ratios shown in Table 6.6.

As can be seen,

$$H_0\!:\omega_2 = 0 \quad \text{versus} \quad H_1\!:\omega_2 \neq 0$$

can be rejected based on the $F_{1,25}$ distribution with $0.025 < P < 0.05$. From this it can be concluded that the average life span for filters I, II, III, and V together differ from that of filter IV. Also

$$H_0\!:\omega_4 = 0 \quad \text{versus} \quad H_1\!:\omega_4 \neq 0$$

can be rejected with $0.01 < P < 0.025$. It can be concluded that the average life span of filter III differs from that of filter V. The null hypotheses $H_0{:}\omega_1 = 0$ and $H_0{:}\omega_3 = 0$ cannot be rejected.

TABLE 6.6 **Orthogonal contrasts among filter types**

Source of Variation	Degrees of Freedom	Sum of Squares	Mean Squares	F
Regression	5	2,298,271		
Reduced model	1	2,095,106.1		
Hypothesis	4	203,164.9	50,791.225	3.3187
ω_1	1	12,330.6485	12,330.6485	0.8057
ω_2	1	64,960.4961	64,960.4961	4.2446
ω_3	1	11,408.3457	11,408.3457	0.7454
ω_4	1	114,465.2943	114,465.2943	7.4793
Residual	25	382,605	15,304.2	
Total (uncorrected)	30	2,680,876		

Orthogonal contrasts are used when the experimenter wants to partition the variation due to treatments into $k-1$ independent components. These contrasts can be chosen rather arbitrarily. However, the researcher usually has certain contrasts that are of particular interest from a physical standpoint. These contrasts will be included in those tested; others that are chosen simply must be orthogonal to those of interest in order for the partitioning to take place. Contrasts of interest can, of course, be tested without finding a full partitioning of $SS_{\text{Reg(Hypothesis)}}$.

6.4

TWO-FACTOR DESIGN WITHOUT INTERACTION: FIXED EFFECTS

Up to this point one-factor problems have been emphasized. Many times in practice the researcher wants to study the effects of two factors in the same experiment. For example, a chemist might want to study the effect of both pressure and temperature on the viscosity of an adhesive; an engineer might study the effect of engine speed and oil type on the life span of a piston ring; a medical researcher might study the effect of an exercise regimen and diet on blood sugar levels in diabetics. Several designs can be used to accomplish this. Two designs that are probably familiar to you from earlier courses in applied statistics are the *two-factor balanced design without interaction* and the *two-factor balanced design with interaction.*

To understand the difference between these two-factor designs, it is necessary to explain *interaction*. An example should illustrate the idea. Suppose that in studying the effect of pressure and temperature on viscosity, pressure is studied at four levels and temperature is studied at two levels. Suppose that the *theoretical* mean responses for the eight pressure-temperature combinations are

		Pressure (Factor I)			
		1	2	3	4
Temperature (Factor II)	1	4	6	4	3
	2	8	2	7	5

A sketch of these means is shown in Figure 6.1. Notice the inconsistent behavior of the response to pressure changes for the two temperatures as exhibited by the crossover in the sketch. At pressures 1, 3, and 4 the average response for temperature 2 exceeds that of temperature 1; however, the opposite is true at pressure 2. An inconsistency such as this indicates an "interaction" between pressure and temperature. If no interaction exists, then changing from temperature 1 to temperature 2 has exactly the same effect on viscosity at each pressure level. For example, if the viscosity for temperature 2 exceeds that of temperature 1 by four units at pressure 1, then it will do so at each of the other pressures also. This results in a sketch in which the line segments for the two temperatures are *parallel*.

FIGURE 6.1 **Line crossovers indicate interaction between pressure and temperature**

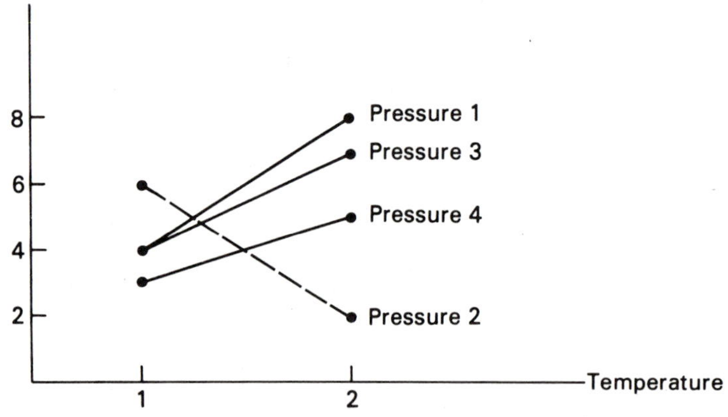

The idea just illustrated can be generalized. If interaction exists, then as we change from one level of factor I to another we notice an inconsistency in the amount of change exhibited by the population means as we move across levels of factor II. This inconsistency could result in a crossover as in the above example or it could be less pronounced. It could result in line segments that do not cross but are nevertheless not parallel. If no interaction is present, the line segments for the levels of factor I will be parallel. Figure 6.2 illustrates the idea.

FIGURE 6.2 (a) A two-factor balanced design with interaction as evidenced by a crossover effect; (b) A two-factor balanced design with interaction as evidenced by nonparallel lines with no crossover; (c) A two-factor balanced design with no interaction as evidenced by parallel line segments

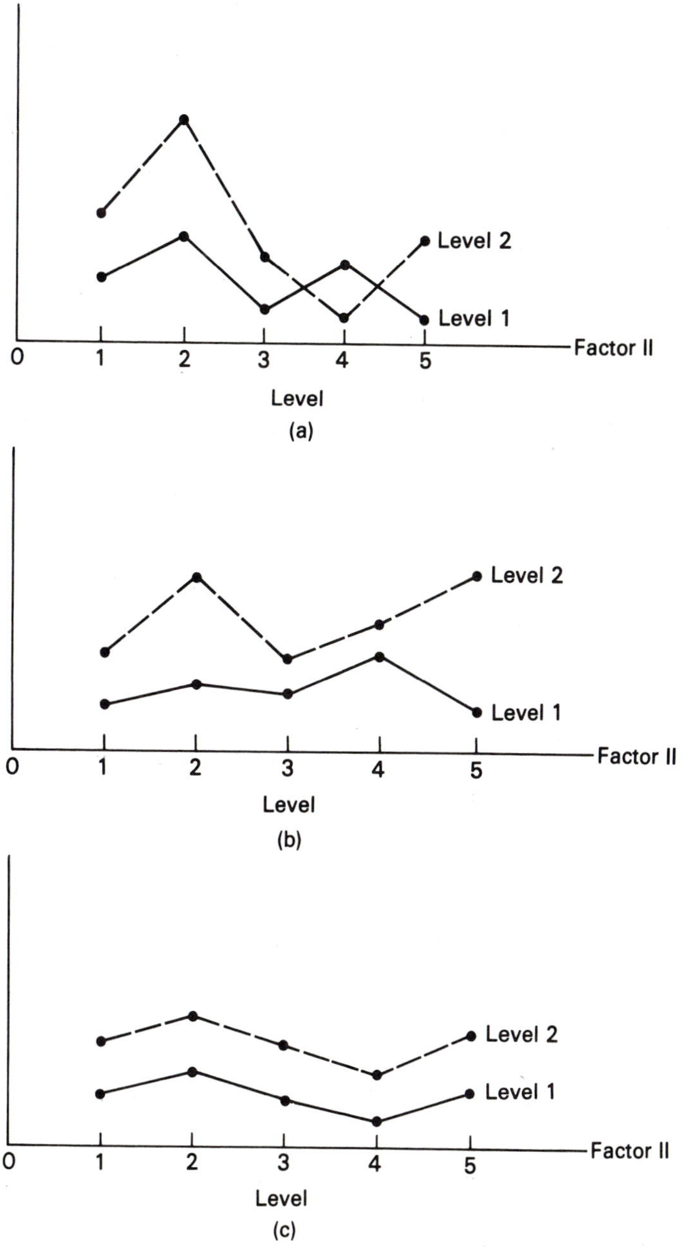

THE ADDITIVE MODEL

When no interaction exists, the model is said to be *additive*. Mathematically, this means that the response can be expressed as the sum of an effect due to factor I, an effect due to factor II, and an effect due to random influences. Here it is assumed that factor I is studied at a levels, factor II at b levels, and that exactly one response is measured at each of the ab treatment combinations. The model for this two-factor design with no interaction is

$$y_{ij} = \mu + \tau_i + \beta_j + \varepsilon_{ij} \qquad \begin{array}{l} i = 1, 2, \ldots, a \\ j = 1, 2, \ldots, b \end{array}$$

Here τ_i denotes the effect due to the fact that the response is measured at the ith level of factor I; β_j denotes the effect due to the fact that the response is measured at the jth level of factor II; ε_{ij} denotes the effect due to random influences on the response. In expanded form, this model becomes

$$\begin{aligned}
y_{11} &= \mu + \tau_1 + \beta_1 + \varepsilon_{11} \\
y_{12} &= \mu + \tau_1 + \beta_2 + \varepsilon_{12} \\
&\vdots \\
y_{1b} &= \mu + \tau_1 + \beta_b + \varepsilon_{1b} \\
y_{21} &= \mu + \tau_2 + \beta_1 + \varepsilon_{21} \\
y_{22} &= \mu + \tau_2 + \beta_2 + \varepsilon_{22} \\
&\vdots \\
y_{2b} &= \mu + \tau_2 + \beta_b + \varepsilon_{2b} \\
&\vdots \\
y_{a1} &= \mu + \tau_a + \beta_1 + \varepsilon_{a1} \\
y_{a2} &= \mu + \tau_a + \beta_2 + \varepsilon_{a2} \\
&\vdots \\
y_{ab} &= \mu + \tau_a + \beta_b + \varepsilon_{ab}
\end{aligned}$$

The system is expressed in matrix form as

$$\mathbf{y} = X\boldsymbol{\beta} + \boldsymbol{\varepsilon}$$

where

$$
\mathbf{y} = \begin{bmatrix} y_{11} \\ y_{12} \\ \vdots \\ y_{1b} \\ -- \\ y_{21} \\ y_{22} \\ \vdots \\ y_{2b} \\ -- \\ \vdots \\ -- \\ y_{a1} \\ y_{a2} \\ \vdots \\ y_{ab} \end{bmatrix}, \quad
\boldsymbol{\beta} = \begin{bmatrix} \mu \\ -- \\ \tau_1 \\ \tau_2 \\ \vdots \\ \tau_a \\ -- \\ \beta_1 \\ \beta_2 \\ \vdots \\ \beta_b \end{bmatrix}, \quad
\boldsymbol{\varepsilon} = \begin{bmatrix} \varepsilon_{11} \\ \varepsilon_{12} \\ \vdots \\ \varepsilon_{1b} \\ -- \\ \varepsilon_{21} \\ \varepsilon_{22} \\ \vdots \\ \varepsilon_{2b} \\ -- \\ \vdots \\ -- \\ \varepsilon_{a1} \\ \varepsilon_{a2} \\ \vdots \\ \varepsilon_{ab} \end{bmatrix},
$$

and

$$
X = \begin{array}{c|cccc|cccc}
\mu & \tau_1 & \tau_2 & \cdots & \tau_a & \beta_1 & \beta_2 & \cdots & \beta_b \\
\hline
1 & 1 & 0 & & 0 & 1 & 0 & \cdots & 0 \\
1 & 1 & 0 & & 0 & 0 & 1 & & 0 \\
\vdots & \vdots & \vdots & & \vdots & \vdots & \vdots & & \vdots \\
1 & 1 & 0 & & 0 & 0 & 0 & & 1 \\
\hline
1 & 0 & 1 & & 0 & 1 & 0 & & 0 \\
1 & 0 & 1 & & 0 & 0 & 1 & & 0 \\
\vdots & \vdots & \vdots & & \vdots & \vdots & \vdots & & \vdots \\
1 & 0 & 1 & & 0 & 0 & 0 & & 1 \\
\hline
\vdots & & & & & & & & \\
\hline
1 & 0 & 0 & & 1 & 1 & 0 & \cdots & 0 \\
1 & 0 & 0 & & 1 & 0 & 0 & & 0 \\
\vdots & \vdots & \vdots & & \vdots & \vdots & \vdots & & \vdots \\
1 & 0 & 0 & & 1 & 0 & 0 & & 1
\end{array}
$$

As usual, it is assumed that $\boldsymbol{\varepsilon}$ is a normally distributed random vector with mean $\mathbf{0}$ and variance $\sigma^2 I$.

TESTABLE HYPOTHESES

The matrix X above is of dimension $ab \times (a+b+1)$. Note that columns 2 through $a+1$ are linearly independent and sum to column 1 and that columns $a+2$ through $a+b+1$ are linearly independent and sum to column 1. Hence X is of rank $(a+b+1)-2=a+b-1$. It is less than full rank. Testable hypotheses assume the form

$$H_0{:}C\boldsymbol{\beta}=\mathbf{0}$$

where C is of rank $m \leqslant a+b-1$.

Two hypotheses are most important.

$H_0{:}\tau_1=\tau_2=\cdots=\tau_a$ (No differences among the effects associated with the levels of factor I)

$H_0'{:}\beta_1=\beta_2=\cdots=\beta_b$ (No differences among the effects associated with the levels of factor II)

To see that these are each testable, it is first shown that contrasts in the τ's or the β's are each estimable. To do so, the matrix $X'X$ is needed. Multiplication will show that

$$X'X=\begin{array}{c} \\ \mu \\ \tau_1 \\ \tau_2 \\ \vdots \\ \tau_a \\ \beta_1 \\ \beta_2 \\ \vdots \\ \beta_b \end{array}\begin{array}{c} \begin{array}{ccccccccc} \mu & \tau_1 & \tau_2 & & \tau_a & \beta_1 & \beta_2 & & \beta_b \end{array} \\ \left[\begin{array}{c|cccc|cccc} ab & b & b & \cdots & b & a & a & \cdots & a \\ \hline b & b & 0 & & 0 & 1 & 1 & \cdots & 1 \\ b & 0 & b & & 0 & 1 & 1 & & 1 \\ \vdots & \vdots & \vdots & & \vdots & \vdots & \vdots & & \vdots \\ b & 0 & 0 & & b & 1 & 1 & & 1 \\ \hline a & 1 & 1 & \cdots & 1 & a & 0 & & 0 \\ a & 1 & 1 & & 1 & 0 & a & & 0 \\ \vdots & \vdots & \vdots & & \vdots & & & & \vdots \\ a & 1 & 1 & & 1 & 0 & & \cdots & a \end{array}\right] \end{array}$$

Notice that this matrix has the same two dependencies among columns as X. Hence it is an $(a+b+1)\times(a+b+1)$ matrix of rank $a+b-1$. The matrix is used to establish the estimability of contrasts in the τ's.

Theorem 6.4.1 In the two-factor design with no interaction, every contrast in the τ's is estimable.

Proof

Let

$$\begin{aligned}\omega &= a_1\tau_1 + a_2\tau_2 + \cdots + a_a\tau_a \\ &= [0 \quad a_1 \quad a_2 \quad \cdots \quad a_a \quad 0 \quad 0 \quad \cdots \quad 0]\boldsymbol{\beta} \\ &= \mathbf{a}'\boldsymbol{\beta}\end{aligned}$$

denote a contrast. By Theorem 5.4.1, $\mathbf{a}'\boldsymbol{\beta}$ is estimable if there exists a solution to the system $(X'X)\mathbf{z} = \mathbf{a}$. By Theorem 5.3.1, the system has a solution if

$$r[X'X \mid \mathbf{a}] = r(X'X)$$

Consider

$$[X'X \mid \mathbf{a}] = \left[\begin{array}{c|cccc|cccc|c} ab & b & b & \cdots & b & a & a & \cdots & a & 0 \\ \hline b & b & 0 & \cdots & 0 & 1 & 1 & \cdots & 1 & a_1 \\ b & 0 & b & & 0 & 1 & 1 & & 1 & a_2 \\ \vdots & \vdots & \vdots & & \vdots & \vdots & \vdots & & \vdots & \vdots \\ b & 0 & 0 & & b & 1 & 1 & & 1 & a_a \\ \hline a & 1 & 1 & \cdots & 1 & a & 0 & \cdots & 0 & 0 \\ a & 1 & 1 & & 1 & 0 & a & & 0 & 0 \\ \vdots & \vdots & \vdots & & \vdots & \vdots & \vdots & & \vdots & \vdots \\ a & 1 & 1 & & 1 & 0 & 0 & & a & 0 \end{array}\right]$$

Since ω is a contrast, $\Sigma_{i=1}^{a} a_i = 0$. Hence rows 2 through $a+1$ sum to row 1, as do rows $a+2$ through $a+b+1$. Hence the rank of $[X'X \mid \mathbf{a}]$ is $a+b+1-2 = a+b-1 = r[X'X \mid \mathbf{a}]$ and the proof is complete. ■

A similar argument will show that every contrast in the β's is estimable. To see that H_0 is testable, note that it is equivalent to

$$\begin{aligned}\tau_1 - \tau_2 &= 0 \\ \tau_1 + \tau_2 - 2\tau_3 &= 0 \\ \tau_1 + \tau_2 + \tau_3 - 3\tau_4 &= 0 \\ &\vdots \\ \tau_1 + \tau_2 + \tau_3 + \cdots + \tau_{a-1} - (a-1)\tau_a &= 0\end{aligned}$$

This is a system of orthogonal contrasts that are linearly independent as well as estimable. By Definition 6.1.1, H_0 is testable. In matrix form, we are testing

$$H_0 : C\boldsymbol{\beta} = \mathbf{0}$$

where C is the $(a-1) \times (a+b+1)$ matrix given by

$$C = \begin{array}{c} \\ \end{array}\begin{array}{cccccccccccc} \mu & \tau_1 & \tau_2 & \tau_3 & & \tau_{a-1} & \tau_a & \beta_1 & \beta_2 & & \beta_b \end{array}$$

$$C = \begin{bmatrix} 0 & 1 & -1 & 0 & \cdots & 0 & 0 & 0 & 0 & \cdots & 0 \\ 0 & 1 & 1 & -2 & \cdots & 0 & 0 & 0 & 0 & \cdots & 0 \\ \vdots & & & & & & & & & & \vdots \\ 0 & 1 & 1 & 1 & \cdots & 1 & a-1 & 0 & 0 & & 0 \end{bmatrix}$$

As in the one-factor setting, testable hypotheses can be handled in a variety of ways. An F ratio of the form developed in Section 6.1 can be used; the model can be reparameterized to the full rank case and the methods of Chapter 4 employed; or the technique used in Chapter 4 to derive tests on a subvector of $\boldsymbol{\beta}$ can be applied. The third technique will be used here. In particular, the regression sum of squares for the full model is found; the regression sum of squares for the model that results by assuming that $\tau_1 = \tau_2 = \cdots = \tau_a$ is found; the difference between the two, $SS_{\text{Reg(Hypothesis)}}$, is used to test H_0.

Recall that the regression sum of squares for the full model is

$$SS_{\text{Reg(Full)}} = \mathbf{y}'X(X'X)^c X'\mathbf{y} = \mathbf{b}'X'\mathbf{y}$$

Since the regression sum of squares is invariant to the choice of the solution to the normal equations, any solution to the system

$$(X'X)\mathbf{b} = X'\mathbf{y}$$

can be used to compute $S_{\text{Reg(Full)}}$. Theoretically, a solution can be found by computing

$$(X'X)^c X'\mathbf{y}$$

where $(X'X)^c$ is any conditional inverse for $X'X$. SAS and other statistical packages find estimates for $\boldsymbol{\beta}$ in this way. Since it is not easy to find an explicit expression for $(X'X)^c$ for this design, a different approach will be taken here. Constraints will be applied to solve the system of normal equations. A *constraint* is a restriction placed on the elements of the solution vector $\mathbf{b}$ that is imposed solely for the purpose of expediting the solution of the normal equations. In general, the number of constraints required is $p - r$ where p is the number of columns of X and r is its rank. Here the number of constraints required is $(a + b + 1) - (a + b - 1) = 2$. As you shall see, convenient constraints are

$$\sum_{i=1}^{a} \hat{\tau}_i = \sum_{j=1}^{b} \hat{\beta}_j = 0$$

To see why these constraints are useful, note that $X'\mathbf{y}$ is given by

$$\begin{array}{c} \mu \\ \tau_1 \\ \tau_2 \\ \vdots \\ \tau_a \\ \\ \beta_1 \\ \beta_2 \\ \vdots \\ \beta_b \end{array}
\left[\begin{array}{cccc|cccc|c|cccc}
1 & 1 & \cdots & 1 & 1 & 1 & \cdots & 1 & \cdots & 1 & 1 & \cdots & 1 \\
1 & 1 & \cdots & 1 & 0 & 0 & \cdots & 0 & & 0 & 0 & \cdots & 0 \\
0 & 0 & \cdots & 0 & 1 & 1 & \cdots & 1 & & 0 & 0 & \cdots & 0 \\
\vdots & & & & & & & & & & & & \\
0 & 0 & \cdots & 0 & 0 & 0 & \cdots & 0 & & 1 & 1 & \cdots & 1 \\
\hline
1 & 0 & \cdots & 0 & 1 & 0 & \cdots & 0 & & 1 & 0 & \cdots & 0 \\
0 & 1 & \cdots & 0 & 0 & 1 & \cdots & 0 & & 0 & 1 & \cdots & 0 \\
\vdots & & & & & & & & & & & & \\
0 & 0 & \cdots & 1 & 0 & 0 & \cdots & 1 & & 0 & 0 & \cdots & 1
\end{array}\right]
\left[\begin{array}{c} y_{11} \\ y_{12} \\ \vdots \\ y_{1b} \\ \hline y_{21} \\ y_{22} \\ \vdots \\ y_{2b} \\ \hline \vdots \\ \hline y_{a1} \\ y_{a2} \\ \vdots \\ y_{ab} \end{array}\right]
=
\left[\begin{array}{c} y_{..} \\ y_{1.} \\ y_{2.} \\ \vdots \\ y_{a.} \\ \hline y_{.1} \\ y_{.2} \\ \vdots \\ y_{.b} \end{array}\right]$$

Remember that $y_{..}$ denotes the sum of all responses, $y_{i.}$ denotes the sum of the responses to level i of factor I, and $y_{.j}$ denotes the sum of the responses to the jth level of factor II.

We want to solve the system $(X'X)\mathbf{b} = X'\mathbf{y}$. The left side of this equation is given by

$$\left[\begin{array}{c|cccc|cccc}
ab & b & b & \cdots & b & a & a & \cdots & a \\
\hline
b & b & & 0 & & 1 & 1 & \cdots & 1 \\
b & & b & & & 1 & 1 & & 1 \\
\vdots & & & \ddots & & \vdots & \vdots & & \vdots \\
b & & 0 & & b & 1 & 1 & & 1 \\
\hline
a & 1 & 1 & \cdots & 1 & a & & 0 & \\
a & 1 & 1 & & 1 & & a & & \\
\vdots & \vdots & \vdots & & \vdots & & & \ddots & \\
a & 1 & 1 & & 1 & & 0 & & a
\end{array}\right]
\left[\begin{array}{c} \hat{\mu} \\ \hat{\tau}_1 \\ \hat{\tau}_2 \\ \vdots \\ \hat{\tau}_a \\ \hat{\beta}_1 \\ \hat{\beta}_2 \\ \vdots \\ \hat{\beta}_b \end{array}\right]
=
\left[\begin{array}{c} ab\hat{\mu} + b\sum_{i=1}^{a} \hat{\tau}_i + a\sum_{j=1}^{b} \hat{\beta}_j \\ b\hat{\mu} + b\hat{\tau}_1 \\ b\hat{\mu} + b\hat{\tau}_2 \\ \vdots \\ b\hat{\mu} + b\hat{\tau}_a \\ a\hat{\mu} + a\hat{\beta}_1 \\ a\hat{\mu} + a\hat{\beta}_2 \\ \vdots \\ a\hat{\mu} + a\hat{\beta}_b \end{array}\right]$$

If we apply the constraints, then we obtain the system

$$\begin{aligned} ab\hat{\mu} &= y_{..} \\ b\hat{\mu} + b\hat{\tau}_i &= y_{i.} \\ a\hat{\mu} + a\hat{\beta}_j &= y_{.j} \end{aligned}$$

This reduces to

$$\hat{\mu} = \bar{y}_{..}$$
$$\hat{\tau}_i = \bar{y}_{i.} - \bar{y}_{..}$$
$$\hat{\beta}_j = \bar{y}_{.j} - \bar{y}_{..}$$

Therefore, let us define $\hat{\tau}_i$ and $\hat{\beta}_j$ by

$$\hat{\mu} = \bar{y}_{..} \quad \hat{\tau}_i = \bar{y}_{i.} - \bar{y}_{..} \quad \hat{\beta}_j = \bar{y}_{.j} - \bar{y}_{..}$$

Exercise 30 shows that for this choice of estimators,

$$\sum_{i=1}^{a} \hat{\tau}_i = \sum_{j=1}^{b} \hat{\beta}_j = 0$$

Hence,

$$\mathbf{b} = \begin{bmatrix} \bar{y}_{..} \\ \bar{y}_{1.} - \bar{y}_{..} \\ \bar{y}_{2.} - \bar{y}_{..} \\ \vdots \\ \bar{y}_{a.} - \bar{y}_{..} \\ \bar{y}_{.1} - \bar{y}_{..} \\ \bar{y}_{.2} - \bar{y}_{..} \\ \vdots \\ \bar{y}_{.b} - \bar{y}_{..} \end{bmatrix}$$

is a convenient solution to the normal equations (see Exercise 31). This solution is used to find $SS_{\text{Reg(Full)}}$. In particular,

$$SS_{\text{Reg(Full)}} = \mathbf{b}'X'\mathbf{y} = y_{..}^2/ab + \sum_{i=1}^{a} y_{i.}(\bar{y}_{i.} - \bar{y}_{..}) + \sum_{j=1}^{b} y_{.j}(\bar{y}_{.j} - \bar{y}_{..})$$

This can be reduced algebraically to the following simplified expression (see Exercise 32):

$$SS_{\text{Reg(Full)}} = \sum_{i=1}^{a} y_{i.}^2/b + \sum_{j=1}^{b} y_{.j}^2/a - y_{..}^2/ab$$

To find the reduced model, it is assumed that $\tau_1 = \tau_2 = \cdots = \tau_a = \tau$. The resulting model is

$$y_{ij} = \mu^* + \beta_j + \varepsilon_{ij} \qquad \begin{aligned} i &= 1, 2, \ldots, a \\ j &= 1, 2, \ldots, b \end{aligned}$$

where $\mu^* = \mu + \tau$. This is simply a one-way classification model with $N = ab$, $n_j = a$, and $k = b$. In Section 6.2 it was shown that the regression sum of squares for such a model is

$$SS_{\text{Reg(Reduced)}} = \sum_{j=1}^{b} y_{.j}^2/a$$

The sum of squares associated with the hypothesis

$$H_0:\tau_1 = \tau_2 = \cdots = \tau_a$$

is

$$\begin{aligned} SS_{\text{Reg(Hypothesis)}} &= SS_{\text{Reg(Full)}} - SS_{\text{Reg(Reduced)}} \\ &= \sum_{i=1}^{a} y_{i.}^2/b + \sum_{j=1}^{b} y_{.j}^2/a - y_{..}^2/ab - \sum_{j=1}^{b} y_{.j}^2/a \\ &= \sum_{i=1}^{a} y_{i.}^2/b - y_{..}^2/ab \end{aligned}$$

The number of degrees of freedom associated with the full model is the same as the rank of X, namely $a + b - 1$; the number of degrees of freedom associated with the reduced model is b; thus, the number of degrees of freedom associated with the hypothesis on levels of factor I is $(a + b - 1) - b = a - 1$. The regression sum of squares has $ab - (a + b - 1) = ab - a - b + 1 = (a - 1)(b - 1)$ associated degrees of freedom. The F ratio used to test H_0 is

$$F_{a-1,(a-1)(b-1)} = \frac{\left[\sum_{i=1}^{a} y_{i.}^2/b - y_{..}^2/ab\right]\Big/(a-1)}{SS_{\text{Res}}/(a-1)(b-1)}$$

This information is summarized in Table 6.7.

A similar argument is used to show that the F ratio used to test

$$H_0':\beta_1 = \beta_2 = \cdots = \beta_b$$

is

$$F_{b-1,(a-1)(b-1)} = \frac{\left[\sum_{j=1}^{b} y_{.j}^2/a - y_{..}^2/ab\right]\Big/(b-1)}{SS_{\text{Res}}/(a-1)(b-1)}$$

If we adjust the total sum of squares by subtracting the correction factor as was done in Section 6.2, then all of this information is summarized in Table 6.8. This table is similar to those usually found in elementary courses in applied statistics.

Example 6.4.1 illustrates this model in a simple setting.

Example 6.4.1 A study of the solubility of two of the most commonly encapsulated enzyme preparations is conducted. The purpose of the study is to determine the effect of capsule type and biological fluid on the time required for the capsule to dissolve. Two biological fluids, gastric and duodenal, and two capsule types, A and B, are used in the study. Four identical samples of the preparation are obtained. Two are randomly selected for encapsulation in capsule type A; the others are encapsulated

TABLE 6.7
ANOVA table for the two-factor design with no interaction and one response measured at each of the *ab* treatment combinations

Source of Variation	Degrees of Freedom	Sum of Squares	Mean Square	F
Regression	$a+b-1$	$\sum_{i=1}^{a} y_{i.}^2/b + \sum_{j=1}^{b} y_{.j}^2/a - y_{..}^2/ab$	$SS_{\text{Reg}}/(a+b-1)$	
Reduced model	b	$\sum_{j=1}^{b} y_{.j}^2/a$		
Hypothesis	$a-1$	$\sum_{i=1}^{a} y_{i.}^2/b - y_{..}^2/ab$	$SS_{\text{Reg(Hypothesis)}}/(a-1)$	$\dfrac{SS_{\text{Reg(Hypothesis)}}/(a-1)}{SS_{\text{Res}}/(a-1)(b-1)}$
Residual	$(a-1)(b-1)$	$\sum_{i=1}^{a}\sum_{j=1}^{b} y_{ij}^2 - SS_{\text{Reg}}$		
Total (uncorrected)	ab	$\sum_{i=1}^{a}\sum_{j=1}^{b} y_{ij}^2$		

TABLE 6.8
ANOVA table for the two-factor design with no interaction and one response measured at each of the *ab* treatment combinations based on corrected totals

Source of Variation	Degrees of Freedom	Sum of Squares	Mean Square	F
Reg (Hypothesis I)	$a-1$	$\sum_{i=1}^{a} y_{i.}^2/b - y_{..}^2/ab$	$SS_{\text{Reg(Hypothesis I)}}/(a-1)$	$\dfrac{SS_{\text{Reg(Hypothesis I)}}/(a-1)}{SS_{\text{Res}}/(a-1)(b-1)}$
Reg (Hypothesis II)	$b-1$	$\sum_{j=1}^{b} y_{.j}^2/a - y_{..}^2/ab$	$SS_{\text{Reg(Hypothesis II)}}/(b-1)$	$\dfrac{SS_{\text{Reg(Hypothesis II)}}/(b-1)}{SS_{\text{Res}}/(a-1)(b-1)}$
Residual	$(a-1)(b-1)$	$\sum_{i=1}^{a}\sum_{j=1}^{b} y_{ij}^2 - \sum_{i=1}^{a} y_{i.}^2/b - \sum_{j=1}^{b} y_{.j}^2/a + y_{..}^2/ab$	$SS_{\text{Res}}/(a-1)(b-1)$	
Corrected total	$ab-1$	$\sum_{i=1}^{a}\sum_{j=1}^{b} y_{ij}^2 - y_{..}^2/ab$		

in type B. One of each type capsule is then randomly selected and dissolved in gastric juices; the others are dissolved in duodenal juices. These data are obtained:

		Fluid Type (Factor I) Gastric	Duodenal	
Capsule Type (Factor II)	A	39.5	31.2	$y_{.1} = 70.7$
	B	47.4	44.0	$y_{.2} = 91.4$
		$y_{1.} = 86.9$	$y_{2.} = 75.2$	$y_{..} = 162.1$

For these data, $a = b = 2$ and

$$\sum_{i=1}^{2} y_{i.}^2/b = 6603.325 \qquad \sum_{j=1}^{2} y_{.j}^2/a = 6676.225$$

$$\sum_{i=1}^{2}\sum_{j=1}^{2} y_{ij}^2 = 6716.45 \qquad y_{..}^2/ab = 6569.1025$$

Null hypotheses to be tested are

$H_0: \tau_1 = \tau_2$ (No difference in response between the two biological fluids)

$H_0': \beta_1 = \beta_2$ (No difference in response between the two capsule types)

The ANOVA following the format of Table 6.8 is shown in Table 6.9.

The critical point for an $\alpha = 0.1$ level test based on the $F_{1,1}$ distribution is 39.86. Since neither of the F ratios obtained exceeds this value, neither H_0 nor H_0' can be rejected based on these data.

TABLE 6.9 **ANOVA for solubility data**

Source of Variation	Degrees of Freedom	Sum of Squares	Mean Square	F
Reg (Hypothesis I)	1	34.2225	34.2225	5.70
Reg (Hypothesis II)	1	107.1225	107.1225	17.846
Residual	1	6.0025	6.0025	
Corrected total	3	147.3475		

This example is presented primarily to illustrate the computations required to analyze the two-factor model with no interaction by hand. In practice, an experiment of this sort would not be very practical for two reasons. First, the overall sample size is so small that very few degrees of freedom are associated with either the numerator or the denominator of the F ratios used to test H_0 and H'_0. This makes it very difficult to detect differences even if they do exist. As you probably suspect, this difficulty can be overcome by taking $n > 1$ observations per treatment combination. The analysis of such a design parallels that presented here. The theoretical development of the analysis is outlined in Exercise 38. The second practical problem is not solved as easily. In particular, it is assumed a priori that no interaction exists between capsule type and fluid type. A researcher who is thoroughly familiar with the problem might have sound biological reasons for believing that this is true. However, the presence or absence of interaction is often not known prior to experimentation. As a result, the experimental data are used to detect interaction. To do this, more than one observation must be available for each of the ab treatment combinations. Models of this sort are presented in Section 6.6.

Contrasts among levels of factor I or factor II can be tested via a t test or an F test in a manner analogous to that developed in the one-way classification setting. The statistics required are demonstrated in Exercises 39 and 40.

One other point is important. In the two-factor design without interaction, randomization is a *two-way process.* We randomly assign a level of factor I to each experimental unit and then randomly assign a level of factor II. Thus there are ab possible treatment combinations, each of which in effect is randomly assigned to exactly one experimental unit. The fact that randomization is done with respect to both factors makes it possible to test for differences among levels of each factor by means of the F tests developed here.

6.5

RANDOMIZED COMPLETE BLOCK DESIGN: FIXED EFFECTS

One of the tests usually presented in elementary statistics courses is the paired t test. This is a test used to compare two treatment means in the presence of an extraneous variable. An *extraneous* variable is a variable that is present and that can affect the response but that is not of direct interest to the researcher. For example, in comparing the life span of two paints, the location in which the paint is used can be considered an extraneous variable. Location is not the focus of the study, but it certainly affects the life span of paint and its effect needs to be controlled. In studying the ability of two different sunscreens to protect an individual from burning, the variable "skin type" might be viewed as an extraneous variable.

When comparing a treatments with $a \geqslant 2$ in the presence of an extraneous variable, the *randomized complete block design* is useful. The extraneous variable is used to form b blocks where a *block* is a group of a experimental units that are as nearly alike as possible relative to the extraneous or blocking variable. The a treatments are randomly assigned to the experimental units within each block. The end result is ab possible treatment-block combinations, each associated with exactly one experimental unit. The model for such a design is

$$y_{ij} = \mu + \tau_i + \beta_j + \varepsilon_{ij} \qquad \begin{array}{l} i = 1, 2, \ldots, a \\ j = 1, 2, \ldots, b \end{array}$$

Here τ_i represents a treatment effect and β_j a block effect. On the surface this model appears to be identical to the one presented in the last section. However, there is an important difference. Here randomization occurs only *once*. We randomly assign treatments to experimental units within blocks; we do *not* randomly assign blocks to experimental units. Each unit falls naturally into exactly one of the b blocks depending on its value of the extraneous variable. As a result, the only null hypothesis that can be tested is

$$H_0: \tau_1 = \tau_2 = \cdots = \tau_a$$

This hypothesis is tested exactly as in the two-factor design without interaction. There is no valid test for determining the effectiveness of blocking. That is,

$$H_0': \beta_1 = \beta_2 = \cdots = \beta_b$$

cannot be tested using an F test [**12**] [**2**]. However, the usefulness of blocking can be assessed. Usually this is done by computing the relative efficiency of the randomized complete block design as compared to the one-way classification design. A common measure of this efficiency is developed in [**12**] and is given by

$$RE = \frac{SS_{\text{Blocks}} + b(a-1)s^2}{(ab-1)s^2}$$

where s^2 is the residual mean square obtained from the original ANOVA and

$$SS_{\text{Blocks}} = \sum_{j=1}^{b} \frac{y_{.j}^2}{a} - \frac{y_{..}^2}{ab}$$

In the two-factor design without interaction this sum of squares is called $SS_{\text{Reg(Hypotheses II)}}$. Any value of RE that exceeds 1 indicates that blocking was effective to some extent. However, there is an easier way to assess the effectiveness of blocking. A "pseudo" F ratio that takes the same form as that used to test for differences among levels of factor II in the two-factor design with no interaction can be formed:

$$F_{\text{pseudo}} = \frac{SS_{\text{Blocks}}/(b-1)}{SS_{\text{Res}}/(a-1)(b-1)}$$

Arnold, Lentner, and Hinkleman [**2**] have shown that F_{pseudo} and RE are linearly related via the equation

$$RE = c + (1 - c)F_{\text{pseudo}}$$

where $c = b(a-1)/(ab-1)$. It is easy to see that $c \leqslant 1$ and that

$$\text{If } F_{\text{pseudo}} < 1, \text{ then } RE < 1$$
$$\text{If } F_{\text{pseudo}} = 1, \text{ then } RE = 1$$
$$\text{If } F_{\text{pseudo}} > 1, \text{ then } RE > 1$$

Hence even though the ratio F_{pseudo} cannot be used to conduct a formal F test on the differences among blocks, it can be used to assess the effectiveness of blocking in exactly the same way as the usual relative efficiency measure.

This model assumes that there is no interaction between blocks and treatments. It is illustrated in Example 6.5.1.

Example 6.5.1 The highway department is studying four different types of paving for possible use on interstate highways. Because it is known that location within the state can influence results due to differences in weather and traffic patterns, location is treated as an extraneous variable. Three sections of highway in different parts of the state are chosen for experimentation. Each section constitutes a block. Each is divided into four strips, and the four paving types are randomly assigned to strips within each block. The idea is illustrated in Figure 6.3.

One year after paving, the amount of wear for each strip is ascertained. We want to test

$$H_0: \tau_1 = \tau_2 = \tau_3 = \tau_4 \qquad \text{(No differences in wear among the four paving types)}$$

These data are obtained (higher values represent greater wear):

		Paving Type			
		1	2	3	4
	1	42.7	39.3	48.5	32.8
Location	2	50.0	38.0	49.7	40.2
(Block)	3	51.9	46.3	53.5	51.1

Computationally these data are handled exactly as in the two-factor design with no interaction. The ANOVA is shown in Table 6.10.

Based on the $F_{3,6}$ distribution, H_0 can be rejected with $0.025 < P < 0.05$. Differences among paving types have been detected. The observed value of F_{pseudo} is 8.47. Since this value exceeds 1, it can be concluded that blocking is effective. Location tends to affect wear among paving types.

As in the two-factor design with no interaction, contrasts among paving type can be tested; however, there is no test for contrasts among locations.

FIGURE 6.3 **Paving types are randomly assigned to strips within each section of highway**

Block 1	Block 2	Block 3
Type 1	Type 2	Type 3
Type 4	Type 3	Type 1
Type 2	Type 1	Type 2
Type 3	Type 4	Type 4

TABLE 6.10 **ANOVA for paving data**

Source of Variation	Degrees of Freedom	Sum of Squares	Mean Square	F
Treatments	3	205.28	68.43	5.81
Blocks	2	199.46	99.73	8.47 (pseudo)
Residual	6	70.69	11.78	
Total (corrected)	11	475.43		

6.6

TWO-FACTOR DESIGN WITH INTERACTION: FIXED EFFECTS

In a two-factor design with interaction, two factors are studied in the same experiment and an interaction term is included in the model. Recall that, intuitively speaking, interaction is the failure of levels of factor I to behave consistently across levels of factor II, and vice versa. Consider, for example, the theoretical cell means given in Table 6.11 for an experiment with two levels of each of two factors.

TABLE 6.11 **Theoretical cell means for an experiment with two levels of each of two factors**

		Factor I 1	Factor I 2
Factor II	1	3	6
	2	6	3

There is obvious interaction exhibited here since in row 1 the first level of factor I has a smaller average value than that of the second level, whereas the situation is reversed in row 2. In other words, the effect of factor I differs depending upon the level of factor II under consideration. The presence of interaction often can mask the effect of a factor due to averaging. In the situation described above, a test on levels of factor I would probably not detect any differences because on the *average* the two levels are identical. However, they are quite different when considered row by row. The implication of this is that in a two-factor design we first test for the presence of interaction. Subsequent hypothesis-testing strategy depends on the outcome of this initial test.

THE GENERAL TWO-FACTOR MODEL

The model for a two-factor design with interaction is

$$y_{ijk} = \mu + \tau_i + \beta_j + (\tau\beta)_{ij} + \varepsilon_{ijk} \qquad \begin{aligned} i &= 1, 2, \ldots, a \\ j &= 1, 2, \ldots, b \\ k &= 1, 2, \ldots, n \end{aligned}$$

Here τ_i denotes the effect of the ith level of factor I, β_j denotes the effect of the jth level of factor II, and $(\tau\beta)_{ij}$ denotes the interaction between the ith level of factor I and the jth level of factor II. These hypotheses are to be tested:

H_0: No interaction

H_0': No differences among the effects of levels of factor I

H_0'': No differences among the effects of levels of factor II

To see how to test these hypotheses, we consider in detail a design with two levels of each factor and two observations per treatment combination. The results obtained in this special case are extended easily to the general setting in which there are a levels of factor I, b levels of factor II, and n observations per treatment combination.

The data layout for the 2×2 design is given in Table 6.12. In matrix notation the model is

$$\mathbf{y} = X\boldsymbol{\beta} + \boldsymbol{\varepsilon}$$

TABLE 6.12 **Data layout for a two-factor design with $a = b = n = 2$**

		Factor I: 1	Factor I: 2
Factor II	1	y_{111}, y_{112}	y_{211}, y_{212}
	2	y_{121}, y_{122}	y_{221}, y_{222}

where

$$
X = \begin{array}{c}
\begin{array}{ccccccccc}
\mu & \tau_1 & \tau_2 & \beta_1 & \beta_2 & (\tau\beta)_{11} & (\tau\beta)_{12} & (\tau\beta)_{21} & (\tau\beta)_{22}
\end{array} \\
\left[\begin{array}{ccccccccc}
1 & 1 & 0 & 1 & 0 & 1 & 0 & 0 & 0 \\
1 & 1 & 0 & 1 & 0 & 1 & 0 & 0 & 0 \\
1 & 1 & 0 & 0 & 1 & 0 & 1 & 0 & 0 \\
1 & 1 & 0 & 0 & 1 & 0 & 1 & 0 & 0 \\
1 & 0 & 1 & 1 & 0 & 0 & 0 & 1 & 0 \\
1 & 0 & 1 & 1 & 0 & 0 & 0 & 1 & 0 \\
1 & 0 & 1 & 0 & 1 & 0 & 0 & 0 & 1 \\
1 & 0 & 1 & 0 & 1 & 0 & 0 & 0 & 1
\end{array} \right]
\end{array},
$$

$$
\beta = \begin{bmatrix} \mu \\ \tau_1 \\ \tau_2 \\ \beta_1 \\ \beta_2 \\ (\tau\beta)_{11} \\ (\tau\beta)_{12} \\ (\tau\beta)_{21} \\ (\tau\beta)_{22} \end{bmatrix}, \quad
y = \begin{bmatrix} y_{111} \\ y_{112} \\ y_{121} \\ y_{122} \\ y_{211} \\ y_{212} \\ y_{221} \\ y_{222} \end{bmatrix}, \quad \text{and} \quad
\varepsilon = \begin{bmatrix} \varepsilon_{111} \\ \varepsilon_{112} \\ \varepsilon_{121} \\ \varepsilon_{122} \\ \varepsilon_{211} \\ \varepsilon_{212} \\ \varepsilon_{221} \\ \varepsilon_{222} \end{bmatrix}.
$$

The matrix is of dimension 8×9. To determine its rank, let c_i denote the ith column of X. Note that each column of X can be expressed as a linear combination of c_6 through c_9, the columns that code the interaction terms, and that these columns are linearly independent (see Exercise 44). Hence the rank of X is four.

TESTING FOR INTERACTION

Our first task is to develop a test for detecting the presence of interaction. It is tempting to assume that if no interaction exists then each of the interaction effects, $(\tau\beta)_{ij}$, is zero and that at least one of these is nonzero otherwise. However, the situation is a bit more complex than this. The trouble arises from the fact that the interaction effects are not estimable. Hence the hypothesis

$$H_0 : C\beta = 0$$

where

$$
C = \begin{bmatrix}
0 & 0 & 0 & 0 & 0 & 1 & 0 & 0 & 0 \\
0 & 0 & 0 & 0 & 0 & 0 & 1 & 0 & 0 \\
0 & 0 & 0 & 0 & 0 & 0 & 0 & 1 & 0 \\
0 & 0 & 0 & 0 & 0 & 0 & 0 & 0 & 1
\end{bmatrix}
$$

is not testable (see Exercise 45). Example 6.6.1 illustrates the problem numerically.

Example 6.6.1 Consider the following hypothetical data and assume that in each case $\varepsilon_{ij} = 0$.

		Factor I	
		1	2
Factor II	1	6	5
	2	6	5

Consider these possible parameter sets:

Parameter Set I		*Parameter Set II*		
$\mu = 0$	$(\tau\beta)_{11} = 1$	$\mu = 0$	$(\tau\beta)_{ij} = 0$	$i = 1, 2$
$\tau_1 = 2$	$(\tau\beta)_{12} = 2$	$\tau_1 = 5$		$j = 1, 2$
$\tau_2 = 1$	$(\tau\beta)_{21} = 1$	$\tau_2 = 4$		
$\beta_1 = 3$	$(\tau\beta)_{22} = 2$	$\beta_1 = 1$		
$\beta_2 = 2$		$\beta_2 = 1$		

Note that for each parameter set

$$y_{ij} = \mu + \tau_i + \beta_j + (\tau\beta)_{ij} \qquad i = 1, 2 \quad j = 1, 2$$

Thus either parameter set could have generated the data. It is obvious that the data exhibit no interaction. Nonetheless, it is possible that $(\tau\beta)_{ij}$ could be nonzero, as evidenced by parameter set I. Parameter set II demonstrates that $(\tau\beta)_{ij}$ can be zero and still have no interaction. Thus the values of the interaction terms alone cannot be used to detect the presence or absence of interaction.

To understand the theoretical definition of *no interaction* let us reparameterize the 2×2 model by defining

$$\mu_{ij} = \mu + \tau_i + \beta_j + (\tau\beta)_{ij} \qquad i = 1, 2 \quad j = 1, 2$$

In reparameterized form the model becomes

$$y_{ijk} = \mu_{ij} + \varepsilon_{ijk} \qquad i = 1, 2 \quad j = 1, 2 \quad k = 1, 2$$

The model expresses the idea that within each cell or treatment combination the responses vary randomly about the cell mean μ_{ij}. We want to express mathematically the idea that levels of factor I behave consistently across levels of factor II, and vice versa. We want to express the idea that the cell by cell difference in means from one column to another is the same for each row and that the cell by cell difference in means from one row to another is the same for each column. This idea is demonstrated in Example 6.6.2.

Example 6.6.2 The 2×2 design has four cell means:

		Factor I 1	Factor I 2
Factor II	1	μ_{11}	μ_{21}
	2	μ_{12}	μ_{22}

For the levels of factor I to behave consistently across levels of factor II, the difference in cell means between levels 1 and 2 of factor I must be the same in each row. That is, we must have

$$\mu_{11} - \mu_{21} = \mu_{12} - \mu_{22}$$

or

$$(\mu_{11} - \mu_{21}) - (\mu_{12} - \mu_{22}) = 0$$

Similarly, for the levels of factor II to behave consistently across levels of factor I, the difference in cell means between levels 1 and 2 of factor II must be the same in each column. This means that

$$\mu_{11} - \mu_{12} = \mu_{21} - \mu_{22}$$

or

$$(\mu_{11} - \mu_{12}) - (\mu_{21} - \mu_{22}) = 0$$

For no interaction to exist, each of these equations must hold. Note, however, that these are algebraically equivalent so that, in fact, no interaction is obtained in the 2×2 case whenever

$$\mu_{11} + \mu_{22} = \mu_{12} + \mu_{21}$$

The sum of the means along the main diagonal of the table of cell means is equal to the sum of the off-diagonal means. It is easy to apply this criterion to the cell means of Table 6.11 (on p. 275) to see that interaction exists here.

In general, a two-factor design has a levels of factor I, b levels of factor II, and ab cells. Whenever a or b exceed 2, the mathematical definition of no interaction becomes more complex even though the underlying idea of consistency of cell means across rows and columns remains the same. Definition 6.6.1 formalizes the idea in the general setting.

Definition 6.6.1

Consider the two-factor model

$$y_{ijk} = \mu + \tau_i + \beta_j + (\tau\beta)_{ij} + \varepsilon_{ijk} \qquad \begin{aligned} i &= 1, 2, \ldots, a \\ j &= 1, 2, \ldots, b \\ k &= 1, 2, \ldots, n \end{aligned}$$

Let

$$\mu_{ij} = \mu + \tau_i + \beta_j + (\tau\beta)_{ij}$$

There is no interaction if and only if

$$(\mu_{ij} - \mu_{ij'}) - (\mu_{i'j} - \mu_{i'j'}) = 0$$

for all i, i', j, j'.

It is easy to verify by substitution that the definition of no interaction can be rephrased in terms of the original model parameters. In particular, no interaction exists if and only if

$$[(\tau\beta)_{ij} - (\tau\beta)_{ij'}] - [(\tau\beta)_{i'j} - (\tau\beta)_{i'j'}] = 0$$

for all i, i', j, and j' (see Exercise 47). In general, this criterion generates $ab(a-1)(b-1)$ equations of which all but $(a-1)(b-1)$ are redundant. For example, in the 2×2 case, the definition generates four equations that are all algebraically equivalent to a single equation; in the 3×3 case, 36 equations are generated, of which only four are unique. From this it can be seen that to test for no interaction based on the original model parameters, we test a null hypothesis of the form

$$H_0\colon C\boldsymbol{\beta} = \mathbf{0}$$

where C is an appropriately chosen matrix of ones and zeros of dimension $(a-1)(b-1) \times (a+b+ab+1)$. As an example, let us again consider the case in which $a = b = n = 2$.

Example 6.6.3 In a two-factor design with $a = b = n = 2$, the null hypothesis of no interaction is expressed as

$$H_0\colon [(\tau\beta)_{11} - (\tau\beta)_{12}] - [(\tau\beta)_{21} - (\tau\beta)_{22}] = 0$$

In matrix form, this becomes

$$H_0: C\boldsymbol{\beta} = 0$$

where

$$C = [0 \quad 0 \quad 0 \quad 0 \quad 0 \quad 1 \quad -1 \quad -1 \quad 1] \quad \text{and} \quad \boldsymbol{\beta} = \begin{bmatrix} \mu \\ \tau_1 \\ \tau_2 \\ \beta_1 \\ \beta_2 \\ (\tau\beta)_{11} \\ (\tau\beta)_{12} \\ (\tau\beta)_{21} \\ (\tau\beta)_{22} \end{bmatrix}$$

Exercise 52 shows that this null hypothesis is testable.

As the number of levels of factors I and II increases, the system of equations needed to express the notion of no interaction becomes more complex. (This idea is demonstrated in Exercise 53.) For this reason, it is convenient to reparameterize the model in such a way that interaction can be expressed more simply by means of the new parameters. In particular, we want to reparameterize in such a way that the reparameterized "interaction" terms will be estimable. By so doing, interaction can be tested by considering the numerical values of these terms directly. Again, the 2×2 case provides a guide to the general technique used. Consider the two-factor model

$$y_{ijk} = \mu + \tau_i + \beta_j + (\tau\beta)_{ij} + \varepsilon_{ijk} \qquad \begin{aligned} i &= 1, 2 \\ j &= 1, 2 \\ k &= 1, 2 \end{aligned}$$

Define $\bar{\tau}$, $\bar{\beta}$, $\bar{\mu}_{..}$, $\overline{(\tau\beta)}$, $\overline{(\tau\beta)}_{i.}$, and $\overline{(\tau\beta)}_{.j}$ by

$$\bar{\tau} = \sum_{i=1}^{2} \tau_i/2 \qquad \overline{(\tau\beta)} = \sum_{i=1}^{2} \sum_{j=1}^{2} (\tau\beta)_{ij}/4$$

$$\bar{\beta} = \sum_{j=1}^{2} \beta_j/2 \qquad \overline{(\tau\beta)}_{i.} = \sum_{j=1}^{2} (\tau\beta)_{ij}/2$$

$$\bar{\mu}_{..} = \sum_{i=1}^{2} \sum_{j=1}^{2} \mu_{ij}/4 \qquad \overline{(\tau\beta)}_{.j} = \sum_{i=1}^{2} (\tau\beta)_{ij}/2$$

The model can be expressed via these parameters as

$$\begin{aligned} y_{ijk} = {} & [\mu + \bar{\tau} + \bar{\beta} + \overline{(\tau\beta)}] + [\tau_i - \bar{\tau} + \overline{(\tau\beta)}_{i.} - \overline{(\tau\beta)}] \\ & + [\beta_j - \bar{\beta} + \overline{(\tau\beta)}_{.j} - \overline{(\tau\beta)}] \\ & + [(\tau\beta)_{ij} - \overline{(\tau\beta)}_{i.} - \overline{(\tau\beta)}_{.j} + \overline{(\tau\beta)}] + \varepsilon_{ijk} \end{aligned}$$

or

$$y_{ijk} = \mu^* + \tau_i^* + \beta_j^* + (\tau\beta)_{ij}^* + \varepsilon_{ijk}$$

where

$$\mu^* = \mu + \bar{\tau} + \bar{\beta} + (\overline{\tau\beta})$$
$$\tau_i^* = \tau_i - \bar{\tau} + (\overline{\tau\beta})_{i.} - (\overline{\tau\beta})$$
$$\beta_j^* = \beta_j - \bar{\beta} + (\overline{\tau\beta})_{.j} - (\overline{\tau\beta})$$
$$(\tau\beta)_{ij}^* = (\tau\beta)_{ij} - (\overline{\tau\beta})_{i.} - (\overline{\tau\beta})_{.j} + (\overline{\tau\beta})$$

It can be shown that each of these new parameters is estimable (see Exercise 54). Hence it is reasonable to expect that the null hypothesis of no interaction can be expressed simply in terms of these redefined parameters. In particular, it can be shown that no interaction exists if and only if $(\tau\beta)_{ij}^* = 0$ for each i and j. Theorem 6.6.1 presents this result in the general setting. Notationally, in general, we define

$$\bar{\tau} = \sum_{i=1}^{a} \tau_i/a \qquad \bar{\mu}_{..} = \sum_{i=1}^{a}\sum_{j=1}^{b} \mu_{ij}/ab$$
$$\bar{\beta} = \sum_{j=1}^{b} \beta_j/b \qquad (\overline{\tau\beta}) = \sum_{i=1}^{a}\sum_{j=1}^{b} (\tau\beta)_{ij}/ab$$
$$(\overline{\tau\beta})_{i.} = \sum_{j=1}^{b} (\tau\beta)_{ij}/b$$
$$(\overline{\tau\beta})_{.j} = \sum_{i=1}^{a} (\tau\beta)_{ij}/a$$

and define μ^*, τ_i^*, β_j^*, and $(\tau\beta)_{ij}^*$ as before.

Theorem 6.6.1 Consider the two-factor model reparameterized as

$$y_{ijk} = \mu^* + \tau_i^* + \beta_j^* + (\tau\beta)_{ij}^* + \varepsilon_{ijk} \qquad \begin{aligned} i &= 1, 2, \ldots, a \\ j &= 1, 2, \ldots, b \\ k &= 1, 2, \ldots, n \end{aligned}$$

where

$$\mu^* = \mu + \bar{\tau} + \bar{\beta} + (\overline{\tau\beta})$$
$$\tau_i^* = \tau_i - \bar{\tau} + (\overline{\tau\beta})_{i.} - (\overline{\tau\beta})$$
$$\beta_j^* = \beta_j - \bar{\beta} + (\overline{\tau\beta})_{.j} - (\overline{\tau\beta})$$
$$(\tau\beta)_{ij}^* = (\tau\beta)_{ij} - (\overline{\tau\beta})_{i.} - (\overline{\tau\beta})_{.j} + (\overline{\tau\beta})$$

No interaction exists if and only if $(\tau\beta)_{ij}^* = 0$ for each i and j.

Proof

Assume that no interaction exists. It has been shown that

$$[(\tau\beta)_{ij} - (\tau\beta)_{ij'}] - [(\tau\beta)_{i'j} - (\tau\beta)_{i'j'}] = 0$$

for all i, j, i', j'. Hence,

$$\begin{aligned}0 &= \sum_{i'=1}^{a} \sum_{j'=1}^{b} [(\tau\beta)_{ij} - (\tau\beta)_{ij'} - (\tau\beta)_{i'j} + (\tau\beta)_{i'j'}] \\ &= \sum_{i'=1}^{a} \left[b(\tau\beta)_{ij} - b(\overline{\tau\beta})_{i.} - b(\tau\beta)_{i'j} + \sum_{j'=1}^{b} (\tau\beta)_{i'j'} \right] \\ &= ab(\tau\beta)_{ij} - ab(\overline{\tau\beta})_{i.} - ab(\overline{\tau\beta})_{.j} + ab(\overline{\tau\beta})\end{aligned}$$

By dividing by ab, it can be concluded that if no interaction exists, then

$$(\tau\beta)_{ij} - (\overline{\tau\beta})_{i.} - (\overline{\tau\beta})_{.j} + (\overline{\tau\beta}) = (\tau\beta)^*_{ij} = 0$$

as desired. To prove the converse, assume that $(\tau\beta)^*_{ij} = 0$ for all i and j. Then

$$(\tau\beta)^*_{ij} - (\tau\beta)^*_{i'j} - (\tau\beta)^*_{ij'} + (\tau\beta)^*_{i'j'} = 0$$

for all i, i', j, and j'. Substituting,

$$\begin{aligned}&[(\tau\beta)_{ij} - (\overline{\tau\beta})_{i.} - (\overline{\tau\beta})_{.j} + (\overline{\tau\beta})] - [(\tau\beta)_{i'j} - (\overline{\tau\beta})_{i'.} - (\overline{\tau\beta})_{.j} + (\overline{\tau\beta})] \\ &\quad - [(\tau\beta)_{ij'} - (\overline{\tau\beta})_{i.} - (\overline{\tau\beta})_{.j'} + (\overline{\tau\beta})] \\ &\quad + [(\tau\beta)_{i'j'} - (\overline{\tau\beta})_{i'.} - (\overline{\tau\beta})_{.j'} + (\overline{\tau\beta})] \\ &\quad = [(\tau\beta)_{ij} - (\tau\beta)_{i'j}] - [(\tau\beta)_{ij'} - (\tau\beta)_{i'j'}] \\ &\quad = 0 \text{ for all } i, i', j, \text{ and } j'\end{aligned}$$

By definition, no interaction exists. ■

In beginning courses in applied statistics, the null hypothesis of no interaction in a two-factor design is often presented as

$$H_0 : (\tau\beta)_{ij} = 0 \quad \text{for each } i \text{ and } j$$

As you can see from Theorem 6.6.1, this is not incorrect provided that it is understood that the underlying model is the reparameterized model just developed; the $(\tau\beta)_{ij}$ referred to is our $(\tau\beta)^*_{ij}$. In particular, it is important to understand that the interaction term is not a parameter that is associated only with the ijth treatment combination. Rather, it is a composite measure that contains information originally associated with the ijth cell, $(\tau\beta)_{ij}$; the ith column, $(\overline{\tau\beta})_{i.}$; the jth row, $(\overline{\tau\beta})_{.j}$, and the overall design, $(\overline{\tau\beta})$.

One other subtle point is pertinent. By defining τ^*_i, β^*_j, and $(\tau\beta)^*_{ij}$ as has been done, certain restrictions have been induced. In particular, it can be shown that

$$\sum_{i=1}^{a} \tau^*_i = 0 \qquad \sum_{i=1}^{a} \sum_{j=1}^{b} (\tau\beta)^*_{ij} = 0 \qquad \sum_{j=1}^{b} (\tau\beta)^*_{.j} = 0$$

$$\sum_{j=1}^{b} \beta^*_j = 0 \qquad \sum_{i=1}^{a} (\tau\beta)^*_{i.} = 0$$

(see Exercise 57). These constraints make it easy to solve the normal equations for the reparameterized model.

To derive the test for interaction, it is constructive to consider a relatively small model in detail. It will be easy then to see how to extend the results to the general case. To this end, consider the two-factor model

$$y_{ijk} = \mu^* + \tau_i^* + \beta_j^* + (\tau\beta)_{ij}^* + \varepsilon_{ijk} \qquad \begin{array}{l} i = 1, 2, 3 \\ j = 1, 2 \\ k = 1, 2, \ldots, n \end{array}$$

The design matrix for this model is a $6n \times 12$ matrix of ones and zeros. (You are asked to find this matrix in Exercise 58.) Multiplication will show that $X'X$ is the 12×12 matrix given by

$$X'X = \left[\begin{array}{c|ccc|cc|cccccc}
6n & 2n & 2n & 2n & 3n & 3n & n & n & n & n & n & n \\
\hline
2n & 2n & 0 & 0 & n & n & n & n & 0 & 0 & 0 & 0 \\
2n & 0 & 2n & 0 & n & n & 0 & 0 & n & n & 0 & 0 \\
2n & 0 & 0 & 2n & n & n & 0 & 0 & 0 & 0 & n & n \\
\hline
3n & n & n & n & 3n & 0 & n & 0 & n & 0 & n & 0 \\
3n & n & n & n & 0 & 3n & 0 & n & 0 & n & 0 & n \\
\hline
n & n & 0 & 0 & n & 0 & n & 0 & 0 & 0 & 0 & 0 \\
n & n & 0 & 0 & 0 & n & 0 & n & 0 & 0 & 0 & 0 \\
n & 0 & n & 0 & n & 0 & 0 & 0 & n & 0 & 0 & 0 \\
n & 0 & n & 0 & 0 & n & 0 & 0 & 0 & n & 0 & 0 \\
n & 0 & 0 & n & n & 0 & 0 & 0 & 0 & 0 & n & 0 \\
n & 0 & 0 & n & 0 & n & 0 & 0 & 0 & 0 & 0 & n
\end{array}\right]$$

The vector $X'\mathbf{y}$ is given by

$$X'\mathbf{y} = \begin{bmatrix} y_{\cdots} \\ y_{1\cdot\cdot} \\ y_{2\cdot\cdot} \\ y_{3\cdot\cdot} \\ y_{\cdot 1\cdot} \\ y_{\cdot 2\cdot} \\ y_{\cdot 11} \\ y_{\cdot 12} \\ y_{\cdot 21} \\ y_{\cdot 22} \\ y_{\cdot 31} \\ y_{\cdot 32} \end{bmatrix}$$

where

$$y_{...} = \sum_{i=1}^{3} \sum_{j=1}^{2} \sum_{k=1}^{n} y_{ijk}$$

$$y_{i..} = \sum_{j=1}^{2} \sum_{k=1}^{n} y_{ijk}$$

$$y_{.j.} = \sum_{i=1}^{3} \sum_{k=1}^{n} y_{ijk}$$

$$y_{ij.} = \sum_{k=1}^{n} y_{ijk}$$

The normal equations are given by

$$(X'X)\mathbf{b} = X'\mathbf{y}$$

or

$$y_{...} = 6n\hat{\mu}^* + 2n \sum_{i=1}^{3} \hat{\tau}_i^* + 3n \sum_{j=1}^{2} \hat{\beta}_j^* + n \sum_{i=1}^{3} \sum_{j=1}^{2} \widehat{(\tau\beta)}_{ij}^*$$

$$y_{i..} = 2n\hat{\mu}^* + 2n\hat{\tau}_i^* + n \sum_{j=1}^{2} \hat{\beta}_j^* + n \sum_{i=1}^{2} \widehat{(\tau\beta)}_{ij}^*$$

$$y_{.j.} = 3n\hat{\mu}^* + n \sum_{i=1}^{3} \hat{\tau}_i^* + 3n\hat{\beta}_j^* + n \sum_{i=1}^{3} \widehat{(\tau\beta)}_{ij}^*$$

$$y_{ij.} = n\hat{\mu}^* + n\hat{\tau}_i^* + n\hat{\beta}_j^* + n\widehat{(\tau\beta)}_{ij}^*$$

If we place constraints on the estimators that parallel those induced on the parameters, then the first equation yields

$$y_{...} = 6n\hat{\mu}^*$$

or

$$\hat{\mu}^* = \bar{y}_{...}$$

The second equation then reduces to

$$y_{i..} = 2n\bar{y}_{...} + 2n\hat{\tau}_i^*$$

or

$$\hat{\tau}_i^* = \bar{y}_{i..} - \bar{y}_{...}$$

In a similar manner, the third equation yields

$$\hat{\beta}_j^* = \bar{y}_{.j.} - \bar{y}_{...}$$

By substitution into the fourth equation, it can be seen that

$$\widehat{(\tau\beta)}_{ij}^* = \bar{y}_{ij.} - \bar{y}_{i..} - \bar{y}_{.j.} + \bar{y}_{...}$$

(You are asked to verify the last two results in Exercise 59.)

In summary, the estimators for μ^*, τ_i^*, β_j^*, and $(\tau\beta)_{ij}^*$ are

$$\hat{\mu}^* = \bar{y}_{...}$$
$$\hat{\tau}_i^* = \bar{y}_{i..} - \bar{y}_{...}$$
$$\hat{\beta}_j^* = \bar{y}_{.j.} - \bar{y}_{...}$$
$$\widehat{(\tau\beta)}_{ij}^* = \bar{y}_{ij.} - \bar{y}_{i..} - \bar{y}_{.j.} + \bar{y}_{...}$$

These results should not be surprising since they parallel the theoretical results found in Exercise 54. A numerical example will clarify these ideas.

Example 6.6.4 Assume that after conducting the experiment described in Example 6.4.1, it is decided to repeat the experiment using five observations for each treatment combination. In this way, differences might be detected that were not detected earlier and a test for interaction can be conducted. These data are obtained:

		Fluid Type (Factor I)	
		Gastric	Duodenal
Capsule Type (Factor II)	A	39 49 63 45 50	31 36 38 33 42
	B	47 39 41 43 36	44 47 42 41 45

For these data

$$\bar{y}_{11.} = 49.2 \qquad \bar{y}_{1..} = 45.2 \qquad \bar{y}_{...} = 42.55$$
$$\bar{y}_{12.} = 41.2 \qquad \bar{y}_{2..} = 39.9$$
$$\bar{y}_{21.} = 36.0 \qquad \bar{y}_{.1.} = 42.6$$
$$\bar{y}_{22.} = 43.8 \qquad \bar{y}_{.2.} = 42.5$$

Estimates of the parameters $(\tau\beta)_{ij}^*$ are

$$\widehat{(\tau\beta)}_{11}^* = \bar{y}_{11.} - \bar{y}_{1..} - \bar{y}_{.1.} + \bar{y}_{...} = 3.95$$
$$\widehat{(\tau\beta)}_{12}^* = \bar{y}_{12.} - \bar{y}_{1..} - \bar{y}_{.2.} + \bar{y}_{...} = -3.95$$
$$\widehat{(\tau\beta)}_{21}^* = \bar{y}_{21.} - \bar{y}_{2..} - \bar{y}_{.1.} + \bar{y}_{...} = -3.95$$
$$\widehat{(\tau\beta)}_{22}^* = \bar{y}_{22.} - \bar{y}_{2..} - \bar{y}_{.2.} + \bar{y}_{...} = 3.95$$

Since these are all rather far removed from zero, the data suggest the presence of interaction. This can also be seen by considering the cell means. The interaction or inconsistency in behavior of the capsule types across the fluids is pictured in Figure 6.4.

FIGURE 6.4 **Lines are not close to parallel, indicating possible interaction between fluid type and capsule type**

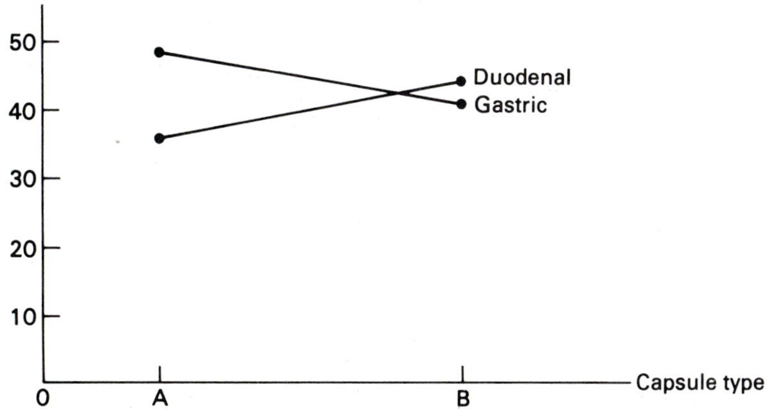

To detect the presence of interaction analytically, it is necessary to test

$$H_0:(\tau\beta)^*_{ij}=0$$

The procedure used to develop the test for interaction is the same as that used on several other occasions. The regression sum of squares for the full model, $SS_{\text{Reg(Full)}}$, is found; a reduced model is found by assuming that no interaction exists, and the regression sum of squares for the reduced model, $SS_{\text{Reg(Reduced)}}$, is found; the difference between these two sums of squares, $SS_{\text{Reg(Hypothesis)}}$, is used to test

$$H_0:(\tau\beta)^*_{ij}=0 \qquad \text{(No interaction)}$$

Most of the work needed to accomplish this has already been done. The regression sum of squares for the full model is

$$SS_{\text{Reg(Full)}}=\mathbf{b}^{*\prime}X'\mathbf{y}$$

where $\mathbf{b}^{*\prime}=[\hat{\mu}^* \quad \hat{\tau}^{*\prime}_i \quad \hat{\beta}^{*\prime}_j \quad \widehat{(\tau\beta)}^{*\prime}_{ij}]$.

Multiplication shows that

$$\begin{aligned}
SS_{\text{Reg(Full)}} &= \frac{y_{\cdots}^2}{abn}+\sum_{i=1}^{a} y_{i\cdot\cdot}(\bar{y}_{i\cdot\cdot}-\bar{y}_{\cdots})+\sum_{j=1}^{b} y_{\cdot j\cdot}(\bar{y}_{\cdot j\cdot}-\bar{y}_{\cdots})\\
&\quad+\sum_{i=1}^{a}\sum_{j=1}^{b} y_{ij\cdot}(\bar{y}_{ij\cdot}-\bar{y}_{i\cdot\cdot}-\bar{y}_{\cdot j\cdot}+\bar{y}_{ij\cdot})\\
&=\frac{y_{\cdots}^2}{abn}+\sum_{i=1}^{a}\frac{y_{i\cdot\cdot}^2}{bn}-\frac{y_{\cdots}^2}{abn}+\sum_{j=1}^{b}\frac{y_{\cdot j\cdot}^2}{an}-\frac{y_{\cdots}^2}{abn}\\
&\quad+\sum_{i=1}^{a}\sum_{j=1}^{b}\frac{y_{ij\cdot}^2}{n}-\sum_{i=1}^{a}\frac{y_{i\cdot\cdot}^2}{bn}-\sum_{j=1}^{b}\frac{y_{\cdot j\cdot}^2}{an}+\frac{y_{\cdots}^2}{abn}\\
&=\sum_{i=1}^{a}\sum_{j=1}^{b}\frac{y_{ij\cdot}^2}{n}
\end{aligned}$$

In the reduced model it is assumed that there is no interaction. The regression sum of squares for this model was derived in Section 6.4. It is given by

$$SS_{\text{Reg(Reduced)}} = \sum_{i=1}^{a} y_{i..}^2/bn + \sum_{j=1}^{b} y_{.j.}^2/an - y_{...}^2/abn$$

Subtracting, $SS_{\text{Reg(Hypothesis)}}$ is found to be

$$\begin{aligned} SS_{\text{Reg(Hypothesis)}} &= SS_{\text{Reg(Full)}} - SS_{\text{Reg(Reduced)}} \\ &= \sum_{i=1}^{a}\sum_{j=1}^{b} \frac{y_{ij.}^2}{n} - \sum_{i=1}^{a} \frac{y_{i..}^2}{bn} - \sum_{j=1}^{b} \frac{y_{.j.}^2}{an} + \frac{y_{...}^2}{abn} \end{aligned}$$

The degrees of freedom associated with these sums of squares are ab, $a+b-1$, and $ab-a-b+1=(a-1)(b-1)$, respectively. The F ratio used to test H_0 is

$$F_{(a-1)(b-1),abn-ab} = \frac{SS_{\text{Reg(Hypothesis)}/(a-1)(b-1)}}{SS_{\text{Res}/(abn-ab)}}$$

The ANOVA table is given in Table 6.13.

To illustrate, the data of Example 6.6.4 are used to test for interaction between fluid type and capsule type.

Example 6.6.5 To test for interaction, these row, column, and cell totals are needed:

$$\begin{array}{lll} y_{11.} = 246 & y_{1..} = 452 & y_{...} = 851 \\ y_{12.} = 206 & y_{2..} = 399 & \\ y_{21.} = 180 & y_{.1.} = 426 & \\ y_{22.} = 219 & y_{.2.} = 425 & \end{array}$$

Here

$$SS_{\text{Total(Uncorrected)}} = \sum_{i=1}^{2}\sum_{j=1}^{2}\sum_{k=1}^{5} y_{ijk}^2 = 37141$$

$$SS_{\text{Reg(Full)}} = \sum_{i=1}^{2}\sum_{j=1}^{2} \frac{y_{ij.}^2}{5} = 36{,}662.6$$

$$\begin{aligned} SS_{\text{Reg(Reduced model)}} &= \sum_{i=1}^{2} \frac{y_{i..}^2}{10} + \sum_{j=1}^{2} \frac{y_{.j.}^2}{10} - \frac{y_{...}^2}{20} \\ &= 36350.55 \end{aligned}$$

The ANOVA is given in Table 6.14.

Note that the null hypothesis of no interaction can be rejected based on the $F_{1,16}$ distribution with $P < 0.01$. As expected from our examination of cell means and our estimated interactions, interaction between fluid types and capsule types does appear to exist.

TABLE 6.13
ANOVA table used to test for interaction in a two-factor design

Source of Variation	Degrees of Freedom	Sum of Squares	Mean Square	F
Regression	ab	$\sum_{i=1}^{a}\sum_{j=1}^{b}\frac{y_{ij.}^2}{n}$		
Reduced model	$a+b-1$	$\sum_{i=1}^{a}\frac{y_{i..}^2}{bn}+\sum_{j=1}^{b}\frac{y_{.j.}^2}{an}-\frac{y_{...}^2}{abn}$		
Hypothesis	$(a-1)(b-1)$	$\sum_{i=1}^{a}\sum_{j=1}^{b}\frac{y_{ij.}^2}{n}-\sum_{i=1}^{a}\frac{y_{i..}^2}{bn}-\sum_{j=1}^{b}\frac{y_{.j.}^2}{an}+\frac{y_{...}^2}{abn}$	$\frac{SS_{\mathrm{Reg(Hypothesis)}}}{(a-1)(b-1)}$	$\frac{SS_{\mathrm{Reg(Hypothesis)}}/(a-1)(b-1)}{SS_{\mathrm{Res}}/(abn-ab)}$
Residual	$abn-ab$	$\sum_{i=1}^{a}\sum_{j=1}^{b}\sum_{k=1}^{n}y_{ijk}^2-\sum_{i=1}^{a}\sum_{j=1}^{b}\frac{y_{ij.}^2}{n}$	$\frac{SS_{\mathrm{Res}}}{abn-ab}$	
Total	abn	$\sum_{i=1}^{a}\sum_{j=1}^{b}\sum_{k=1}^{n}y_{ijk}^2$		

TABLE 6.14
ANOVA testing for interaction in solubility study

Source of Variation	Degrees of Freedom	Sum of Squares	Mean Square	F
Regression	4	36,662.6		
Reduced model	3	36,350.55		
Hypothesis	1	312.05	312.05	10.44
Residual	16	478.4	29.9	
Total (uncorrected)	20	37,141		

If interaction is detected in a two-factor design, it is suggested that factors be compared row by row or column by column using the one-way classification procedure. To illustrate, we continue the analysis of the data of Example 6.6.4.

Example 6.6.6 Let us continue our study by testing to see if there is a difference in the average time required for these two types of capsules to dissolve in gastric juices. To do so, these totals are needed:

$$\sum_{j=1}^{2}\sum_{k=1}^{5} y_{1jk}^2 = 20{,}972 \qquad y_{1..} = 452$$

$$y_{11.} = 246 \qquad \sum_{j=1}^{2} \frac{y_{1j.}^2}{5} = 20{,}590.4$$

$$y_{12.} = 206$$

The ANOVA for the design is given in Table 6.2. The ANOVA for these data is found in Table 6.15.

Based on the $F_{1,8}$ distribution, $P > 0.1$ (critical point 3.46). Since this P value is large, it cannot be concluded that there is a difference in the average time required for these two types of capsules to dissolve in gastric juices. The analysis of the data for duodenal juices is left as an exercise (see Exercise 60).

TABLE 6.15 **One-way ANOVA for comparing capsules in gastric juices**

Source of Variation	Degrees of Freedom	Sum of Squares	Mean Square	F
Regression	2	20,590.4		
Reduced model	1	20,430.4		
Hypothesis	1	160.0	160.0	3.35
Residual	8	381.6	47.7	
Total (uncorrected)	10	20,972.0		

TESTS ON MAIN EFFECTS

If no interaction is detected in a two-factor design, then tests can be performed to compare levels of factor I and those of factor II overall rather than row by row or column by column. Tests of this sort are called tests for *main* effects. The null hypotheses being tested are

$H'_0: \tau_1^* = \tau_2^* = \cdots = \tau_a^* = 0$ (On average, no differences among the levels of factor I)

$H''_0: \beta_1^* = \beta_2^* = \cdots = \beta_b^* = 0$ (On average, no differences among the levels of factor II)

The setting is now almost the same as the one presented in Section 6.4 with $n \geqslant 1$. The only difference is the manner in which the residual sum of squares is handled. In the earlier model, it is assumed that there is no interaction. If this assumption is false, any variation due to interaction is ignored and included as part of the residual sum of squares. Here we have isolated some variation due to interaction but deemed it negligible. It is, nonetheless, kept apart in the analysis so that the residual sum of squares truly represents random or unexplained variation in response. The ANOVA table for testing for interaction as well as main effects is presented in Table 6.16.

Example 6.6.7 illustrates the procedure.

Example 6.6.7 A study of the decomposition of leaf packs is conducted. Twenty-four leaf packs are prepared and randomly assigned to four different environments over three different exposure periods. The study yields these data with the response being the weight loss of the leaf pack in grams (cell totals are given in parentheses):

		Exposure Time (Factor I)			
		1 month	2 months	3 months	$y_{.j.}$
Environment (Factor II)	1	1.09 1.06 (2.15)	1.35 1.53 (2.88)	1.60 1.40 (3.00)	8.03
	2	1.16 1.03 (2.19)	1.38 1.35 (2.73)	2.18 1.77 (3.95)	8.87
	3	1.01 1.04 (2.05)	1.63 1.51 (3.14)	1.66 1.98 (3.64)	8.83
	4	0.90 1.03 (1.93)	1.60 1.72 (3.32)	1.73 1.76 (3.49)	8.74
	$y_{i..}$	8.32	12.07	14.08	$y_{...} = 34.47$

TABLE 6.16
ANOVA table for the general two-factor design with interaction

Source of Variation	Degrees of Freedom	Sum of Squares	Mean Square (MS)	F
Regression (full)	ab	$\sum_{i=1}^{a}\sum_{j=1}^{b} y_{ij.}^2/n$		
Mean	1	$y_{...}^2/abn$		
Factor I	$a-1$	$\sum_{i=1}^{a} y_{i..}^2/bn - y_{...}^2/abn$	$SS_{\text{Factor I}}/(a-1)$	$MS_{\text{Factor I}}/MS_{\text{Res}}$
Factor II	$b-1$	$\sum_{j=1}^{b} y_{.j.}^2/an - y_{...}^2/abn$	$SS_{\text{Factor II}}/(b-1)$	$MS_{\text{Factor II}}/MS_{\text{Res}}$
Interaction	$(a-1)(b-1)$	$SS_{\text{Reg(Full)}} - SS_{\text{Mean}} - SS_{\text{Factor I}} - SS_{\text{Factor II}}$	$SS_{\text{Interaction}}/(a-1)(b-1)$	$MS_{\text{Interaction}}/MS_{\text{Res}}$
Residual	$abn-ab$	$SS_{\text{Total}} - SS_{\text{Reg(Full)}}$	$SS_{\text{Res}}/(abn-ab)$	
Total (uncorrected)	abn	$\sum_{i=1}^{a}\sum_{j=1}^{b}\sum_{k=1}^{n} y_{ijk}^2$		

Let us first test the null hypothesis of no interaction at the $\alpha = 0.05$ level. For these data,

$$\sum_{i=1}^{3}\sum_{j=1}^{4} y_{ij.}^2/n = 104.01/2 = 52.005$$

$$y_{...}^2/abn = (34.47)^2/24 = 49.507$$

$$\sum_{i=1}^{3} y_{i..}^2/bn = 413.154/8 = 51.644$$

$$\sum_{j=1}^{4} y_{.j.}^2/an = 297.514/6 = 49.586$$

$$\sum_{i=1}^{3}\sum_{j=1}^{4}\sum_{k=1}^{2} y_{ijk}^2 = 52.208$$

Based on these sums, the full analysis is as shown in Table 6.17.

The critical point for an $\alpha = 0.05$ level test based on the $F_{6,12}$ distribution is 3.00. Since $2.7692 < 3.00$, the null hypothesis of no interaction is not rejected. We now test for main effects. The null hypothesis of no differences among the levels of factor I can be rejected ($f = 63.2544$ and $P < 0.01$). However, no differences are detected among levels of factor II ($f = 1.5799$, $P > 0.1$). It should be evident that presetting the α level on the test for interaction made interpreting these data easier. In reality, things are not so clearcut. The P value for the interaction test lies between 0.05 and 0.10. Some statisticians would probably interpret this as an indication of the presence of interaction and interpret the data on a row by row basis. Computing Supplement J illustrates the use of SAS in this setting.

TABLE 6.17 **ANOVA table for the leaf decomposition data of Example 6.6.7**

Source of Variation	Degrees of Freedom	Sum of Squares	Mean Square (MS)	F
Regression (full)	12	52.005		
Mean	1	49.507		
Factor I (time)	2	2.137	1.0685	63.2248
Factor II (environment)	3	0.079	0.0263	1.5562
Interaction	6	0.2820	0.0470	2.7810
Residual	12	0.2030	0.0169	
Total (uncorrected)	24	52.208		

A few closing remarks are in order. First, in the two-factor design that has been presented in this section, it is assumed that there are n observations per cell. That is, cell sizes are identical. This is an essential part of the design. If this is not true then the analysis is quite different from that presented here. Second, in this chapter we have introduced only the basic concepts underlying the field of linear models. Many more linear models are used in applied statistics. The sums of squares, mean squares, and F ratios used to analyze other models can be derived by applying the techniques developed in Section 6.1 for the general linear model. The detailed analysis of other commonly encountered models is found in most texts on experimental design [**6**] [**12**]. Third, all the designs considered to this point are *fixed effects* models. It is assumed that the levels of the factors studied are deliberately chosen by the researcher because they are of particular interest. Models in which the levels are randomly selected from a larger set of possible levels are considered briefly in Chapter 7.

EXERCISES

Section 6.1

1. Consider the one-way classification model with fixed effects and $k = 3$. Check each of these null hypotheses for testability.
   a. $H_0: \tau_1 = \tau_2$
   b. $H_0: \tau_1 - 2\tau_2 + \tau_3 = 0$
   c. $H_0: \tau_1 = 0$
   d. $H_0: \tau_1 = \tau_3$ and $\tau_1 - 2\tau_2 + \tau_3 = 0$
   e. $H_0: \tau_1 = \tau_2$ and $\tau_1 = \tau_3$ and $2\tau_1 - \tau_2 - \tau_3 = 0$

2. Consider the one-way classification model with fixed effects where $k = 3$ and $N = n_1 + n_2 + n_3$. Show that the hypothesis of Example 6.1.1 is testable by verifying that $C(X'X)^c X'X = C$ where

$$(X'X) = \begin{bmatrix} N & n_1 & n_2 & n_3 \\ n_1 & n_1 & 0 & 0 \\ n_2 & 0 & n_2 & 0 \\ n_3 & 0 & 0 & n_3 \end{bmatrix} \quad \text{and} \quad (X'X)^c = \begin{bmatrix} 0 & 0 & 0 & 0 \\ 0 & \frac{1}{n_1} & 0 & 0 \\ 0 & 0 & \frac{1}{n_2} & 0 \\ 0 & 0 & 0 & \frac{1}{n_3} \end{bmatrix}$$

*3. Let $C\boldsymbol{\beta} = \mathbf{0}$ be testable. Prove that $C(X'X)^c C'$ is unique. (*Hint*: Let $(X'X)_1^c$ denote another conditional inverse for $X'X$. Show that $C(X'X)^c C' = C(X'X)_1^c C'$. Remember that since $C\boldsymbol{\beta}$ is testable, $C(X'X)_1^c(X'X) = C$.)

*4. In this exercise you will show that $C(X'X)^c C'$ is nonsingular. In the less than full rank model, X is $n \times p$ of rank $r < p$. Hence X has exactly r linearly independent rows. Since elements of X are estimable, there are exactly r linearly independent estimable functions of the form $\mathbf{x}_i'\boldsymbol{\beta}$ where $\mathbf{x}_i'$ denotes a row of X. Any other estimable function must be a linear combination of these functions. That is, if $\mathbf{c}'\boldsymbol{\beta}$ is estimable then there exists a vector $\mathbf{a}'$ such that $\mathbf{c}' = \mathbf{a}'X$. The vector $\mathbf{a}'$ can in turn be expressed in the form $\mathbf{a}' = \mathbf{k}'X'$ for some vector $\mathbf{k}'$. Combining these results, $\mathbf{c}'\boldsymbol{\beta} = \mathbf{k}'X'X$. Consider the testable hypothesis $C\boldsymbol{\beta} = \mathbf{0}$. Since each element of $C\boldsymbol{\beta}$ is estimable, the above discussion implies that there exists a matrix K such that $C = K(X'X)$.

a. What are the dimensions of C? of K?
b. What is the rank of C?
c. Argue that $r(K) \leqslant m$.
d. Argue that $m \leqslant r(K)$.
e. What is the rank of K?
f. Show that $C(X'X)^c C'$ can be expressed in the form $P'P$ where $P = XK'$.
g. What are the dimensions of P?
h. Argue that $r(P) \leqslant m$.
i. Use the fact that $C = P'X$ to argue that $m \leqslant r(P)$.
j. What is the rank of P?
k. What are the dimensions of $P'P$?
l. Argue that $C(X'X)^c C'$ is nonsingular as claimed.

***5.** In this exercise you will argue that the F ratio developed to test the testable hypothesis $H_0\!:\!C\boldsymbol{\beta} = \mathbf{0}$ results in a right-tailed test.

a. Use Theorem 2.2.1 to find

$$E\{(C\mathbf{b})'[C(X'X)^c C']^{-1}(C\mathbf{b})/m\}$$

b. Argue that if H_0 is true, then the expected value of part a is equal to σ^2.
c. Argue that $[C(X'X)^c C']^{-1}$ is positive definite. (*Hint*: See part f of Exercise 4, Theorem 4.1.4, and Exercise 17c in Chapter 4.)
d. Argue that if H_0 is not true, then the expected value of part a exceeds σ^2.
e. Argue that F values near 1 indicate that H_0 is probably true and that values substantially larger than 1 support the contention that H_0 is false.

6. Find the numerator of the F ratio needed to test the null hypothesis of part a in Exercise 1.

7. Find the numerator of the F ratio needed to test the null hypothesis of part b in Exericse 1.

8. Find a general expression for the numerator of the F ratio needed to test a contrast in the one-way classification model with $k = 3$.

9. A study of sleep deprivation is conducted. Twenty-four subjects are randomly divided into three groups of size eight each. After differing amounts of sleep deprivation, all subjects are asked to perform a task that requires manual dexterity. A score from 0 (poor) to 10 (excellent) is recorded. These data result:

Group I (16 hours deprivation)		Group II (20 hours deprivation)		Group III (24 hours deprivation)	
8.95	6.48	7.70	8.04	5.99	5.78
8.04	7.81	5.81	5.96	6.79	7.60
7.72	7.50	6.61	7.30	6.43	5.78
6.21	6.90	6.07	7.46	5.85	6.00

Assuming that the one-way classification model with fixed effects is appropriate, use these data to test $H_0:\tau_1=\tau_2=\tau_3$.

Section 6.2

10. Consider the one-way classification model with Z given by

$$Z=\begin{bmatrix}1&0&0\\0&1&0\\0&1&0\\0&0&1\\0&0&1\\0&0&1\end{bmatrix}\quad\text{and}\quad Z_2=\begin{bmatrix}1\\1\\1\\1\\1\\1\end{bmatrix}$$

Here $k=3$, $n_1=1$, $n_2=2$, $n_3=3$, and $N=6$. In this exercise you will show that

$$r([Z(Z'Z)^{-1}Z'-Z_2(Z_2'Z_2)^{-1}Z_2'])=2=k-1$$

thus justifying the use of the Cochran–Fisher theorem in this special case. The argument is generalized in Exercise 11.

a. Verify that

$$Z(Z'Z)^{-1}=\begin{bmatrix}1&0&0\\0&\frac{1}{2}&0\\0&\frac{1}{2}&0\\0&0&\frac{1}{3}\\0&0&\frac{1}{3}\\0&0&\frac{1}{3}\end{bmatrix}$$

b. Verify that

$$Z(Z'Z)^{-1}Z'=\begin{bmatrix}1&0&0&0&0&0\\0&\frac{1}{2}&\frac{1}{2}&0&0&0\\0&\frac{1}{2}&\frac{1}{2}&0&0&0\\0&0&0&\frac{1}{3}&\frac{1}{3}&\frac{1}{3}\\0&0&0&\frac{1}{3}&\frac{1}{3}&\frac{1}{3}\\0&0&0&\frac{1}{3}&\frac{1}{3}&\frac{1}{3}\end{bmatrix}$$

c. Verify that

$$Z_2(Z_2'Z_2)^{-1}Z_2'=\begin{bmatrix}\frac{1}{6}&\frac{1}{6}&\frac{1}{6}&\frac{1}{6}&\frac{1}{6}&\frac{1}{6}\\\frac{1}{6}&\frac{1}{6}&\frac{1}{6}&\frac{1}{6}&\frac{1}{6}&\frac{1}{6}\\\frac{1}{6}&\frac{1}{6}&\frac{1}{6}&\frac{1}{6}&\frac{1}{6}&\frac{1}{6}\\\frac{1}{6}&\frac{1}{6}&\frac{1}{6}&\frac{1}{6}&\frac{1}{6}&\frac{1}{6}\\\frac{1}{6}&\frac{1}{6}&\frac{1}{6}&\frac{1}{6}&\frac{1}{6}&\frac{1}{6}\\\frac{1}{6}&\frac{1}{6}&\frac{1}{6}&\frac{1}{6}&\frac{1}{6}&\frac{1}{6}\end{bmatrix}$$

d. Verify that

$$Z_2(Z_2'Z_2)^{-1}Z_2'Z(Z'Z)^{-1}Z'=Z_2(Z_2'Z_2)^{-1}Z_2'$$

e. Find the general expression for

$$[Z(Z'Z)^{-1}Z'-Z_2(Z_2'Z_2)^{-1}Z_2']^2$$

and use the result of part d to verify that in this special case

$$Z(Z'Z)^{-1}Z' - Z_2(Z_2'Z_2)^{-1}Z_2'$$

is idempotent.

f. Find $\mathrm{tr}[Z(Z'Z)^{-1}Z']$, $\mathrm{tr}[Z_2(Z_2'Z_2)^{-1}Z_2']$, and $\mathrm{tr}([Z(Z'Z)^{-1}Z' - Z_2(Z_2'Z_2)^{-1}Z_2'])$.

g. Use Theorem 1.5.2 to show that

$$r([Z(Z'Z)^{-1}Z' - Z_2(Z_2'Z_2)^{-1}Z_2'] = 2$$

as claimed.

11. Consider the general one-way classification model with fixed effects. Parallel each of the steps in Exercise 10 to show that

$$r([Z(Z'Z)^{-1}Z' - Z_2(Z_2'Z_2)^{-1}Z_2']) = k - 1$$

thus justifying the use of the Cochran–Fisher theorem in the derivation of the test statistic used to test $H_0{:}\mu_1 = \mu_2 = \cdots = \mu_k$.

12. Show that, in general, the noncentrality parameter for

$$\frac{SS_{\mathrm{Reg(Full)}} - SS_{\mathrm{Reg(Reduced)}}}{\sigma^2}$$

in the one-way classification model is

$$\begin{aligned}\lambda &= (1/2\sigma^2)(Z\boldsymbol{\alpha})'[Z(Z'Z)^{-1}Z' - Z_2(Z_2'Z_2)^{-1}Z_2'](Z\boldsymbol{\alpha}) \\ &= (1/2\sigma^2)[\boldsymbol{\alpha}'Z'Z\boldsymbol{\alpha} - \boldsymbol{\alpha}'Z'Z_2(Z_2'Z_2)^{-1}Z_2'Z\boldsymbol{\alpha}]\end{aligned}$$

13. Consider the one-way classification model of Exercise 10.

a. Show that when $\mu_1 = \mu_2 = \mu_3 = \mu$ then $\boldsymbol{\alpha}'Z'Z\boldsymbol{\alpha} = 6\mu^2$.

b. Show that when $\mu_1 = \mu_2 = \mu_3 = \mu$ then

$$\boldsymbol{\alpha}'Z'Z_2(Z_2'Z_2)^{-1}Z_2'Z\boldsymbol{\alpha} = 6\mu^2$$

c. Show that when $\mu_1 = \mu_2 = \mu_3 = \mu$ then $\lambda = 0$, thus verifying that when H_0 is true the random variable

$$\frac{\mathbf{y}'[Z(Z'Z)^{-1}Z' - Z_2(Z_2'Z_2)^{-1}Z_2']\mathbf{y}}{\sigma^2}$$

follows a chi-squared distribution with $k - 1 = 2$ degrees of freedom.

14. Generalize the argument given in Exercise 13 to any one-way classification model with fixed effects.

15. In this exercise you will justify the use of a right-tailed F test in testing $H_0{:}\mu_1 = \mu_2 = \cdots = \mu_k$ in the one-way classification model.

a. Consider the statistic

$$\left[\sum_{i=1}^{k} y_{i.}^2/n_i - y_{..}^2/N\right]\Big/(k-1) = \frac{\mathbf{y}'[Z(Z'Z)^{-1}Z' - Z_2(Z_2'Z_2)^{-1}Z_2']\mathbf{y}}{k-1} = Q$$

Use Theorem 2.2.1 to show that the expected value of Q is

$$\sigma^2 + \left[\sum_{i=1}^{k} n_i\mu_i^2 - \left(\sum_{i=1}^{k} n_i\mu_i\right)^2\Big/N\right]\Big/(k-1)$$

b. Let $\bar{\mu} = (\sum_{i=1}^{k} n_i\mu_i)/N$. Show that

$$\sum_{i=1}^{k} n_i\mu_i^2 - \left(\sum_{i=1}^{k} n_i\mu_i\right)^2 \Big/ N = \sum_{i=1}^{k} n_i(\mu_i - \bar{\mu})^2 \geq 0$$

thus showing that $E[Q] \geq \sigma^2$.

c. Show that if $\mu_1 = \mu_2 = \cdots = \mu_k$ then $E[Q] = \sigma^2$.

16. Construct the ANOVA table based on the corrected total sum of squares for the data of Example 6.2.1.

17. Use the reparameterized model to construct the ANOVA table for the data of Exercise 9 based on the uncorrected total sum of squares. Test $H_0{:}\mu_1 = \mu_2 = \mu_3$ and verify that the F statistic obtained via the reparameterized model is identical to the one found in Exercise 9.

18. Construct the ANOVA table for the data of Exercise 9 based on the reparameterized model and the corrected total sum of squares.

19. Consider the usual one-way classification model with fixed effects given by

$$y_{ij} = \mu + \tau_i + \varepsilon_{ij} \qquad \begin{array}{l} i = 1, 2, 3 \\ j = 1, 2 \end{array}$$

Let

$$\boldsymbol{\alpha} = \begin{bmatrix} \mu + \dfrac{\tau_1 + \tau_2 + \tau_3}{3} \\ \dfrac{\tau_1 - \tau_2}{2} \\ \dfrac{\tau_1 + \tau_2 - 2\tau_3}{6} \end{bmatrix}$$

and let

$$Z = \begin{bmatrix} 1 & 1 & 1 \\ 1 & 1 & 1 \\ 1 & -1 & 1 \\ 1 & -1 & 1 \\ 1 & 0 & -2 \\ 1 & 0 & -2 \end{bmatrix}$$

a. Show that Z is of full rank.

b. Show that $X\boldsymbol{\beta} = Z\boldsymbol{\alpha}$.

c. Show that α_1, α_2, and α_3 are each estimable.

d. Show that α_1, α_2, and α_3 are independent, thus showing that $Z\boldsymbol{\alpha}$ is a reparameterization of the original model.

e. Show that testing $H_0{:}\tau_1 = \tau_2 = \tau_3$ is equivalent to testing $H_0{:}\alpha_2 = \alpha_3 = 0$.

f. Find $Z'Z$, $(Z'Z)^{-1}$, and $Z'\mathbf{y}$.

g. Let $y_{..}$, $y_{1.}$, $y_{2.}$, and $y_{3.}$ denote the overall total and the totals for samples 1, 2, and 3, respectively. Let $\bar{y}_{..}$, $\bar{y}_{1.}$, $\bar{y}_{2.}$, and $\bar{y}_{3.}$ denote the corresponding sample means. Show that the least squares estimators for α_1, α_2, and α_3 are

$$\hat{\alpha}_1 = \bar{y}_{..} \qquad \hat{\alpha}_2 = \frac{\bar{y}_{1.} - \bar{y}_{2.}}{2} \qquad \hat{\alpha}_3 = \frac{\bar{y}_{1.} + \bar{y}_{2.} - 2\bar{y}_{3.}}{6}$$

h. Show that

$$SS_{\text{Reg(Full)}} = \frac{y_{..}^2}{6} + \frac{(y_{1.} - y_{2.})^2}{4} + \frac{(y_{1.} + y_{2.} - 2y_{3.})^2}{12}$$

i. Show that

$$SS_{\text{Reg(Reduced)}} = \frac{y_{..}^2}{6}$$

j. Find $SS_{\text{Reg(Hypothesis)}}$.
k. Find the ANOVA table used to test H_0 based on this reparameterization of the model $\mathbf{y} = X\boldsymbol{\beta} + \boldsymbol{\varepsilon}$.
l. Show that

$$SS_{\text{Reg(Hypothesis)}} = \sum_{i=1}^{3} \frac{y_{i.}^2}{2} - \frac{\left(\sum_{i=1}^{3} y_{i.}\right)^2}{6}$$

thus showing that $SS_{\text{Reg(Hypothesis)}}$ for this reparameterization is identical to the one given in Table 6.2.
m. Show that $SS_{\text{Reg(Full)}}$ for this reparameterization is identical to the one given in Table 6.2.
n. Show that the F ratio used to test $H_0\!:\tau_1 = \tau_2 = \tau_3$ via this reparameterization is identical to the one in Table 6.2.

20. The reparameterization given in Exercise 19 assumes that $n_1 = n_2 = n_3 = 2$ and $N = 6$ for convenience. For samples of size n, this reparameterization leads to these sums of squares:

$$SS_{\text{Reg(Full)}} = \frac{y_{..}^2}{N} + \frac{(y_{1.} - y_{2.})^2}{2n} + \frac{(y_{1.} + y_{2.} - 2y_{3.})^2}{6n}$$

$$SS_{\text{Reg(Reduced)}} = \frac{y_{..}^2}{N}$$

$$SS_{\text{Reg(Hypothesis)}} = SS_{\text{Reg(Full)}} - SS_{\text{Reg(Reduced)}}$$

Use the data of Example 6.2.1 to verify that these sums of squares and the ANOVA based on the reparameterization are identical to those found earlier and reported in Table 6.3.

Section 6.3

21. Construct the ANOVA for the data of Example 6.3.1 based on the corrected total sum of squares.

22. Relative to Example 6.3.1, design a contrast that allows ball performance on asphalt to be compared to that on soft surfaces. Find the contrast sum of squares and conduct the appropriate two-tailed test.

23. An industrial psychologist is investigating absenteeism among production-line workers. He is experimenting with three types of work hours: (I) 4-day work week with a 10-hour day; (II) 5-day work week with an 8-hour day on flextime; (III) 5-day work week with a structured 8-hour day. These data are obtained on the average number of days missed per worker under each plan:

I	II	III
$\bar{y}_{1.} = 9$	$\bar{y}_{2.} = 6.2$	$\bar{y}_{3.} = 10.1$
$n_1 = 100$	$n_2 = 85$	$n_3 = 90$
$\sum_{i=1}^{3} \sum_{j=1}^{n_i} y_{ij}^2 = 50{,}511$		

Test $H_0{:}\mu_1 = \mu_2 = \mu_3$ at the $\alpha = 0.05$ level.

24. Consider the information in Exercise 23. Form a contrast to compare flextime with the other two schedules combined. Can it be claimed that flextime reduces absenteeism? Explain by testing the appropriate contrast.

25. Consider the contrasts ω_1 and ω_2 of Example 6.3.2. Show that if $n_1 = n_5 = 2$, $n_2 = n_3 = 3$, and $n_4 = 5$, then ω_1 and ω_2 are orthogonal.

26. Consider the model

$$y_{ij} = \mu + \tau_i + \varepsilon_{ij} \qquad i = 1, 2, 3$$

in which $n_1 = 1$, $n_2 = 2$, and $n_3 = 3$. Show that the contrasts $\omega_1 = \mu_1 - \mu_2$ and $\omega_2 = \mu_1 + 2\mu_2 - 3\mu_3$ are orthogonal. Would these contrasts be orthogonal if sample sizes were equal?

27. Verify the computations of Example 6.3.3.

28. In a small study of inflation, the following categories of foodstuffs are considered:

I: Staples (sugar, salt, flour)
II: Fresh fruits and vegetables
III: Canned fruits and vegetables
IV: Meat
V: Dairy products

The measured variable is the percentage increase in cost over a two-month period. These summary data result: $n_1 = n_2 = n_3 = n_4 = n_5 = 10$.

$$y_{1.} = 1 \qquad \sum_{i=1}^{5} \sum_{j=1}^{10} y_{ij}^2 = 40$$

$$y_{2.} = 5$$

$$y_{3.} = 6$$

$$y_{4.} = 10$$

$$y_{5.} = 3$$

a. Test $H_0{:}\mu_1 = \mu_2 = \mu_3 = \mu_4 = \mu_5$ at the $\alpha = 0.10$ level.
b. Use single degree of freedom F tests to test these orthogonal contrasts:

$$\omega_1 = \tfrac{1}{2}\mu_2 + \tfrac{1}{2}\mu_3 - \tfrac{1}{3}\mu_1 - \tfrac{1}{3}\mu_4 - \tfrac{1}{3}\mu_5$$

$$\omega_2 = \mu_2 - \mu_3$$

$$\omega_3 = \mu_1 - \mu_4$$

Interpret your results in a practical sense.

c. There exists a fourth contrast that is orthogonal to those of part b. Find the sum of squares associated with this contrast.

Section 6.4

29. Each table below shows the theoretical means for a two-factor experiment. Sketch these means following the format of Figure 6.1. In each case, decide whether or not there is interaction between levels of factors I and II.

a. Factor II

	Factor I: 1	2	3	4
1	1	3	4	0
2	4	6	7	3
3	2	4	5	1

b. Factor II

	Factor I: 1	2	3	4
1	1	3	0	0
2	4	6	5	3
3	2	4	5	1

c. Factor II

	Factor I: 1	2	3	4
1	1	3	4	0
2	4	5	7	3
3	2	4	5	1

30. Verify that $\Sigma_{i=1}^{a} \hat{\tau}_i = \Sigma_{j=1}^{b} \hat{\beta}_j = 0$ where

$$\hat{\tau}_i = \bar{y}_{i.} - \bar{y}_{..} \quad \text{and} \quad \hat{\beta}_j = \bar{y}_{.j} - \bar{y}_{..}$$

31. Consider the two-factor design with no interaction. Verify that

$$\mathbf{b} = \begin{bmatrix} \bar{y}_{..} \\ \bar{y}_{1.} - \bar{y}_{..} \\ \bar{y}_{2.} - \bar{y}_{..} \\ \vdots \\ \bar{y}_{a.} - \bar{y}_{..} \\ \bar{y}_{.1} - \bar{y}_{..} \\ \bar{y}_{.2} - \bar{y}_{..} \\ \vdots \\ \bar{y}_{.b} - \bar{y}_{..} \end{bmatrix}$$

is a solution to the system $(X'X)\mathbf{b} = X'\mathbf{y}$ as claimed.

32. Show that $SS_{\text{Reg(Full)}}$ can be written as

$$SS_{\text{Reg(Full)}} = \sum_{i=1}^{a} y_{i.}^2/b + \sum_{j=1}^{b} y_{.j}^2/a - y_{..}^2/ab$$

33. The life span of paints used on automobiles is studied for four types of undercoating and three paint formulas. The response is the elapsed time between painting and the onset of surface breakdown of the paint in years. These data are obtained:

		Formula (Factor I)		
		1	2	3
Undercoat (Factor II)	1	1.09	1.35	1.60
	2	1.16	1.38	2.18
	3	1.01	1.63	1.66
	4	0.90	1.60	1.73

Assume that the two-factor design with no interaction is appropriate. Test for differences among levels of factor I and for differences among levels of factor II.

In Exercises 34–37 you will derive the expectation for $SS_{\text{Reg(Full)}}$, $SS_{\text{Reg(Reduced)}}$, and $SS_{\text{Reg(Hypothesis)}}$ for the two-factor design with no interaction and $n = 1$. Via these expectations you will justify the use of a right-tailed F test in testing H_0 and H_0'.

34. Consider the two-factor model without interaction given as

$$y_{ij} = \mu + \tau_i + \beta_j + \varepsilon_{ij} \qquad \begin{array}{l} i = 1, 2, \ldots, a \\ j = 1, 2, \ldots, b \end{array}$$

Define τ_i^*, β_j^*, $\bar{\tau}$, $\bar{\beta}$, and μ^* by

$$\bar{\tau} = \sum_{i=1}^{a} \tau_i/a \qquad \tau_i^* = \tau_i - \bar{\tau} \qquad \mu^* = \mu + \bar{\tau} + \bar{\beta}$$

$$\bar{\beta} = \sum_{j=1}^{b} \beta_j/b \qquad \beta_j^* = \beta_j - \bar{\beta}$$

a. Show that the model can be rewritten as

$$y_{ij} = \mu^* + \tau_i^* + \beta_j^* + \varepsilon_{ij}$$

b. Let

$$\boldsymbol{\gamma}^* = \begin{bmatrix} \mu^* \\ \tau_1^* \\ \tau_2^* \\ \vdots \\ \tau_a^* \\ \beta_1^* \\ \beta_2^* \\ \vdots \\ \beta_b^* \end{bmatrix}$$

Note that the reparameterized model can be expressed as

$$\mathbf{y} = X\boldsymbol{\gamma}^* + \boldsymbol{\varepsilon}$$

where X is the design matrix for the original model and that $\mathbf{y}$ is normally distributed with mean $X\gamma^*$ and variance $\sigma^2 I$. Use Theorem 2.2.1 to show that

$$E[SS_{\text{Reg(Full)}}] = \sigma^2 \operatorname{tr}[X(X'X)^c X'] + \gamma^{*\prime}X'X\gamma^*$$

c. Show that $\Sigma_{i=1}^{a} \tau_i^* = \Sigma_{j=1}^{b} \beta_j^* = 0$.

d. Show that

$$\gamma^{*\prime}X'X\gamma^* = (\mu^*)^2 ab + b\sum_{i=1}^{a}(\tau_i^*)^2 + a\sum_{j=1}^{b}(\beta_j^*)^2$$

e. Show that

$$E[SS_{\text{Reg(Full)}}] = \sigma^2(a+b-1) + ab(\mu^*)^2 + b\sum_{i=1}^{a}(\tau_i^*)^2 + a\sum_{j=1}^{b}(\beta_j^*)^2$$

35. Consider the hypothesis

$$H_0: \tau_1 = \tau_2 = \cdots = \tau_a = \tau$$

and the reduced model

$$y_{ij} = (\mu + \tau) + \beta_j + \varepsilon_{ij} \qquad j = 1, 2, \ldots, b$$

a. Show that under H_0, $\bar{\tau} = \Sigma_{i=1}^{a} \tau_i / a = \tau$.

b. Use the result of part a to show that the reduced model can be expressed as

$$y_{ij} = \mu^* + \beta_j^* + \varepsilon_{ij} \qquad \begin{array}{l} i = 1, 2, \ldots, a \\ j = 1, 2, \ldots, b \end{array}$$

where

$$\mu^* = \mu + \bar{\tau} + \bar{\beta} \quad \text{and} \quad \beta_j^* = \beta_j - \bar{\beta}$$

This is a one-way classification model with $n_j = a$ for $j = 1, 2, \ldots, b$; $k = b$, and $n = ab$.

c. Find X and $X'X$ for the model of part b. What is $r(X'X)$?

d. Use Theorem 2.2.1 to show that

$$E[SS_{\text{Reg(Reduced)}}] = b\sigma^2 + ab(\mu^*)^2 + a\sum_{j=1}^{b}(\beta_j^*)^2$$

36. Recall that

$$E[SS_{\text{Reg(Hypothesis)}}] = SS_{\text{Reg(Full)}} - SS_{\text{Reg(Reduced)}}$$

a. Use this relationship to show that

$$SS_{\text{Reg(Hypothesis)}} = (a-1)\sigma^2 + b\sum_{i=1}^{a}(\tau_i^*)^2$$

where $H_0: \tau_1 = \tau_2 = \cdots = \tau_a = \tau$.

b. The F ratio used to test H_0 is

$$\frac{SS_{\text{Reg(Hypothesis)}}/(a-1)}{SS_{\text{Res}}/(a-1)(b-1)}$$

Argue that if H_0 is true, ideally its value should be close to 1 and that it should exceed 1 otherwise. This, in effect, implies that the F test employed to test H_0 is a right-tailed test.

37. Find $E[SS_{\text{Reg(Hypothesis)}}]$ used in testing

$$H_0':\beta_1 = \beta_2 = \cdots = \beta_b = \beta$$

38. a. Consider a two-factor design with no interaction with $a = 2$, $b = 3$, and $n > 1$ observations per treatment combination. Find the design matrix and $X'X$ for such a model.

b. Find $X'X$ for a general two-factor design with no interaction and $n > 1$ observations per treatment combination.

c. What changes must be made in Table 6.7 to adjust to this new setting?

39. Let $\Sigma_{i=1}^{a} a_i \tau_i$ and $\Sigma_{j=1}^{b} b_j \beta_j$ denote contrasts among levels of factors I and II, respectively, in a two-factor design with $n = 1$ and without interaction.

a. Argue that the t statistic for testing

$$H_0: \sum_{i=1}^{a} a_i \tau_i = 0$$

is

$$t_{(a-1)(b-1)} = \frac{\sum_{i=1}^{a} a_i \bar{y}_{i.}}{s\sqrt{\sum_{i=1}^{a} a_i^2/b}}$$

b. Use the data of Exercise 33 to test

$$H_0:\tau_1 + \tau_2 - 2\tau_3 = 0 \quad \text{versus} \quad H_1:\tau_1 + \tau_2 - 2\tau_3 < 0$$

at the $\alpha = 0.05$ level.

c. What is the t statistic used to test

$$H_0': \sum_{j=1}^{b} b_j \beta_j = 0$$

40. Let $\Sigma_{i=1}^{a} a_i \tau_i$ and $\Sigma_{j=1}^{b} b_j \beta_j$ denote contrasts among levels of factors I and II, respectively, in a two-factor design without interaction and $n = 1$.

a. What is the F ratio used to test

$$H_0': \sum_{j=1}^{b} b_j \beta_j = 0$$

b. Use the data of Exercise 33 to test

$$H_0':\beta_1 + \beta_2 - \beta_3 - \beta_4 = 0$$

at the $\alpha = 0.05$ level.

Section 6.5

41. Verify that $RE = c + (1 - c)F_{\text{pseudo}}$ where $c = b(a - 1)/(ab - 1)$.

42. Find the relative efficiency for the experiment described in Example 6.5.1.

43. A study is conducted to compare the speed with which four private carriers can deliver mail. Since the destination of the mail can affect delivery time, this variable is treated as an extraneous variable. Identical packages are shipped to eight destinations and the following data on the time in hours required for delivery are obtained:

		Carrier			
		1	2	3	4
Destination (Block)	1	8	10	9	11
	2	12	11	9	14
	3	3	5	4	6
	4	24	26	23	25
	5	9	8	10	7
	6	6	7	5	9
	7	16	17	16	18
	8	18	19	18	20

a. Test the null hypothesis that there is no difference in delivery time among the four carriers.
b. Is there evidence that blocking is effective?
c. State the contrast used to compare the average delivery time for carrier 4 to that of the other three combined.
d. Test the contrast of part c via a single degree of freedom F test.
e. Find RE for these data via its definition. Verify that, apart from roundoff error, $RE = c + (1 - c)F_{\text{pseudo}}$.

Section 6.6

44. Consider the two-factor design with $a = b = n = 2$.
a. Show that c_6 through c_9 are linearly independent.
b. Express each of the columns c_1 through c_5 in terms of c_6 through c_9.

45. Consider the two-factor design with $a = b = n = 2$.
a. Verify that $X'X$ is given by

$$X'X = \left[\begin{array}{ccccc|cccc} 8 & 4 & 4 & 4 & 4 & 2 & 2 & 2 & 2 \\ 4 & 4 & 0 & 2 & 2 & 2 & 2 & 0 & 0 \\ 4 & 0 & 4 & 2 & 2 & 0 & 0 & 2 & 2 \\ 4 & 2 & 2 & 4 & 0 & 2 & 0 & 2 & 0 \\ 4 & 2 & 2 & 0 & 4 & 0 & 2 & 0 & 2 \\ \hline 2 & 2 & 0 & 2 & 0 & 2 & 0 & 0 & 0 \\ 2 & 2 & 0 & 0 & 2 & 0 & 2 & 0 & 0 \\ 2 & 0 & 2 & 2 & 0 & 0 & 0 & 2 & 0 \\ 2 & 0 & 2 & 0 & 2 & 0 & 0 & 0 & 2 \end{array}\right] \leftarrow M$$

b. Find the conditional inverse for $X'X$ based on the 4×4 minor M shown above.

c. Verify that $H = (X'X)^c X'X$ is given by

$$H = \begin{bmatrix} 0 & 0 & 0 & 0 & 0 & 0 & 0 & 0 & 0 \\ 0 & 0 & 0 & 0 & 0 & 0 & 0 & 0 & 0 \\ 0 & 0 & 0 & 0 & 0 & 0 & 0 & 0 & 0 \\ 0 & 0 & 0 & 0 & 0 & 0 & 0 & 0 & 0 \\ 0 & 0 & 0 & 0 & 0 & 0 & 0 & 0 & 0 \\ 1 & 1 & 0 & 1 & 0 & 1 & 0 & 0 & 0 \\ 1 & 1 & 0 & 0 & 1 & 0 & 1 & 0 & 0 \\ 1 & 0 & 1 & 1 & 0 & 0 & 0 & 1 & 0 \\ 1 & 0 & 1 & 0 & 1 & 0 & 0 & 0 & 1 \end{bmatrix}$$

d. Let

$$C = \begin{bmatrix} 0 & 0 & 0 & 0 & 0 & 1 & 0 & 0 & 0 \\ 0 & 0 & 0 & 0 & 0 & 0 & 1 & 0 & 0 \\ 0 & 0 & 0 & 0 & 0 & 0 & 0 & 1 & 0 \\ 0 & 0 & 0 & 0 & 0 & 0 & 0 & 0 & 1 \end{bmatrix}$$

Show that $H_0 : C\boldsymbol{\beta} = \mathbf{0}$ is not testable by showing that $CH \neq C$.

46. Consider these theoretical cell means for 2×2 designs. Which exhibit no interaction? Check using the criterion developed in Example 6.6.2.

a. $\mu_{11} = 3$ $\quad \mu_{21} = 1$
$\mu_{12} = 5$ $\quad \mu_{22} = 3$

b. $\mu_{11} = 3$ $\quad \mu_{21} = 5$
$\mu_{12} = 2$ $\quad \mu_{22} = 6$

c. $\mu_{11} = -2$ $\quad \mu_{21} = 0$
$\mu_{12} = 5$ $\quad \mu_{22} = 7$

***47.** Verify by substitution into Definition 6.6.1 that no interaction exists if and only if

$$[(\tau\beta)_{ij} - (\tau\beta)_{ij'}] - [(\tau\beta)_{i'j} - (\tau\beta)_{i'j'}] = 0$$

48. Rephrase the no interaction criterion developed in Example 6.6.2 in terms of the original model parameters $(\tau\beta)_{ij}$. Verify that each parameter set in Example 6.6.1 satisfies this criterion.

49. Verify that in the 2×2 case, Definition 6.6.1 leads to four nontrivial equations, each of which is equivalent to the criterion for no interaction developed in Example 6.6.2.

50. Consider these theoretical means for two-factor designs. Which exhibit no interaction? Explain graphically.

a. $\mu_{11} = 4$ $\quad \mu_{21} = 5$ $\quad \mu_{31} = 1$
$\mu_{12} = 6$ $\quad \mu_{22} = 7$ $\quad \mu_{32} = 2$

b. $\mu_{11} = 3$ $\quad \mu_{21} = 1$ $\quad \mu_{31} = 8$
$\mu_{12} = 5$ $\quad \mu_{22} = 3$ $\quad \mu_{32} = 10$

c. $\mu_{11} = 5$ $\quad \mu_{21} = 4$ $\quad \mu_{31} = 9$
$\mu_{12} = 4$ $\quad \mu_{22} = 3$ $\quad \mu_{32} = 8$
$\mu_{13} = 7$ $\quad \mu_{23} = 6$ $\quad \mu_{33} = 11$

51. Consider these theoretical means for a two-factor design.

$$\begin{array}{lll} \mu_{11}=8 & \mu_{21}=0 & \mu_{31}=10 \\ \mu_{12}=6 & \mu_{22}=? & \mu_{32}=? \\ \mu_{13}=4 & \mu_{23}=? & \mu_{33}=? \end{array}$$

Fill in the missing values in such a way that no interaction exists.

52. Show that the null hypothesis stated in Example 6.6.3 is testable by showing that $CH = C$ where $H = (X'X)^c X'X$ (see Exercise 45).

53. Verify by substitution that in a two-factor design with $a = 3$ and $b = 2$ Definition 6.6.1 generates 12 equations, only two of which are unique. Find the matrix C used to test the null hypothesis of no interaction for such a model. (*Hint*: Let $i = j = 1$ and $i' = 2$ to obtain one equation. Let $i = 2, j = 1$, and $i' = 3$ to obtain another.)

54. Consider the two-factor design with $a = b = n = 2$.

a. Show that for each i and j, $\mu_{ij} = \mu + \tau_i + \beta_j + (\tau\beta)_{ij}$ is estimable (see Exercise 45).
b. Use Theorem 5.5.2 to show that $\bar{\mu}_{..}$ is estimable. Verify that $\bar{\mu}_{..} = \Sigma_{i=1}^{a} \Sigma_{j=1}^{b} \mu_{ij}/ab = \mu^*$, thus showing that μ^* is estimable.
c. Show that for each i, $\bar{\mu}_{i.} = \Sigma_{j=1}^{b} \mu_{ij}/b$ is estimable. Argue that $\bar{\mu}_{i.} - \bar{\mu}_{..}$ is estimable. Verify that $\tau_i^* = \bar{\mu}_{i.} - \bar{\mu}_{..}$, thus showing that τ_i^* is estimable for $i = 1, 2$.
d. Show that for each j, $\bar{\mu}_{.j} = \Sigma_{i=1}^{a} \mu_{ij}/a$ and $\bar{\mu}_{.j} - \bar{\mu}_{..}$ are estimable. Verify that $\beta_j^* = \bar{\mu}_{.j} - \bar{\mu}_{..}$, thus showing that β_j^* is estimable for $j = 1, 2$.
e. Verify that $(\tau\beta)_{ij}^* = \mu_{ij} - \bar{\mu}_{i.} - \bar{\mu}_{.j} + \bar{\mu}_{..}$ and argue that this parameter is estimable.

55. Consider these theoretical means in a two-factor model:

$$\begin{array}{ll} \mu_{11}=2 & \mu_{12}=4 \\ \mu_{21}=4 & \mu_{22}=6 \end{array}$$

a. Find $\bar{\mu}_{1.}$, $\bar{\mu}_{2.}$, $\bar{\mu}_{.1}$, $\bar{\mu}_{.2}$, and $\bar{\mu}_{..}$.
b. Find $(\tau\beta)_{11}^*$, $(\tau\beta)_{12}^*$, $(\tau\beta)_{21}^*$, and $(\tau\beta)_{22}^*$. (*Hint*: See Exercise 54, part e.)
c. Is interaction present? Explain.

56. Consider these theoretical means in a two-factor model:

$$\begin{array}{lll} \mu_{11}=2 & \mu_{21}=4 & \mu_{31}=0 \\ \mu_{12}=1 & \mu_{22}=3 & \mu_{32}=0 \end{array}$$

Show that interaction exists by finding an interaction term $(\tau\beta)_{ij}^*$ that is nonzero.

57. Verify that in the reparameterized model,

$$\sum_{i=1}^{a} \tau_i^* = \sum_{j=1}^{b} \beta_j^* = \sum_{i=1}^{a} \sum_{j=1}^{b} (\tau\beta)_{ij}^* = 0$$

as claimed. Verify that this guarantees that

$$\sum_{i=1}^{a} (\tau\beta)_{i.}^* = \sum_{j=1}^{b} (\tau\beta)_{.j}^* = 0$$

58. Let

$$y_{ijk} = \mu^* + \tau_i^* + \beta_j^* + (\tau\beta)_{ij}^* + \varepsilon_{ijk} \qquad \begin{array}{l} i = 1, 2, 3 \\ j = 1, 2 \\ k = 1, 2, \ldots, n \end{array}$$

Find the design matrix for this model. Use this to verify that $X'X$ and $X'\mathbf{y}$ are as claimed on page 284.

59. Verify that the estimators for β_j^* and $(\tau\beta)_{ij}^*$ are

$$\hat{\beta}_j^* = \bar{y}_{.j.} - \bar{y}_{...}$$

and

$$\widehat{(\tau\beta)}_{ij}^* = \bar{y}_{ij.} - \bar{y}_{i..} - \bar{y}_{.j.} + \bar{y}_{...}$$

as claimed.

60. Use a one-way analysis to compare the behavior of capsule A to that of capsule B when in duodenal juices based on the data of Example 6.6.4.

61. Verify the sums and the ANOVA given in Example 6.6.7.

62. Ozonation as a secondary treatment for effluent following absorption by ferrous chloride was studied for three reaction times and three PH levels. These data are obtained on effluent decline:

Reaction Time in Minutes (Factor II)		PH Level (Factor I): 7	9	10.5
	20	23	16	14
		21	18	13
		22	15	16
	40	20	14	12
		22	13	11
		19	12	10
	60	21	13	11
		20	12	13
		19	12	12

Derive the complete ANOVA table for these data and interpret the findings.

COMPUTING SUPPLEMENT G (Testing Contrasts)

PROC GLM is used to test hypotheses in the less than full rank case. The data of Example 6.1.2 are used as an illustration.

The output of this program is shown in Figure 6.5. Several things in this output are important. First, the estimates for β_0, β_1, β_2, and β_3 shown by (1)–(4), respectively, are different from those of Example 6.1.2. This should not be surprising since, as the SAS "NOTE" indicates, parameter estimates are not unique

Statement	Purpose
`DATA TAR;`	Names the data set
`INPUT Y X1 X2 X3;`	Names the variable
`CARDS;`	Signals that the data follow

`34.6 1 0 0`
`35.1 1 0 0`
`.`
`.`
`.`
`37.7 1 0 0`
`38.8 0 1 0`
`39.0 0 1 0`
`.`
`.`
`.`
`53.6 0 1 0`
`26.7 0 0 1`
`26.7 0 0 1`
`.`
`.`
`31.2 0 0 1`
	Data lines

`;`	Signals ends of data
`PROC GLM;`	Calls for the general linear models procedure
`MODEL Y=X1 X2 X3;`	Identifies the independent variables as X1, X2, and X3; names Y as the response
`CONTRAST 'EQUAL MEANS' X1 1 X2 -1 X3 0,` `X1 1 X2 0 X3 -1;`	Asks GLM to test $H_0{:}C\boldsymbol{\beta}=\mathbf{0}$; the values listed after the variable names form the last three columns of the matrix C

in the less than full rank model. The estimates found by SAS are based on a conditional inverse that differs from the one given in Example 6.1.2. The sum of squares associated with the hypothesis $C\boldsymbol{\beta}=\mathbf{0}$ is

$$(C\mathbf{b})'[C(X'X)^c C']^{-1}(C\mathbf{b}) = 1251.533$$

This sum of squares is shown by (5). The residual sum of squares and s^2 are given at (6) and (7), respectively. The F ratio used to test $H_0{:}C\boldsymbol{\beta}=\mathbf{0}$ is given by (8). Note that this value agrees with that found by hand in Example 6.1.2. The P value for the test ($P < 0.0001$) is given by (9). Any testable hypothesis can be tested via an appropriately chosen CONTRAST statement.

FIGURE 6.5 **Testing contrasts using the data of Example 6.1.2**

SAS 1

GENERAL LINEAR MODELS PROCEDURE

DEPENDENT VARIABLE: Y

SOURCE	DF	SUM OF SQUARES	MEAN SQUARE	F VALUE
MODEL	2	1251.53266667	625.76633333	60.63
ERROR	27	278.66100000 ⑥	10.32077778 ⑦	PR > F
CORRECTED TOTAL	29	1530.19366667		0.0001

R-SQUARE	C.V.	ROOT MSE	Y MEAN
0.817892	8.8885	3.21259673	36.14333333

SOURCE	DF	TYPE I SS	F VALUE	PR > F
X1	1	0.17066667	0.02	0.8986
X2	1	1251.36200000	121.25	0.0001
X3	0	0.00000000	.	.

SOURCE	DF	TYPE III SS	F VALUE	PR > F
X1	0	0.00000000	.	.
X2	0	0.00000000	.	.
X3	0	0.00000000	.	.

CONTRAST	DF	SS	F VALUE	PR > F
equal means	2	1251.53266667 ⑤	60.63 ⑧	0.0001 ⑨

PARAMETER	ESTIMATE	T FOR H0: PARAMETER=0	PR > |T|	STD ERROR OF ESTIMATE
INTERCEPT	28.18000000 B ①	27.74	0.0001	1.01591229
X1	8.07000000 B ②	5.62	0.0001	1.43671694
X2	15.82000000 B ③	11.01	0.0001	1.43671694
X3	0.00000000 B ④	.	.	.

NOTE: THE X'X MATRIX HAS BEEN DEEMED SINGULAR AND A GENERALIZED INVERSE HAS BEEN EMPLOYED TO SOLVE THE NORMAL EQUATIONS. THE ABOVE ESTIMATES REPRESENT ONLY ONE OF MANY POSSIBLE SOLUTIONS TO THE NORMAL EQUATIONS. ESTIMATES FOLLOWED BY THE LETTER B ARE BIASED AND DO NOT ESTIMATE THE PARAMETER BUT ARE BLUE FOR SOME LINEAR COMBINATION OF PARAMETERS (OR ARE ZERO). THE EXPECTED VALUE OF THE BIASED ESTIMATORS MAY BE OBTAINED FROM THE GENERAL FORM OF ESTIMABLE FUNCTIONS. FOR THE BIASED ESTIMATORS, THE STD ERR IS THAT OF THE BIASED ESTIMATOR AND THE T VALUE TESTS H0: E(BIASED ESTIMATOR) = 0. ESTIMATES NOT FOLLOWED BY THE LETTER B ARE BLUE FOR THE PARAMETER.

COMPUTING SUPPLEMENT H (One-Way Classification)

A one-way ANOVA with fixed effects based on the reparameterized model developed in Section 6.2 can be run easily on SAS via either PROC GLM or PROC ANOVA. The appropriate GLM program for analyzing the data of Example 6.2.1 is shown below.

Statement	Purpose
`DATA TAR;`	Names the data set
`INPUT METHOD $ REMOVE;`	Names the variables; $ indicates that the method of tar removal will be coded alphanumerically
`CARDS;`	Signals that the data follow
`AF 34.6`	Data lines
`AF 35.1`	
.	
.	
.	
`AF 37.7`	
`FS 38.8`	
`FS 39.0`	
.	
.	
.	
`FS 53.6`	
`FCC 26.7`	
`FCC 26.7`	
.	
.	
.	
`FCC 31.2`	
`;`	Signals the end of data
`PROC GLM;`	Asks for the general linear models procedure
`CLASSES METHOD;`	Indicates that the data are grouped according to the values of the variable METHOD;
`MODEL REMOVE= METHOD;`	Identifies the variable REMOVE as the response variable
`TITLE COAL-TAR DATA;`	Titles the output

The output of this program is shown in Figure 6.6. The F ratio used to test $H_0{:}\mu_1 = \mu_2 = \mu_3$ is shown by (1) and its P value is given by (2). The ANOVA table is based on the corrected total sum of squares. The sum of squares that SAS calls "Model" is our $SS_{\text{Reg(Hypothesis)}}$. This sum of squares is shown by (3). The SAS "error" sum of squares shown by (4) is what we called SS_{Res}.

FIGURE 6.6 **Running a one-way ANOVA on the data of Example 6.2.1**

```
                              COAL-TAR DATA                                        2

                    GENERAL LINEAR MODELS PROCEDURE

DEPENDENT VARIABLE: REMOVE

SOURCE                  DF      SUM OF SQUARES          MEAN SQUARE        F VALUE

MODEL                    2      1251.53266667 (3)      625.76633333     (1) 60.63

ERROR                   27       278.66100000 (4)       10.32077778        PR > F

CORRECTED TOTAL         29      1530.19366667                          (2) 0.0001

R-SQUARE              C.V.            ROOT MSE          REMOVE MEAN

0.817892            8.8885          3.21259673          36.14333333

SOURCE                  DF         TYPE I SS    F VALUE      PR > F

METHOD                   2     1251.53266667      60.63      0.0001

SOURCE                  DF       TYPE III SS    F VALUE      PR > F

METHOD                   2     1251.53266667      60.63      0.0001
```

COMPUTING SUPPLEMENT I (Two-Factor Design, No Interaction or Randomized Blocks)

A two-factor design with no interaction or a randomized complete block design can be analyzed on SAS via PROC GLM or PROC ANOVA. The code is illustrated by analyzing the data of Example 6.5.1.

Statement	Purpose
`DATA HIGHWAY;`	Names the data set
`INPUT LOCATION TYPE WEAR;`	Names the variables
`CARDS;`	Signals that the data follow
`1 1 42.7`	Data lines
`1 2 39.3`	
`1 3 48.5`	
`1 4 32.8`	
`2 1 50.0`	
`2 2 38.0`	
`2 3 49.7`	
`2 4 40.2`	
`3 1 51.9`	
`3 2 46.3`	
`3 3 53.5`	
`3 4 51.1`	
`;`	Signals end of data
`PROC ANOVA;`	Calls for the analysis of variance procedure
`CLASSES LOCATION TYPE;`	Indicates that the responses are identified by both a location number and a paving type number
`MODEL WEAR=LOCATION TYPE;`	Identifies the variable WEAR as the response variable
`TITLE RANDOMIZED BLOCKS;`	Titles output

The output of this program is shown in Figure 6.7. The following quantities are given:

(1) SS_{Blocks}

(2) $SS_{\text{Reg(Hypothesis)}}$

(3) F ratio used to test H_0

(4) P value for the test on H_0

(5) F_{pseudo}

(6) s^2

FIGURE 6.7 **Running a randomized complete block design using the data of Example 6.5.1**

```
                        RANDOMIZED BLOCKS                              2

                  ANALYSIS OF VARIANCE PROCEDURE

DEPENDENT VARIABLE: WEAR

SOURCE              DF      SUM OF SQUARES          MEAN SQUARE      F VALUE

MODEL                5        404.72500000          80.94500000         6.87

ERROR                6         70.70166667          11.78361111(6)    PR > F

CORRECTED TOTAL     11        475.42666667                            0.0181

R-SQUARE           C.V.           ROOT MSE            WEAR MEAN

0.851288         7.5722         3.43272648          45.33333333

SOURCE              DF          ANOVA SS   F VALUE      PR > F

LOCATION             2   199.45166667(1)   8.46(5)   0.0179
TYPE                 3   205.27333333(2)   5.81(3)   0.0330(4)
```

COMPUTING SUPPLEMENT J (Two-Factor Design with Interaction)

Either PROC ANOVA or PROC GLM can be used to analyze data in a two-factor design with interaction when cell sizes are equal. When cell sizes are not equal, PROC ANOVA is not applicable. The data of Example 6.6.7 are used as an illustration.

Statement	Purpose
`DATA LEAF;`	Names the data set
`INPUT TIME ENV GRAMS;`	Names the variables
`CARDS;`	Signals that the data follow
`1 1 1.09`	Data lines
`1 1 1.06`	
`2 1 1.35`	
`2 1 1.53`	
.	
.	

Statement	Purpose
`.`	Data Lines
`3 4 1.76`	
`;`	Signals end of data
`PROC ANOVA;`	Asks for the analysis of variance procedure
`CLASSES TIME ENV;`	Indicates that the data are grouped according to the values of the two variables TIME and ENV
`MODEL GRAMS=TIME ENV TIME*ENV;`	Identifies the variable GRAMS as the response; TIME*ENV indicates the presence of an interaction term in the model
`TITLE Two-Factor Design with Interaction;`	Titles the output

The output of this program is shown in Figure 6.8. The F ratio for testing for interaction is shown at (1) and its P value is at (2). Apart from roundoff error, these agree with the values found in Example 6.6.7. If the interaction is deemed nonsignificant, the main effects F ratios are shown at (3) and (4) with their respective P values given by (5) and (6).

FIGURE 6.8 **Running a two-factor design with interaction using the data of Example 6.6.7**

```
                 TWO-FACTOR DESIGN WITH INTERACTION                          2

                   ANALYSIS OF VARIANCE PROCEDURE

DEPENDENT VARIABLE: GRAMS

SOURCE                   DF        SUM OF SQUARES          MEAN SQUARE       F VALUE

MODEL                    11            2.49621250           0.22692841         13.31

ERROR                    12            0.20455000           0.01704583       PR > F

CORRECTED TOTAL          23            2.70076250                             0.0001

R-SQUARE               C.V.              ROOT MSE           GRAMS MEAN

0.924262             9.0903            0.13055969           1.43625000

SOURCE                   DF             ANOVA SS    F VALUE      PR > F

TIME                      2           2.13667500    62.67(4)    0.0001(6)
ENV                       3           0.07817917     1.53(3)    0.2576(5)
TIME*ENV                  6           0.28135833     2.75(1)    0.0641(2)
```

C H A P T E R 7

ADDITIONAL TOPICS

In this chapter some special topics are introduced. These topics are important applications of the more general structures presented in Chapters 1–6. Two of these topics, analysis of covariance and components of variance models, are widely used in many fields.

7.1 ANALYSIS OF COVARIANCE

In Chapter 6 we discussed several special cases in which the general linear less than full rank model applies. These special cases include the one-way classification, two-way classification, and randomized complete block designs. Recall that in the randomized complete block model the experimenter allows a block effect for the model in order to systematically account for variation (variation in response due to blocks) that would ordinarily fall into experimental error. The experimenter can exploit other methods of error reduction. At times extraneous variation in the experiment is in the form of *continuous measurements.* For example, a researcher may want to compare the effect of three drugs on some physiological response. Subjects to be used in the experiment are selected at random, with five subjects using each of the three drugs. At this point, the best analytical technique to use would appear to be a one-way classification analysis of variance; the variation among the subjects plays an important role in determining the random error term in the analysis.

If the researcher can determine that certain measurements on the subjects used in the experiment help characterize or describe the variation among subjects, then these variables can be included in the model. For example, if the subject's weight is taken into account, the error variance can be reduced substantially. As a result, we may write the model.

$$y_{ij} = \mu + \tau_i + \beta x_{ij} + \varepsilon_{ij} \qquad \begin{aligned} i &= 1, 2, 3 \\ j &= 1, 2, 3, 4, 5 \end{aligned}$$

where y_{ij} is the response of the jth subject with the ith drug, τ_i is the usual treatment effect, ε_{ij} is the random error, and x_{ij}, the covariate, is the weight of the jth subject using the ith drug. As in the case of blocking, we are accounting for extraneous variation whose source is the experimental unit—the subject. But in this case the form of the "extra information", that is, the subject's weight, is different from that in the randomized block. The above model is called an *analysis of covariance* model. The purpose of the terminology will be apparent when we investigate the method of analysis.

The general analysis of covariance model with a single set of treatments and a *single covariate* is given by

$$y_{ij} = \mu + \tau_i + \beta x_{ij} + \varepsilon_{ij} \qquad \begin{aligned} i &= 1, 2, \dots, t \\ j &= 1, 2, \dots, n \end{aligned}$$

This combination of ANOVA and regression model terms is applied in biology, agriculture, and many engineering fields. It is important to categorize the model in terms of rank so that we can proceed with a discussion of the analysis. If we invoke the usual general linear model

$$\mathbf{y} = X\boldsymbol{\beta} + \boldsymbol{\varepsilon}$$

we find that

$$\mathbf{y} = \begin{bmatrix} y_{11} \\ y_{12} \\ \vdots \\ y_{1n} \\ \hline y_{21} \\ \vdots \\ y_{2n} \\ \hline \vdots \\ \hline y_{t1} \\ \vdots \\ y_{tn} \end{bmatrix}, \quad X = \begin{bmatrix} 1 & 1 & 0 & \cdots & 0 & x_{11} \\ 1 & 1 & 0 & & 0 & x_{12} \\ \vdots & \vdots & \vdots & & \vdots & \vdots \\ 1 & 1 & 0 & & 0 & x_{1n} \\ \hline 1 & 0 & 1 & & 0 & x_{21} \\ 1 & 0 & 1 & & 0 & x_{22} \\ \vdots & \vdots & \vdots & & \vdots & \vdots \\ 1 & 0 & 1 & & 0 & x_{2n} \\ \hline \vdots & \vdots & \vdots & & \vdots & \vdots \\ \hline 1 & 0 & 0 & & 1 & x_{t1} \\ 1 & 0 & 0 & & 1 & x_{t2} \\ \vdots & \vdots & \vdots & & \vdots & \vdots \\ 1 & 0 & 0 & & 1 & x_{tn} \end{bmatrix}, \quad \text{and} \quad \boldsymbol{\beta} = \begin{bmatrix} \mu \\ \tau_1 \\ \tau_2 \\ \vdots \\ \tau_t \\ \beta \end{bmatrix}$$

The X matrix is $tn \times (t+2)$. It is clear that, as in the case of a one-factor ANOVA, the X matrix is less than full rank. In fact the rank of X is $t+1$. As a result, any estimation or hypothesis testing will be approached in the context of the less than full rank model. The primary motivation in the analysis of covariance is to test equality of treatment effects and to estimate treatment means, adjusted for the covariate.

INFERENCE ON TREATMENT EFFECTS

We first consider hypothesis testing on treatment effects. We approach it in much the same fashion as we did in our ANOVA illustration. The hypothesis is written as

$$H_0: \tau_1 = \tau_2 = \cdots = \tau_t \quad \text{versus} \quad H_1: \text{Not all } \tau_i \text{ are equal}$$

Now it becomes convenient to reparameterize the model in order to test H_0. Suppose we rewrite the analysis of covariance model as

$$y_{ij} = \mu + \tau_i^* + \beta(x_{ij} - \bar{x}_{i.}) + \varepsilon_{ij} \qquad \begin{array}{l} i = 1, 2, \ldots, t \\ j = 1, 2, \ldots, n \end{array}$$

where $\bar{x}_{i.} = \Sigma_{j=1}^n x_{ij}/n$ and $\tau_i^* = \tau_i + \beta\bar{x}_{i.}$. As a result, the $X'X$ matrix becomes

$$X'X = \begin{bmatrix} nt & n & n & n & \cdots & n & 0 \\ n & n & 0 & 0 & \cdots & 0 & 0 \\ n & 0 & n & 0 & \cdots & 0 & 0 \\ \vdots & \vdots & \vdots & \vdots & & \vdots & \vdots \\ n & 0 & 0 & 0 & \cdots & n & 0 \\ 0 & 0 & 0 & 0 & \cdots & 0 & E_{xx} \end{bmatrix}$$

where $E_{xx} = \Sigma_{i=1}^t \Sigma_{j=1}^n (x_{ij} - \bar{x}_{i.})^2$ (see Exercise 1). The vector $X'\mathbf{y}$ becomes

$$X'\mathbf{y} = \begin{bmatrix} \sum_i \sum_j y_{ij} \\ \sum_{j=1}^n y_{1j} \\ \sum_{j=1}^n y_{2j} \\ \vdots \\ \sum_{j=1}^n y_{tj} \\ E_{xy} \end{bmatrix} = \begin{bmatrix} y_{..} \\ y_{1.} \\ \vdots \\ y_{t.} \\ E_{xy} \end{bmatrix}$$

where $E_{xy} = \Sigma_{i=1}^t \Sigma_{j=1}^n (x_{ij} - \bar{x}_{i.})\, y_{ij}$. Notice that the reparameterization allows the covariate to be orthogonal to the mean and to all treatment effects (see Exercise 2).

We develop the test of the hypothesis on treatment effects by first considering estimates $\hat{\mu}$, $\hat{\tau}_i^*$, $\hat{\beta}$ under the "full model" and then under a reduced model as was done in Section 6.2. For the full model (reparameterized form), the normal equations can be written

$$(X'X)\mathbf{b} = X'\mathbf{y}$$

$$\begin{bmatrix} nt & n & n & \cdots & n & 0 \\ n & n & 0 & \cdots & 0 & 0 \\ & 0 & & & & \\ \vdots & \vdots & & & \vdots & \vdots \\ n & 0 & & & n & \\ 0 & 0 & & & 0 & E_{xx} \end{bmatrix} \begin{bmatrix} \hat{\mu} \\ \hat{\tau}_1^* \\ \hat{\tau}_2^* \\ \vdots \\ \hat{\tau}_t^* \\ \hat{\beta} \end{bmatrix} = \begin{bmatrix} y_{..} \\ y_{1.} \\ \vdots \\ y_{t.} \\ E_{xy} \end{bmatrix}$$

Notice that the $X'X$ matrix is $(t+2) \times (t+2)$ of rank $t+1$. As in the case of ANOVA, the equations can be solved via the linear constraint

$$\sum_{i=1}^{t} \hat{\tau}_i^* = 0$$

It is quite simple to see that the solutions to the equations are given by

$$\hat{\mu} = \bar{y}_{..}$$
$$\hat{\tau}_i^* = \bar{y}_{i.} - \bar{y}_{..}$$
$$\hat{\beta} = E_{xy}/E_{xx}$$

(see Exercise 3). The estimates of the grand mean μ and the adjusted treatment effect τ_i^* are exactly the same as their counterparts in ANOVA (see Section 6.4). The estimator $\hat{\beta}$ has as its structure a form that is very similar to the slope of a simple regression (see Exercise 4). The difference is that here the sums of squares of the x's and the sum of crossproducts of the x's and y's are corrected for $\bar{x}_{i.}$ rather than $\bar{x}_{..}$ (overall mean) since the presence of treatment effects allows for relevance of different means, one for each treatment. In fact, if we carefully study the analysis of covariance model it becomes clear that it is equivalent to t separate regressions with different intercepts (the ith intercept being $\mu + \tau_i$) and *common slope* (the common slope being β).

Let us now develop the test procedure. As before, we must assume that the ε_{ij} are independent and normally distributed with mean 0 and variance σ^2. Recall that if we follow the technique used for the regression model in Section 4.2 and the ANOVA model in Section 6.2 we require the difference

$$SS_{\text{Reg(Full)}} - SS_{\text{Reg(Reduced)}}$$

The $SS_{\text{Reg(Full)}}$ model is written as

$$\begin{aligned} SS_{\text{Reg(Full)}} &= \mathbf{b}'X'\mathbf{y} \\ &= \bar{y}_{..}(y_{..}) + \sum_{i=1}^{t} (\bar{y}_{i.} - \bar{y}_{..})\, y_{i.} + E_{xy}^2/E_{xx} \\ &= \sum_{i=1}^{t} y_{i.}^2/n + E_{xy}^2/E_{xx} \end{aligned}$$

with $t+1$ degrees of freedom (rank of X) (see Exercise 5).

In dealing with the reduced model, we can use the material presented in Chapter 3. It should be obvious that under $H_0: \tau_1 = \tau_2 = \cdots = \tau_t = \tau$, the analysis of covariance model can be written

$$\begin{aligned} y_{ij} &= (\mu + \tau) + \beta x_{ij} + \varepsilon_{ij} \\ &= \mu^* + \beta x_{ij} + \varepsilon_{ij} \end{aligned}$$

This is merely a simple linear regression model. We need only find the regression sum of squares. This is given by

$$SS_{\text{Reg(Reduced)}} = y_{..}^2/tn + S_{xy}^2/S_{xx}$$

with 2 degrees of freedom (see Exercise 6). Here,

$$S_{xy} = \sum_{i=1}^{t} \sum_{j=1}^{n} (x_{ij} - \bar{x}_{..})(y_{ij} - \bar{y}_{..})$$

$$S_{xx} = \sum_{i=1}^{t} \sum_{j=1}^{n} (x_{ij} - \bar{x}_{..})^2$$

Notice the distinction between the (S_{xy}, S_{xx}) and the (E_{xy}, E_{xx}) quantities. As a result, the appropriate regression sum of squares, denoted by $R(\tau_1, \tau_2, \ldots, \tau_t|\mu, \beta)$ is given by

$$\begin{aligned} R(\tau_1, \tau_2, \ldots, \tau_t|\mu, \beta) &= SS_{\text{Reg(Full)}} - SS_{\text{Reg(Reduced)}} \\ &= \left[\sum_{i=1}^{t} y_{i.}^2/n - \frac{(y_{..})^2}{tn}\right] + \left[E_{xy}^2/E_{xx} - \frac{S_{xy}^2}{S_{xx}}\right] \end{aligned}$$

Let us denote the first bracketed term on the right by B_{yy} because it reflects variation between or among treatments. Thus $R(\tau_1, \tau_2, \ldots, \tau_t|\mu, \beta)$ becomes

$$R(\tau_1, \tau_2, \ldots, \tau_t|\mu, \beta) = B_{yy} + \frac{E_{xy}^2}{E_{xx}} - \frac{S_{xy}^2}{S_{xx}}$$

with $(t + 1) - 2 = t - 1$ degrees of freedom. Notice again that the first portion of the sum of squares is identical to the sum of squares used for the same equal treatment effects hypothesis. The term in brackets is a *covariate adjustment* term.

To develop the error sum of squares, we merely use

$$SS_E = \mathbf{y}'\mathbf{y} = \mathbf{b}'X'\mathbf{y}$$

which results in the following expression:

$$\begin{aligned} SS_E &= \sum_i \sum_j y_{ij}^2 - \left[\sum_{i=1}^{t} \frac{y_{i.}^2}{n} + \frac{E_{xy}^2}{E_{xx}}\right] \\ &= \left[\sum_i \sum_j y_{ij}^2 - \sum_{i=1}^{t} \frac{y_{i.}^2}{n}\right] - \frac{E_{xy}^2}{E_{xx}} \end{aligned}$$

We denote the first bracketed term on the right by E_{yy} because it denotes random variation within treatments. Thus SS_E becomes

$$SS_E = E_{yy} - E_{xy}^2/E_{xx}$$

with degrees of freedom $nt - r(X) = nt - t - 1$. We learned that in the general setting with the less than full rank model, the F test for testing the hypothesis invoked in the reduced model is merely a ratio of mean squares. In this case we have

$$F = \frac{[SS_{\text{Reg(Full)}} - SS_{\text{Reg(Reduced)}}]/(t-1)}{SS_E/(nt-t-1)}$$

THE ANALYSIS OF COVARIANCE TABLE

The procedure for testing treatment effects in the presence of covariate information is straightforward. As in the case of experiments with blocking, the purpose of using the covariate is to reduce σ^2, experimental error. One degree of freedom is lost from experimental error. For this type of analysis, sum of squares results are often summarized using an *analysis of covariance table*, which is given in Table 7.1.

TABLE 7.1
Analysis of covariance table

Source	Sum of Squares of y	Sum of Products	Sum of Squares of x
Between treatments	$B_{yy} = \sum_{i=1}^{t} \frac{y_{i.}^2}{n} - \frac{(y_{..})^2}{nt}$	$B_{xy} = \sum_{i=1}^{t} x_{i.}\, y_{i.} - \frac{(y_{..})(x_{..})}{nt}$	$B_{xx} = \sum_{i=1}^{t} \frac{x_{i.}^2}{n} - \frac{(x_{..})^2}{nt}$
Within treatments	$E_{yy} = \sum \sum y_{ij}^2 - \sum_{i=1}^{t} \frac{y_{i.}^2}{n}$	$E_{xy} = \sum \sum x_{ij}\, y_{ij} - \sum_{i=1}^{y} \frac{y_{i.}\, x_{i.}}{n}$	$E_{xx} = \sum_i \sum_j x_{ij}^2 - \sum_{i=1}^{t} \frac{x_{i.}^2}{n}$
Total	$S_{yy} = \sum \sum y_{ij}^2 - \frac{(y_{..})^2}{nt}$	$S_{xy} = \sum_i \sum_j x_{ij}\, y_{ij} - \frac{(y_{..})(x_{..})}{nt}$	$S_{xx} = \sum_i \sum_j x_{ij}^2 - \frac{(x_{..})^2}{nt}$

Unlike ANOVA, which merely partitions and analyzes variance, the technique of analysis of covariance partitions both variance and covariance, with the "between" and "within" determined by the treatment structure. (An example follows the discussion of tests on β, the slope parameter.)

INFERENCE ON THE SLOPE COEFFICIENT

The point estimate of the slope is $\hat{\beta} = E_{xy}/E_{xx}$. In structure, a test of hypothesis on β resembles a test on the slope in simple linear regression. The orthogonality that was produced by the reparameterization, that is, the use of $x_{ij} - \bar{x}_{i.}$, allows us to write the regression explained by the slope adjusted for the other parameters. You are asked in Exercise 7 to verify that

$$R(\beta|\mu, \tau_1, \ldots, \tau_t) = R(\beta) = \frac{E_{xy}^2}{E_{xx}}$$

This sum of squares has 1 degree of freedom. As a result, the appropriate F test for testing

$$H_0{:}\beta = 0 \quad \text{versus} \quad H_1{:}\beta \neq 0$$

is given by

$$F = \frac{E_{xy}^2/E_{xx}}{[E_{yy} - E_{xy}^2/E_{xx}]/[nt - t - 1]}$$

REVIEW OF ANALYSIS OF COVARIANCE ASSUMPTIONS

We have already indicated that the usual normality, independence, and common error variance are necessary assumptions for analysis of covariance. An additional very important assumption is made, one that is obvious from the model but often ignored in practice. Namely, we assume that we are dealing in t regression models with possibly different intercepts and common slopes. Often, in practice, the mere presence of a covariate automatically suggests the need for analysis of covariance to account for variation due to the covariate. However, it is not always obvious that the impact on the response due to the covariate will be constant across treatments (see Exercise 1). That is, it is not obvious that there is a common slope. In Exercise 8 you are asked to develop a test for testing for equality of slopes.

Example 7.1.1 In an experiment conducted by the Health and Physical Education Department at Virginia Polytechnic Institute and State University to study the effect of exercise on total cholesterol, four groups of subjects were selected for the study and four different forms of exercises were selected as the treatments. One treatment (no exercise) was considered as the control. The additional groups were characterized by the following activities: running, weightlifting, and both running and weightlifting. The number of subjects were 8, 8, 10, and 5, respectively. (Though our development was done assuming equal sample sizes in each group, the obvious changes for the unequal sample sizes should be apparent [see Exercise 9].) In order to take into account the differences among the subjects, the prestudy value of total cholesterol was used as a covariate. The response for each subject is the poststudy value of total cholesterol. The controlled study lasted eight weeks. The data are as follows (the value x is prestudy cholesterol):

Control	y:	75.0	72.5	62.0	60.0	53.0	53.0	65.0	63.5
	x:	75.0	55.0	54.5	54.0	70.5	51.0	76.0	69.0

Running	y:	49.0	53.5	30.0	40.5	51.5	57.5	49.0	74.0		
	x:	49.5	50.0	27.5	38.5	50.0	68.5	48.5	60.5		
Weightlifting	y:	54.5	79.5	64.0	69.0	50.5	58.0	63.5	76.0	55.5	68.0
	x:	55.0	78.0	49.5	58.5	64.0	63.5	54.5	75.0	50.5	66.5
Both	y:	59.0	54.5	50.5	63.0	65.0					
	x:	78.0	73.0	47.0	82.0	62.5					

The analysis of covariance was conducted. The following information produces a test of equality of group means, adjusted for the covariate, that is, a test of H_0: $\tau_1 = \tau_2 = \tau_3 = \tau_4$:

$$R(\tau_1, \tau_2, \ldots, \tau_t | \mu, \beta) = 384.6059 \qquad (3\ df)$$

$$SS_E = 1{,}555.0408 \qquad (26\ df)$$

The F test for equality of treatment effects is given by

$$F = \frac{384.6059/3}{1{,}555.0408/26} = 2.14$$

The P value is 0.1190. As a result, we must say that there is no significant difference in mean cholesterol among the exercise groups.

The test on the covariate was conducted via the hypothesis

$$H_0:\beta = 0 \quad \text{versus} \quad H_1:\beta \neq 0$$

For these data,

$$E_{xy}^2/E_{xx} = 978.0592$$

with 1 degree of freedom. As a result the F statistic is given by

$$F = \frac{978.0592}{1{,}555.0408/26} = 16.35$$

which is significant at the 0.0004 level. Thus we conclude that the pretest cholesterol level has a significant impact on the posttest cholesterol level. Computing Supplement K demonstrates the code needed for the computer analysis of these data.

ESTIMATES OF TREATMENT MEANS

In the previous illustration we were testing to detect differences among treatment effects. Recall that in the one-way analysis of variance, equality of treatment effects is equivalent to equality of *treatment means.* That is, with the model

$$y_{ij} = \mu + \tau_i + \varepsilon_{ij} \qquad i = 1, 2, \ldots, t$$
$$j = 1, 2, \ldots, n$$

the quantity $\mu + \tau_i$ is estimable and is estimated by $\bar{y}_{i.}$, the sample treatment mean. In analysis of covariance, quantities such as μ, τ_i, and τ_i^* are not estimable. However, $\mu + \tau_i$ is estimable and is a relevant parameter that can be referred to as a treatment mean. But what is the estimator for $\mu_i = \mu + \tau_i$? Observe from the normal equations that if we use the constraint $\Sigma_{i=1}^t \hat{\tau}_i^* = 0$, the solutions to the normal equations are

$$\hat{\mu} = \bar{y}_{..}$$
$$\hat{\tau}_i^* = \bar{y}_{i.} - \bar{y}_{..} \qquad (i = 1, 2, \ldots, t)$$

which implies that the estimates of the estimable parameters $\mu_1, \mu_2, \ldots, \mu_t$ are given by

$$\begin{aligned}\hat{\mu}_i &= \hat{\mu} + \hat{\tau}_i^* - \hat{\beta}\bar{x}_{i.}\\ &= \bar{y}_{i.} - \hat{\beta}\bar{x}_{i.} \qquad (i = 1, 2, \ldots, t)\end{aligned}$$

If we want to estimate the mean response for the ith treatment at a given value of x_0 we substitute that value into the estimated regression equation

$$\begin{aligned}\hat{\mu}_{Y|x} &= \hat{\mu}_i + \hat{\beta}x_0\\ &= \bar{y}_{i.} - \hat{\beta}\bar{x}_{i.} + \hat{\beta}x_0\end{aligned}$$

To compare treatments, we need to make the comparison at some common x value. It is convenient to let $x_0 = \bar{x}_{..}$ and evaluate $\hat{\mu}_{Y|\bar{x}}$ for each of the i treatments at this common value. These estimated means are called *adjusted treatment means* or *sample means adjusted for the covariate*. They are given by

$$\hat{\mu}_{i(\text{adjusted})} = \bar{y}_{i.} - \hat{\beta}(\bar{x}_{i.} - \bar{x}_{..})$$

Example 7.1.2 These sample means are computed from the data of Example 7.1.1.

$$\begin{array}{lll}\bar{y}_{1.} = 63 & \bar{x}_{1.} = 63.125 & \bar{x}_{..} = 59.855\\ \bar{y}_{2.} = 50.625 & \bar{x}_{2.} = 49.125 & \\ \bar{y}_{3.} = 63.85 & \bar{x}_{3.} = 61.5 & \\ \bar{y}_{4.} = 58.4 & \bar{x}_{4.} = 68.5 & \end{array}$$

From Computing Supplement K, we see that $\hat{\beta} = 0.52864490$. The adjusted means are

$$\hat{\mu}_{1(\text{adjusted})} = 61.2712$$
$$\hat{\mu}_{2(\text{adjusted})} = 56.2973$$
$$\hat{\mu}_{3(\text{adjusted})} = 62.9803$$
$$\hat{\mu}_{4(\text{adjusted})} = 53.8398$$

If there is no significant difference in treatments then these estimates should be close in value since each of the four estimated regression lines is estimating the common line $\mu_{Y|x} = \mu + \tau + \beta x$. If the treatment effects are different, then we are dealing with more than one regression line. Hence there should be a marked difference in at least two of these estimates.

7.2

RANDOM EFFECTS ANALYSIS OF VARIANCE

The analysis of covariance model and the models considered in Chapters 5 and 6 are all fixed effects models. That is, treatments are chosen a priori by the researcher because they are of particular interest. Thus, model terms that represent treatments, blocks, and interactions are parameters. Occasionally an experiment is encountered in which the levels at which the experiment is conducted are not of interest in themselves. Rather, they represent some of the many levels at which the experiment could have been conducted. The researcher is interested simply in determining whether or not different experimental conditions produce different responses.

Example 7.2.1 Consider a situation in which a chemist is interested in researching a process that produce a certain product. During production various batches of raw materials are used. We are not interested in differences among these particular batches but we would like to know whether or not there is a batch effect. One may visualize a study in which five batches are used. The data may take the following form:

	1	y_{11}	y_{12}	y_{13}	y_{14}
	2	y_{21}	y_{22}	y_{23}	y_{24}
Batch	3	y_{31}	y_{32}	y_{33}	y_{34}
	4	y_{41}	y_{42}	y_{43}	y_{44}
	5	y_{51}	y_{52}	y_{53}	y_{54}

where y_{ij} is the jth yield obtained from the ith batch. An appropriate model is

$$y_{ij} = \mu + T_i + \varepsilon_{ij} \qquad i = 1, 2, \ldots, 5$$
$$j = 1, 2, 3, 4$$

where T_i, the batch effect, is a random variable.

In general, a random effects model assumes the form

$$y_{ij} = \mu + T_i + \varepsilon_{ij} \qquad i = 1, 2, \ldots, t$$
$$j = 1, 2, \ldots, n$$

where T_i are assumed to be independent normally distributed random variables with mean 0 and common variance σ_T^2. The random errors ε_{ij} are, as in the past, assumed to be independent normally distributed random variables with mean 0 and common variance σ^2. It is also assumed that the random variables T_i are independent of ε_{ij}. The hypothesis of differences among treatments is expressed as

$$H_0: \sigma_T^2 = 0$$
$$H_1: \sigma_T^2 \neq 0$$

To conduct the test, we compute MS_{Tr} and MS_E exactly the same as in the fixed effects case. It can be shown that $E[MS_{Tr}] = \sigma^2 + n\sigma_T^2$ and that $E[MSE] = \sigma^2$ [7]. Hence a logical test statistic is MS_{Tr}/MSE. If H_0 is true, the observed value of this ratio should lie close to 1; otherwise it should exceed 1. The theory developed in this text can be applied to show that if H_0 is true, MS_{Tr}/MSE follows an $F_{t-1,nt-t}$ distribution. Thus, the test statistic used to detect treatment effects is exactly the same as that used in the fixed effects setting.

In practice, it is useful to estimate σ^2 and σ_T^2. In this way, we can assess the relative importance of these two sources of variation upon our response. The estimators for these variances are

$$\hat{\sigma}^2 = MSE$$
$$\hat{\sigma}_T^2 = \frac{MS_{Tr} - MSE}{n}$$

Example 7.2.2 A dairy has received complaints that its milk tends to spoil quickly. The dairy receives raw milk for processing in truck shipments from farmers in the surrounding area. It is possible that the raw milk supply may have an important effect on spoilage. To study the effect of the variability due to shipment, the manufacturer selects five shipments at random. After processing each shipment, six cartons are selected at random from each and are stored for 12 days, at which time bacteria counts are made. The basic response used is the square roots of the bacteria counts. (This was accomplished in order to stabilize within treatment variance.*) The following table displays the data:

Shipment	Observations (Square Roots)						Sample Mean
1	24	15	21	28	33	23	23.93
2	14	7	12	17	14	16	13.33
3	11	9	7	13	12	18	11.67
4	7	7	4	7	12	18	9.17
5	19	24	19	15	10	20	17.83

*Robert V. Hogg and Johannes Ledolter, *Engineering Statistics*. p. 205. New York: Macmillan, 1987.

Test $H_0\colon\sigma_T^2=0$ against $H_0\colon\sigma_T^2\neq 0$ to determine if the shipment produces a significant variability in the bacteria count. Also estimate and interpret σ^2 and σ_T^2. The following is the standard analysis of variance table:

Source	*SS*	*df*	*MS*	*F*
Shipments	803.0	4	200.8	9.01
Error	557.2	25	22.3	
Total	1360.2	29		

The F statistic is significant at the 0.01 level, suggesting that we should reject H_0 and conclude that the variance in bacteria is due to different shipments.

In order to quantify the source of variance, consider the estimates $\hat{\sigma}^2$ and $\hat{\sigma}_\tau^2$.

$$\hat{\sigma}^2 = 22.3$$

$$\hat{\sigma}_T^2 = \frac{200.8 - 22.3}{6} = 29.75$$

The shipment-to-shipment variance in bacteria count appears to be somewhat larger than the within-shipment variance.

The random effects model can be extended to more complicated situations, with two-factor ANOVA, two-factor ANOVA with interactions, and multifactor models. Texts such as [7] and [17] provide details for these multifactor models.

7.3

GENERALIZED LINEAR MODELS

In recent years considerable interest has centered on the development of a foundation for a more generalized linear model. In this text and in most texts dealing with linear models, much of the theory and methodology evolves around the assumption of normality of the errors. A body of literature now exists in the area of *generalized linear models* that allows for specific theory and methods for much more general classes of linear models, of which the normal is only a special case. For example, in the biological and physical sciences, there are situations in which there is count data and the error structure is a Poisson distribution. Often there are applications in which the response is binary. As a result, the mean response is in the form of a probability, that is, a binomial parameter. Other distributions accommodated nicely by generalized linear models include the Gamma distribution and the inverse Gaussian.

Details concerning generalized linear models are not covered in this text.

EXERCISES

Section 7.1

1. Write the X matrix for the reparameterized analysis of covariance model and show that $X'X$ is as stated in Section 7.1.

2. Verify that the last column of X in the reparameterized analysis of covariance model is orthogonal to each of the other columns of X, thus verifying the claim that the covariate in the reparameterized model is orthogonal to the mean and to all treatment effects.

3. Verify that by using the constraint $\Sigma_{i=1}^{t} \hat{\tau}_i^* = 0$, a solution to the normal equation in the reparameterized analysis of covariance model is $\hat{\mu} = \bar{y}_{..}$, $\hat{\tau}_i^* = \bar{y}_{i.} - \bar{y}_{..}$, $\hat{\beta} = E_{xy}/E_{xx}$.

4. Verify that in simple linear regression $\hat{\beta}_1 = S_{xy}/S_{xx}$, where $S_{xy} = \Sigma_{i=1}^{n}(x_i - \bar{x})(y_i - \bar{y})$ and $S_{xx} = \Sigma_{i=1}^{n}(x_i - \bar{x})^2$.

5. Verify that $SS_{\text{Reg(Full)}} = \Sigma_{i=1}^{t} y_{i.}^2/n + E_{xy}^2/E_{xx}$ in the reparameterized analysis of covariance model.

6. Verify that $SS_{\text{Reg(Reduced)}} = y_{..}^2/tn + S_{xy}^2/S_{xx}$, where

$$S_{xy} = \sum_{i=1}^{t} \sum_{j=1}^{n} (x_{ij} - \bar{x}_{..})(y_{ij} - \bar{y}_{..})$$

$$S_{xx} = \sum_{i=1}^{t} \sum_{j=1}^{n} (x_{ij} - \bar{x}_{..})^2$$

as claimed in Section 7.1. (*Hint*: Write SS_{Res} as $\Sigma_{i=1}^{t}\Sigma_{j=1}^{n}(y_{ij} - b_0 - b_1 x_{ij})^2$.) Use Exercise 6 of Chapter 3 to rewrite SS_{Res} in terms of b_1. Expand and simplify letting $S_{yy} = \Sigma_{i=1}^{t}\Sigma_{j=1}^{n}(y_{ij} - \bar{y}_{..})^2$. Use the fact that $SS_{\text{Reg}} = SS_{\text{Total}} - SS_{\text{Res}}$ to complete the argument.

7. Verify that in the analysis of covariance,

$$R(\beta|\mu, \tau_1, \tau_2, \ldots, \tau_t) = R(\beta) = \frac{E_{xy}^2}{E_{xx}}$$

(*Hint*: See Section 4.8 and apply the results given there to the X matrix in reparameterized form.)

8. Consider the analysis of covariance model given by

$$y_{ij} = \mu + \tau_i + \beta x_{ij} + \varepsilon_{ij} \qquad i = 1, 2, 3 \quad j = 1, 2, \ldots, n$$

To gain evidence about the appropriateness of the analysis of covariance model, we often postulate the three *separate regression* models

$$y_{ij} = \mu + \tau_i + \beta_i x_{ij} + \varepsilon_{ij} \qquad i = 1, 2, 3 \quad j = 1, 2, \ldots, n$$

and test

$$H_0: \beta_1 = \beta_2 = \beta_3$$

as a prelude to doing analysis of covariance. This is easily tested by using either the full model and reduced model procedures discussed in Section 6.2 or the general linear hypothesis discussed in Section 6.1. The methods give equivalent results. Outline the methodology for testing this hypothesis using both methods.

9. Consider the analysis of covariance model with unequal sample sizes for the treatments

$$y_{ij} = \mu + \tau_i + \beta x_{ij} + \varepsilon_{ij} \qquad \begin{array}{l} i = 1, 2, \ldots, t \\ j = 1, 2, \ldots, n_i \end{array}$$

a. Using the same reparameterized form of the model as in the equal sample size case, develop an expression for $R(\tau_1, \tau_2, \ldots, \tau_t|\mu, \beta)$.
b. Develop an expression for $R(\beta|\mu, \tau_1, \tau_2, \ldots, \tau_t)$.

Section 7.2

10. Verify the computations given in Example 7.2.2.

COMPUTING SUPPLEMENT K (Analysis of Covariance)

PROC GLM can be used to conduct an analysis of covariance. The technique is illustrated using the data of Example 7.1.1.

Statement	Purpose
`DATA ANCOVA;`	Names data set
`INPUT TRT $ Y X ;`	Names variables
`CARDS;`	Signals that data follows
`C    75.0   75.0`	
`C    72.5   55.0`	
`.`	Data lines
`.`	
`.`	
`B    65.0   62.6`	
`;`	Signals end of data
`PROC GLM;`	Calls for the general linear models procedure
`CLASS TRT;`	Identifies the treatment variable
`MODEL Y=TRT X/SOLUTION;`	Identifies X as the covariate and asks for a solution to the normal equations to be printed
`TITLE1 ANALYSIS OF COVARIANCE;`	Titles output

ANALYSIS OF COVARIANCE 2

GENERAL LINEAR MODELS PROCEDURE

DEPENDENT VARIABLE: Y

SOURCE	DF	SUM OF SQUARES	MEAN SQUARE	F VALUE
MODEL	4	1900.65269993	475.16317498	7.94
ERROR	26	1555.04084845 (1)	59.80926340	PR > F
CORRECTED TOTAL	30	3455.69354839		0.0003

R-SQUARE	C.V.	ROOT MSE	Y MEAN
0.550006	13.0331	7.73364490	59.33870968

SOURCE	DF	TYPE I SS	F VALUE	PR > F
TRT	3	922.59354839	5.14	0.0063
X	1	978.05915155	16.35	0.0004

SOURCE	DF	TYPE III SS	F VALUE	PR > F
TRT	3	384.60592753 (2)	2.14 (3)	0.1190 (4)
X	1	978.05915155 (5)	16.35 (6)	0.0004 (7)

PARAMETER		ESTIMATE	T FOR H0: PARAMETER=0	PR > |T|	STD ERROR OF ESTIMATE
INTERCEPT		31.33833845 B	3.73	0.0009	8.40345147
TRT	b	-9.15051432 B	-2.11	0.0445	4.33360918
	c	-1.70904797 B	-0.47	0.6457	3.67453554
	r	-6.68301932 B	-1.67	0.1075	4.00926346
	w	0.00000000 B	.	.	.
X		0.52864490 (8)	4.04	0.0004	0.13072713

NOTE: THE X'X MATRIX HAS BEEN DEEMED SINGULAR AND A GENERALIZED INVERSE HAS BEEN EMPLOYED TO SOLVE THE NORMAL EQUATIONS. THE ABOVE ESTIMATES REPRESENT ONLY ONE OF MANY POSSIBLE SOLUTIONS TO THE NORMAL EQUATIONS. ESTIMATES FOLLOWED BY THE LETTER B ARE BIASED AND DO NOT ESTIMATE THE PARAMETER BUT ARE BLUE FOR SOME LINEAR COMBINATION OF PARAMETERS (OR ARE ZERO). THE EXPECTED VALUE OF THE BIASED ESTIMATORS MAY BE OBTAINED FROM THE GENERAL FORM OF ESTIMABLE FUNCTIONS. FOR THE BIASED ESTIMATORS, THE STD ERR IS THAT OF THE BIASED ESTIMATOR AND THE T VALUE TESTS H0: E(BIASED ESTIMATOR) = 0. ESTIMATES NOT FOLLOWED BY THE LETTER B ARE BLUE FOR THE PARAMETER.

These statistics are given:

(1) SS_E

(2) $R(\tau_1, \tau_2, \ldots \tau_t | \mu, \beta)$

(3) F statistic used to test for equality of treatment means

(4) P value for the test given in 3

(5) $E_{xy}^2 | E_{xx}$

(6) F statistic used to test for common slope

(7) P value for the test given in 6

(8) $\hat{\beta}$

APPENDIX: TABLES

Percentage points, F distribution

Percentage points, Students t distribution

Percentage Points, *F* Distribution $P[F \leqslant t] = .90$

n \ m	1	2	3	4	5	6	7	8	9	10	12	15	20	24	30	40	60	120	∞
1	39.86	49.50	53.59	55.83	57.24	58.20	58.91	59.44	59.86	60.19	60.71	61.22	61.74	62.00	62.26	62.53	62.79	63.06	63.33
2	8.53	9.00	9.16	9.24	9.29	9.33	9.35	9.37	9.38	9.39	9.41	9.42	9.44	9.45	9.46	9.47	9.47	9.48	9.49
3	5.54	5.46	5.39	5.34	5.31	5.28	5.27	5.25	5.24	5.23	5.22	5.20	5.18	5.18	5.17	5.16	5.15	5.14	5.13
4	4.54	4.32	4.19	4.11	4.05	4.01	3.98	3.95	3.94	3.92	3.90	3.87	3.84	3.83	3.82	3.80	3.79	3.78	3.76
5	4.06	3.78	3.62	3.52	3.45	3.40	3.37	3.34	3.32	3.30	3.27	3.24	3.21	3.19	3.17	3.16	3.14	3.12	3.10
6	3.78	3.46	3.29	3.18	3.11	3.05	3.01	2.98	2.96	2.94	2.90	2.87	2.84	2.82	2.80	2.78	2.76	2.74	2.72
7	3.59	3.26	3.07	2.96	2.88	2.83	2.78	2.75	2.72	2.70	2.67	2.63	2.59	2.58	2.56	2.54	2.51	2.49	2.47
8	3.46	3.11	2.92	2.81	2.73	2.67	2.62	2.59	2.56	2.54	2.50	2.46	2.42	2.40	2.38	2.36	2.34	2.32	2.29
9	3.36	3.01	2.81	2.69	2.61	2.55	2.51	2.47	2.44	2.42	2.38	2.34	2.30	2.28	2.25	2.23	2.21	2.18	2.16
10	3.29	2.92	2.73	2.61	2.52	2.46	2.41	2.38	2.35	2.32	2.28	2.24	2.20	2.18	2.16	2.13	2.11	2.08	2.06
11	3.23	2.86	2.66	2.54	2.45	2.39	2.34	2.30	2.27	2.25	2.21	2.17	2.12	2.10	2.08	2.05	2.03	2.00	1.97
12	3.18	2.81	2.61	2.48	2.39	2.33	2.28	2.24	2.21	2.19	2.15	2.10	2.06	2.04	2.01	1.99	1.96	1.93	1.90
13	3.14	2.76	2.56	2.43	2.35	2.28	2.23	2.20	2.16	2.14	2.10	2.05	2.01	1.98	1.96	1.93	1.90	1.88	1.85
14	3.10	2.73	2.52	2.39	2.31	2.24	2.19	2.15	2.12	2.10	2.05	2.01	1.96	1.94	1.91	1.89	1.86	1.83	1.80
15	3.07	2.70	2.49	2.36	2.27	2.21	2.16	2.12	2.09	2.06	2.02	1.97	1.92	1.90	1.87	1.85	1.82	1.79	1.76
16	3.05	2.67	2.46	2.33	2.24	2.18	2.13	2.09	2.06	2.03	1.99	1.94	1.89	1.87	1.84	1.81	1.78	1.75	1.72
17	3.03	2.64	2.44	2.31	2.22	2.15	2.10	2.06	2.03	2.00	1.96	1.91	1.86	1.84	1.81	1.78	1.75	1.72	1.69
18	3.01	2.62	2.42	2.29	2.20	2.13	2.08	2.04	2.00	1.98	1.93	1.89	1.84	1.81	1.78	1.75	1.72	1.69	1.66
19	2.99	2.61	2.40	2.27	2.18	2.11	2.06	2.02	1.98	1.96	1.91	1.86	1.81	1.79	1.76	1.73	1.70	1.67	1.63
20	2.97	2.59	2.38	2.25	2.16	2.09	2.04	2.00	1.96	1.94	1.89	1.84	1.79	1.77	1.74	1.71	1.68	1.64	1.61
21	2.96	2.57	2.36	2.23	2.14	2.08	2.02	1.98	1.95	1.92	1.87	1.83	1.78	1.75	1.72	1.69	1.66	1.62	1.59
22	2.95	2.56	2.35	2.22	2.13	2.06	2.01	1.97	1.93	1.90	1.86	1.81	1.76	1.73	1.70	1.67	1.64	1.60	1.57
23	2.94	2.55	2.34	2.21	2.11	2.05	1.99	1.95	1.92	1.89	1.84	1.80	1.74	1.72	1.69	1.66	1.62	1.59	1.55
24	2.93	2.54	2.33	2.19	2.10	2.04	1.98	1.94	1.91	1.88	1.83	1.78	1.73	1.70	1.67	1.64	1.61	1.57	1.53
25	2.92	2.53	2.32	2.18	2.09	2.02	1.97	1.93	1.89	1.87	1.82	1.77	1.72	1.69	1.66	1.63	1.59	1.56	1.52
26	2.91	2.52	2.31	2.17	2.08	2.01	1.96	1.92	1.88	1.86	1.81	1.76	1.71	1.68	1.65	1.61	1.58	1.54	1.50
27	2.90	2.51	2.30	2.17	2.07	2.00	1.95	1.91	1.87	1.85	1.80	1.75	1.70	1.67	1.64	1.60	1.57	1.53	1.49
28	2.89	2.50	2.29	2.16	2.06	2.00	1.94	1.90	1.87	1.84	1.79	1.74	1.69	1.66	1.63	1.59	1.56	1.52	1.48
29	2.89	2.50	2.28	2.15	2.06	1.99	1.93	1.89	1.86	1.83	1.78	1.73	1.68	1.65	1.62	1.58	1.55	1.51	1.47
30	2.88	2.49	2.28	2.14	2.05	1.98	1.93	1.88	1.85	1.82	1.77	1.72	1.67	1.64	1.61	1.57	1.54	1.50	1.46
40	2.84	2.44	2.23	2.09	2.00	1.93	1.87	1.83	1.79	1.76	1.71	1.66	1.61	1.57	1.54	1.51	1.47	1.42	1.38
60	2.79	2.39	2.18	2.04	1.95	1.87	1.82	1.77	1.74	1.71	1.66	1.60	1.54	1.51	1.48	1.44	1.40	1.35	1.29
120	2.75	2.35	2.13	1.99	1.90	1.82	1.77	1.72	1.68	1.65	1.60	1.55	1.48	1.45	1.41	1.37	1.32	1.26	1.19
∞	2.71	2.30	2.08	1.94	1.85	1.77	1.72	1.67	1.63	1.60	1.55	1.49	1.42	1.38	1.34	1.30	1.24	1.17	1.00

$F = \frac{s_1^2}{s_2^2} = \frac{S_1}{m} \Big/ \frac{S_2}{n}$, where $s_1^2 = S_1/m$ and $s_2^2 = S_2/n$ are independent mean squares estimating a common variance σ^2 and based on m and n degrees of freedom, respectively.

Percentage Points, *F* Distribution $P[F \leqslant t] = .95$

n \ m	1	2	3	4	5	6	7	8	9	10	12	15	20	24	30	40	60	120	∞
1	161.4	199.5	215.7	224.6	230.2	234.0	236.8	238.9	240.5	241.9	243.9	245.9	248.0	249.1	250.1	251.1	252.2	253.3	254.3
2	18.51	19.00	19.16	19.25	19.30	19.33	19.35	19.37	19.38	19.40	19.41	19.43	19.45	19.45	19.46	19.47	19.48	19.49	19.50
3	10.13	9.55	9.28	9.12	9.01	8.94	8.89	8.85	8.81	8.79	8.74	8.70	8.66	8.64	8.62	8.59	8.57	8.55	8.53
4	7.71	6.94	6.59	6.39	6.26	6.16	6.09	6.04	6.00	5.96	5.91	5.86	5.80	5.77	5.75	5.72	5.69	5.66	5.63
5	6.61	5.79	5.41	5.19	5.05	4.95	4.88	4.82	4.77	4.74	4.68	4.62	4.56	4.53	4.50	4.46	4.43	4.40	4.36
6	5.99	5.14	4.76	4.53	4.39	4.28	4.21	4.15	4.10	4.06	4.00	3.94	3.87	3.84	3.81	3.77	3.74	3.70	3.67
7	5.59	4.74	4.35	4.12	3.97	3.87	3.79	3.73	3.68	3.64	3.57	3.51	3.44	3.41	3.38	3.34	3.30	3.27	3.23
8	5.32	4.46	4.07	3.84	3.69	3.58	3.50	3.44	3.39	3.35	3.28	3.22	3.15	3.12	3.08	3.04	3.01	2.97	2.93
9	5.12	4.26	3.86	3.63	3.48	3.37	3.29	3.23	3.18	3.14	3.07	3.01	2.94	2.90	2.86	2.83	2.79	2.75	2.71
10	4.96	4.10	3.71	3.48	3.33	3.22	3.14	3.07	3.02	2.98	2.91	2.85	2.77	2.74	2.70	2.66	2.62	2.58	2.54
11	4.84	3.98	3.59	3.36	3.20	3.09	3.01	2.95	2.90	2.85	2.79	2.72	2.65	2.61	2.57	2.53	2.49	2.45	2.40
12	4.75	3.89	3.49	3.26	3.11	3.00	2.91	2.85	2.80	2.75	2.69	2.62	2.54	2.51	2.47	2.43	2.38	2.34	2.30
13	4.67	3.81	3.41	3.18	3.03	2.92	2.83	2.77	2.71	2.67	2.60	2.53	2.46	2.42	2.38	2.34	2.30	2.25	2.21
14	4.60	3.74	3.34	3.11	2.96	2.85	2.76	2.70	2.65	2.60	2.53	2.46	2.39	2.35	2.31	2.27	2.22	2.18	2.13
15	4.54	3.68	3.29	3.06	2.90	2.79	2.71	2.64	2.59	2.54	2.48	2.40	2.33	2.29	2.25	2.20	2.16	2.11	2.07
16	4.49	3.63	3.24	3.01	2.85	2.74	2.66	2.59	2.54	2.49	2.42	2.35	2.28	2.24	2.19	2.15	2.11	2.06	2.01
17	4.45	3.59	3.20	2.96	2.81	2.70	2.61	2.55	2.49	2.45	2.38	2.31	2.23	2.19	2.15	2.10	2.06	2.01	1.96
18	4.41	3.55	3.16	2.93	2.77	2.66	2.58	2.51	2.46	2.41	2.34	2.27	2.19	2.15	2.11	2.06	2.02	1.97	1.92
19	4.38	3.52	3.13	2.90	2.74	2.63	2.54	2.48	2.42	2.38	2.31	2.23	2.16	2.11	2.07	2.03	1.98	1.93	1.88
20	4.35	3.49	3.10	2.87	2.71	2.60	2.51	2.45	2.39	2.35	2.28	2.20	2.12	2.08	2.04	1.99	1.95	1.90	1.84
21	4.32	3.47	3.07	2.84	2.68	2.57	2.49	2.42	2.37	2.32	2.25	2.18	2.10	2.05	2.01	1.96	1.92	1.87	1.81
22	4.30	3.44	3.05	2.82	2.66	2.55	2.46	2.40	2.34	2.30	2.23	2.15	2.07	2.03	1.98	1.94	1.89	1.84	1.78
23	4.28	3.42	3.03	2.80	2.64	2.53	2.44	2.37	2.32	2.27	2.20	2.13	2.05	2.01	1.96	1.91	1.86	1.81	1.76
24	4.26	3.40	3.01	2.78	2.62	2.51	2.42	2.36	2.30	2.25	2.18	2.11	2.03	1.98	1.94	1.89	1.84	1.79	1.73
25	4.24	3.39	2.99	2.76	2.60	2.49	2.40	2.34	2.28	2.24	2.16	2.09	2.01	1.96	1.92	1.87	1.82	1.77	1.71
26	4.23	3.37	2.98	2.74	2.59	2.47	2.39	2.32	2.27	2.22	2.15	2.07	1.99	1.95	1.90	1.85	1.80	1.75	1.69
27	4.21	3.35	2.96	2.73	2.57	2.46	2.37	2.31	2.25	2.20	2.13	2.06	1.97	1.93	1.88	1.84	1.79	1.73	1.67
28	4.20	3.34	2.95	2.71	2.56	2.45	2.36	2.29	2.24	2.19	2.12	2.04	1.96	1.91	1.87	1.82	1.77	1.71	1.65
29	4.18	3.33	2.93	2.70	2.55	2.43	2.35	2.28	2.22	2.18	2.10	2.03	1.94	1.90	1.85	1.81	1.75	1.70	1.64
30	4.17	3.32	2.92	2.69	2.53	2.42	2.33	2.27	2.21	2.16	2.09	2.01	1.93	1.89	1.84	1.79	1.74	1.68	1.62
40	4.08	3.23	2.84	2.61	2.45	2.34	2.25	2.18	2.12	2.08	2.00	1.92	1.84	1.79	1.74	1.69	1.64	1.58	1.51
60	4.00	3.15	2.76	2.53	2.37	2.25	2.17	2.10	2.04	1.99	1.92	1.84	1.75	1.70	1.65	1.59	1.53	1.47	1.39
120	3.92	3.07	2.68	2.45	2.29	2.17	2.09	2.02	1.96	1.91	1.83	1.75	1.66	1.61	1.55	1.50	1.43	1.35	1.25
∞	3,84	3.00	2.60	2.37	2.21	2.10	2.01	1.94	1.88	1.83	1.75	1.67	1.57	1.52	1.46	1.39	1.32	1.22	1.00

$F = \frac{s_1^2}{s_2^2} = \frac{S_1}{m} \Big/ \frac{S_2}{n}$, where $s_1^2 = S_1/m$ and $s_2^2 = S_2/n$ are independent mean squares estimating a common variance σ^2 and based on m and n degrees of freedom, respectively.

Percentage Points, *F* Distribution $P[F \leq t] = .975$

n \ m	1	2	3	4	5	6	7	8	9	10	12	15	20	24	30	40	60	120	∞
1	647.8	799.5	864.2	899.6	921.8	937.1	948.2	956.7	963.3	968.6	976.7	984.9	993.1	997.2	1001	1006	1010	1014	1018
2	38.51	39.00	39.17	39.25	39.30	39.33	39.36	39.37	39.39	39.40	39.41	39.43	39.45	39.46	39.46	39.47	39.48	39.49	39.50
3	17.44	16.04	15.44	15.10	14.88	14.73	14.62	14.54	14.47	14.42	14.34	14.25	14.17	14.12	14.08	14.04	13.99	13.95	13.90
4	12.22	10.65	9.98	9.60	9.36	9.20	9.07	8.98	8.90	8.84	8.75	8.66	8.56	8.51	8.46	8.41	8.36	8.31	8.26
5	10.01	8.43	7.76	7.39	7.15	6.98	6.85	6.76	6.68	6.62	6.52	6.43	6.33	6.28	6.23	6.18	6.12	6.07	6.02
6	8.81	7.26	6.60	6.23	5.99	5.82	5.70	5.60	5.52	5.46	5.37	5.27	5.17	5.12	5.07	5.01	4.96	4.90	4.85
7	8.07	6.54	5.89	5.52	5.29	5.12	4.99	4.90	4.82	4.76	4.67	4.57	4.47	4.42	4.36	4.31	4.25	4.20	4.14
8	7.57	6.06	5.42	5.05	4.82	4.65	4.53	4.43	4.36	4.30	4.20	4.10	4.00	3.95	3.89	3.84	3.78	3.73	3.67
9	7.21	5.71	5.08	4.72	4.48	4.32	4.20	4.10	4.03	3.96	3.87	3.77	3.67	3.61	3.56	3.51	3.45	3.39	3.33
10	6.94	5.46	4.83	4.47	4.24	4.07	3.95	3.85	3.78	3.72	3.62	3.52	3.42	3.37	3.31	3.26	3.20	3.14	3.08
11	6.72	5.26	4.63	4.28	4.04	3.88	3.76	3.66	3.59	3.53	3.43	3.33	3.23	3.17	3.12	3.06	3.00	2.94	2.88
12	6.55	5.10	4.47	4.12	3.89	3.73	3.61	3.51	3.44	3.37	3.28	3.18	3.07	3.02	2.96	2.91	2.85	2.79	2.72
13	6.41	4.97	4.35	4.00	3.77	3.60	3.48	3.39	3.31	3.25	3.15	3.05	2.95	2.89	2.84	2.78	2.72	2.66	2.60
14	6.30	4.86	4.24	3.89	3.66	3.50	3.38	3.29	3.21	3.15	3.05	2.95	2.84	2.79	2.73	2.67	2.61	2.55	2.49
15	6.20	4.77	4.15	3.80	3.58	3.41	3.29	3.20	3.12	3.06	2.96	2.86	2.76	2.70	2.64	2.59	2.52	2.46	2.40
16	6.12	4.69	4.08	3.73	3.50	3.34	3.22	3.12	3.05	2.99	2.89	2.79	2.68	2.63	2.57	2.51	2.45	2.38	2.32
17	6.04	4.62	4.01	3.66	3.44	3.28	3.16	3.06	2.98	2.92	2.82	2.72	2.62	2.56	2.50	2.44	2.38	2.32	2.25
18	5.98	4.56	3.95	3.61	3.38	3.22	3.10	3.01	2.93	2.87	2.77	2.67	2.56	2.50	2.44	2.38	2.32	2.26	2.19
19	5.92	4.51	3.90	3.56	3.33	3.17	3.05	2.96	2.88	2.82	2.72	2.62	2.51	2.45	2.39	2.33	2.27	2.20	2.13
20	5.87	4.46	3.86	3.51	3.29	3.13	3.01	2.91	2.84	2.77	2.68	2.57	2.46	2.41	2.35	2.29	2.22	2.16	2.09
21	5.83	4.42	3.82	3.48	3.25	3.09	2.97	2.87	2.80	2.73	2.64	2.53	2.42	2.37	2.31	2.25	2.18	2.11	2.04
22	5.79	4.38	3.78	3.44	3.22	3.05	2.93	2.84	2.76	2.70	2.60	2.50	2.39	2.33	2.27	2.21	2.14	2.08	2.00
23	5.75	4.35	3.75	3.41	3.18	3.02	2.90	2.81	2.73	2.67	2.57	2.47	2.36	2.30	2.24	2.18	2.11	2.04	1.97
24	5.72	4.32	3.72	3.38	3.15	2.99	2.87	2.78	2.70	2.64	2.54	2.44	2.33	2.27	2.21	2.15	2.08	2.01	1.94
25	5.69	4.29	3.69	3.35	3.13	2.97	2.85	2.75	2.68	2.61	2.51	2.41	2.30	2.24	2.18	2.12	2.05	1.98	1.91
26	5.66	4.27	3.67	3.33	3.10	2.94	2.82	2.73	2.65	2.59	2.49	2.39	2.28	2.22	2.16	2.09	2.03	1.95	1.88
27	5.63	4.24	3.65	3.31	3.08	2.92	2.80	2.71	2.63	2.57	2.47	2.36	2.25	2.19	2.13	2.07	2.00	1.93	1.85
28	5.61	4.22	3.63	3.29	3.06	2.90	2.78	2.69	2.61	2.55	2.45	2.34	2.23	2.17	2.11	2.05	1.98	1.91	1.83
29	5.59	4.20	3.61	3.27	3.04	2.88	2.76	2.67	2.59	2.53	2.43	2.32	2.21	2.15	2.09	2.03	1.96	1.89	1.81
30	5.57	4.18	3.59	3.25	3.03	2.87	2.75	2.65	2.57	2.51	2.41	2.31	2.20	2.14	2.07	2.01	1.94	1.87	1.79
40	5.42	4.05	3.46	3.13	2.90	2.74	2.62	2.53	2.45	2.39	2.29	2.18	2.07	2.01	1.94	1.88	1.80	1.72	1.64
60	5.29	3.93	3.34	3.01	2.79	2.63	2.51	2.41	2.33	2.27	2.17	2.06	1.94	1.88	1.82	1.74	1.67	1.58	1.48
120	5.15	3.80	3.23	2.89	2.67	2.52	2.39	2.30	2.22	2.16	2.05	1.94	1.82	1.76	1.69	1.61	1.53	1.43	1.31
∞	5.02	3.69	3.12	2.79	2.57	2.41	2.29	2.19	2.11	2.05	1.94	1.83	1.71	1.64	1.57	1.48	1.39	1.27	1.00

$F = \frac{s_1^2}{s_2^2} = \frac{S_1}{m} \Big/ \frac{S_2}{n}$, where $s_1^2 = S_1/m$ and $s_2^2 = S_2/n$ are independent mean squares estimating a common variance σ^2 and based on m and n degrees of freedom, respectively.

Percentage Points, Students *t* Distribution

df \ F	0.60	0.75	0.90	0.95	0.975	0.99	0.995	0.9995
1	0.325	1.000	3.078	6.314	12.706	31.821	63.657	636.619
2	0.289	0.816	1.886	2.920	4.303	6.965	9.925	31.598
3	0.277	0.765	1.638	2.353	3.182	4.541	5.841	12.924
4	0.271	0.741	1.533	2.132	2.776	3.747	4.604	8.610
5	0.267	0.727	1.476	2.015	2.571	3.365	4.032	6.869
6	0.265	0.718	1.440	1.943	2.447	3.143	3.707	5.959
7	0.263	0.711	1.415	1.895	2.365	2.998	3.499	5.408
8	0.262	0.706	1.397	1.860	2.306	2.896	3.355	5.041
9	0.261	0.703	1.383	1.833	2.262	2.821	3.250	4.781
10	0.260	0.700	1.372	1.812	2.228	2.764	3.169	4.587
11	0.260	0.697	1.363	1.796	2.201	2.718	3.106	4.437
12	0.259	0.695	1.356	1.782	2.179	2.681	3.055	4.318
13	0.259	0.694	1.350	1.771	2.160	2.650	3.012	4.221
14	0.258	0.692	1.345	1.761	2.145	2.624	2.977	4.140
15	0.258	0.691	1.341	1.753	2.131	2.602	2.947	4.073
16	0.258	0.690	1.337	1.746	2.120	2.583	2.921	4.015
17	0.257	0.689	1.333	1.740	2.110	2.567	2.898	3.965
18	0.257	0.688	1.330	1.734	2.101	2.552	2.878	3.922
19	0.257	0.688	1.328	1.729	2.093	2.539	2.861	3.883
20	0.257	0.687	1.325	1.725	2.086	2.528	2.845	3.850
21	0.257	0.686	1.323	1.721	2.080	2.518	2.831	3.819
22	0.256	0.686	1.321	1.717	2.074	2.508	2.819	3.792
23	0.256	0.685	1.319	1.714	2.069	2.500	2.807	3.767
24	0.256	0.685	1.318	1.711	2.064	2.492	2.797	3.745
25	0.256	0.684	1.316	1.708	2.060	2.485	2.787	3.725
26	0.256	0.684	1.315	1.706	2.056	2.479	2.779	3.707
27	0.256	0.684	1.314	1.703	2.052	2.473	2.771	3.690
28	0.256	0.683	1.313	1.701	2.048	2.467	2.763	3.674
29	0.256	0.683	1.311	1.699	2.045	2.462	2.756	3.659
30	0.256	0.683	1.310	1.697	2.042	2.457	2.750	3.646
40	0.255	0.681	1.303	1.684	2.021	2.423	2.704	3.551
60	0.254	0.679	1.296	1.671	2.000	2.390	2.660	3.460
120	0.254	0.677	1.289	1.658	1.980	2.358	2.617	3.373
∞	0.253	0.674	1.282	1.645	1.960	2.326	2.576	3.291

REFERENCES

1 Anton, H. 1984. *Elementary linear algebra.* 4th ed. New York: Wiley.

2 Arnold, J., M. Lentner, and K. Hinkleman, 1989. The efficiency of blocking: How to use MS(blocks)/MS(error) correctly. *American Statistician*, 43 (May), 2.

3 Barnett, R., and M. Ziegler, 1987. *Linear algebra: An introduction with applications.* San Francisco: Dellen.

4 Box, G. E. P., and N. R. Napier, 1987. *Empirical model building and response surfaces.* New York: Wiley.

5 Box, G. E. P., W. G. Hunter, and J. S. Hunter, 1978. *Statistics for experimenters.* New York: Wiley.

6 Cochran, W., and G. Cox, 1957. *Experimental designs.* New York: Wiley.

7 Graybill, F. 1976. *An introduction to linear statistical models.* Boston: Duxbury.

8 Halmos, P. 1958. *Finite dimensional vector spaces.* Princeton, N. J.: D. Van Nostrand.

9 Hogg, R., and A. Craig, 1970. *Introduction to mathematical statistics.* New York: Macmillan.

10 Kempthorne, O. 1975. *The design and analysis of experiments.* Huntington, N.Y.: Robert E. Krieger.

11 Kolman, B. 1986. *Elementary linear algebra.* New York: Macmillan.

12 Lentner, M., and T. Bishop, 1986. *Experimental design and analysis.* Blacksburg, Va.: Valley Books.

13 Milton, J. S., and J. Arnold, 1990. *Introduction to probability and statistics: Principles and applications for engineering and the computing sciences.* New York: McGraw-Hill.

14 Mood, A., F. Graybill, and D. Boes, 1963. *Introduction to the theory of statistics.* New York: McGraw-Hill.

15 Mostow, G., J. Sampson, and J. Meyer, 1963. *Fundamental structures of algebra.* New York: McGraw-Hill.

16 Myers, R. 1990. *Classical and modern regression with applications.* Boston: PWS-KENT.

17 Searle, S. 1971. *Linear models.* New York: Wiley.

18 Walpole, R., and R. Myers, 1989. *Probability and statistics for engineers and scientists.* New York: Macmillan.

19 Younger, MarySue, 1979. *A Handbook for Linear Regression.* North Scituate, Ma.: Duxbury.

INDEX